THE POLITICAL ECONOMY
OF AGRICULTURAL,
NATURAL RESOURCE, AND
ENVIRONMENTAL POLICY ANALYSIS

THE POLITICAL ECONOMY
OF AGRICULTURAL,
NATURAL RESOURCE, AND
ENVIRONMENTAL POLICY ANALYSIS

E. WESLEY F. PETERSON

Iowa State University Press / Ames

E. Wesley F. Peterson is a professor of agricultural economics at the University of Nebraska–Lincoln. He holds a master's in public affairs from Princeton as well as a master of arts in economics and a PhD in agricultural economics from Michigan State University. Prior to his current position, he held academic positions in France and at Texas A&M. He also served in the Peace Corps and is the author of numerous publications and papers.

Iowa State University Press
2121 South State Avenue, Ames, Iowa 50014

Orders:	1-800-862-6657
Office:	1-515-292-0140
Fax:	1-515-292-3348
Web site:	www.isupress.com

♾ Printed on acid-free paper in the United States of America

First edition, 2001

Library of Congress Cataloging-in-Publication Data

Peterson, E. Wesley F.
　　The political economy of agricultural, natural resource, and environmental policy analysis / E. Wesley F. Peterson.
　　　　p. cm.
　　Includes bibliographical references and index.
　　ISBN 0-8138-0432-9 (alk. paper)
　　1. Agriculture and state. 2. Natural resources—Management. 3. Environmental policy. I. Title.
　　HD1415 .P377 2001
　　333.7—dc21
　　　　　　　　　　　　　　　　　　　　　　　　　　　2001024285

The last digit is the print number: 9 8 7 6 5 4 3 2 1

This book is dedicated to my parents,
Pete and Marney Peterson, in recognition
of their support and guidance through all
these years.

Contents

Contents

Part II. Analytical Methods and Applications

Preface

This book is intended to serve as a textbook for upper-division, undergraduate courses in agricultural and natural resource policy. Part I includes six chapters describing the theoretical framework that guides the economic analysis of public policies. A major theme in these chapters is the centrality of political and ethical considerations in policy analysis and evaluation. Questions about the best course of action for collective entities such as governments are profoundly ethical and it is a fact of life that decisions about such actions are made in political settings characterized by conflicting values, strategic bargaining and efforts to advance particular sets of interests. These aspects of policy analysis are addressed in Chapters 5 and 6 which follow a detailed discussion of the economics of public policy.

Part II of the book is made up of three core chapters covering basic analytical methods derived from the conceptual foundations in Part I. The core chapters describe benefit-cost analysis (Chapter 7), partial equilibrium analysis (Chapter 10), and analysis based on multisector, systems and economy-wide models (Chapter 13). Each core chapter is followed by one or two case studies of particular policy issues related to food, agriculture, environmental protection and natural resource use. The specific topics of these case studies include a project to rejuvenate the banana industry in a West African country, the problem of nitrate contamination of groundwater in North America and Western Europe, the U.S. sugar program, the conflict between the United States and the European Union over hormones in livestock production and genetically-modified food and the implications of the North American Free Trade Agreement (NAFTA) for agriculture and the environment. The case studies are designed to highlight the interaction of economic analysis with the political and ethical aspects of the problem being addressed by the policy. The final chapter includes a discussion of the place of policy analysis in the broader process of identifying and selecting public policy interventions and a recapitulation of the major themes raised in the book.

Many textbooks on agricultural and natural resource policy emphasize descriptions of particular policies, their origins and history along with discussions of the implications of existing and alternative policies. These topics are also included in this book although in a slightly different format. The emphasis here is on theories and methods for policy analysis with the rich history of government policy intervention in agriculture, natural resources and the environment used to illustrate the application of the analytical principles. Because the details of agricultural and natural resource policies are constantly being revised, descriptions of such policies become dated very rapidly. An emphasis on how to analyze a wide range of government interventions may be of greater long-term use to students than the more narrow focus on specific government programs, many of which may well disappear within the student's professional life. In addition, an understanding of policy analysis is relevant to a wide range of professional demands and many aspects of everyday life. Many students taking courses in agricultural and natural resource policy will end up working in professions for which knowledge of particular policies in these sectors is unnecessary. These students may still find it useful to be able to draw on the principles and methods described in this book as they confront different public policy issues.

Throughout the text, ideas and concepts are presented with commentary on the historical context in which they were developed. The most important of these concepts are highlighted in the text and collected in a list at the end of the chapters, which also include discussion questions and suggestions for further reading. In addition, both domestic and global issues are addressed and every effort has been made to incorporate examples of policy issues from around the world. While the analytical principles discussed have broad applicability, most of the examples and illustrations concern public policy issues in the food, agricultural, environmental and natural resource sectors. There is a natural connection between these topics in that agricultural production depends critically on the natural resource base (land, water and energy) and has significant impacts on the state of the environment (wildlife habitat, biodiversity, wetlands, and clean air and water). In addition, the governments of most countries implement policies designed to regulate food prices, increase agricultural output, insure farm incomes, protect the natural environment, conserve important natural resources, and achieve a variety of other goals related to these sectors of the economy.

Although the presentation of most of the material is based on graphs and non-mathematical explanations, calculus is used in a few places to derive some of the results and there is a brief introduction to the use of matrix algebra for input-output analysis in Chapter 13. In general, students who take advanced courses in policy analysis will have completed a course in calculus but few will have experience in using it in economic analysis and even fewer will have an understanding of mathematical modeling. Mathematical explanations are included in the text as a way to illustrate their usefulness and to provide the few students with mathematical aptitudes or an interest in graduate school a demonstration of how mathematics may be used in economic analysis. The mathematical presentations also illustrate the way economists approach theoretical questions and this can be useful in motivating critical thinking about constrained maximization and other intellectual presuppositions of the discipline. Full understanding of the mathematical intricacies is not needed for grasping the main points so students who are uncomfortable with mathematics can simply skip those parts.

In designing this text, I have tried to include as many potentially relevant topics as possible. Some of these topics are conceptual while others relate to current issues. Discussions of such current issues as food safety, trade policies, agricultural impacts on the environment, and many others are included in the cases studies as well as in the theoretical and applied chapters where they are used to illustrate the principles being derived. The broad range of theoretical topics introduced in Part I includes some, such as Arrow's general possibility theorem, that may be somewhat challenging for undergraduate students. The rationale for including the more challenging topics is that some instructors may wish to alert students to the existence of these very important and influential ideas. Most undergraduates are familiar with such concepts as the prisoners' dilemma and many are capable of working through other complex representations. On the other hand, some students may find this type of material fairly impenetrable and, as with the mathematical presentations, it should be passed over fairly quickly by students and instructors if it does not serve a useful pedagogic purpose. For purposes of undergraduate instruction, I believe that little will be lost if the Arrow and Sen theorems and the theory of the second best in Chapter 5, for example, are left out.

Instructors may also find that they wish to skip the more complex modeling approaches described in Chapter 13 as well as the log-linear displacement model introduced in Chapter 10. Again, these topics are included just in case they may prove useful. Undergraduates often have little experience with the use of models to represent and analyze economic relations. In addition, many have not been exposed to linear regression analysis making it something of a challenge to describe the methods used by economists to analyze public policies. The discussions of these topics in Chapters 10 and 13 and in the case studies are an effort to provide general descriptions of important analytical methods to a non-specialist audience. They are not manuals for applied economic analysis. Many introductory texts on policy analysis do not attempt to go beyond basic benefit-cost analysis and, depending on the nature of the students and the course being offered, that may be a sensible strategy in using this book.

The case studies were selected either because I had done research in that area myself or because I felt that it would be possible to develop a case study based on published work. I have also attempted to include case studies that relate to current issues in food, agriculture, natural resources, and the environment. Some instructors may find it more effective to develop alternative case studies based on their own work and more closely reflecting their methodological and topical interests. It may be that some of the case studies fit well in the particular organization of a course while others do not. Some instructors may find that there is not enough time to work through five case studies although one or two would be of interest. Others may feel that it is important to devote more time to the development of other topics, such as the history of farm programs in the United States, and elect not to use case studies at all. Some of the case studies contain information on such topics as the institutions governing world trade and these sections can be assigned even if the rest of the case study is not used. It is my hope that the various parts of the book, whether entire chapters, sections in a case study or one of the other chapters, or an appendix can be selected and rearranged as needed by instructors to construct a set of reading material that is relevant for a wide variety of policy courses.

Although the primary audience for this text is undergraduate students enrolled in upper-division courses on agricultural and natural resource policy, it could also be used for courses at the master's level. In most graduate courses, the text would need to be complemented by some more advanced readings but it could provide a general foundation, particularly for students with limited backgrounds in policy analysis. In addition, students in broader courses on public policy analysis may find the range of topics covered in this book to be appealing. Food, agriculture, natural resources and the environment are highly relevant for such issues as development in low-income countries or global environmental protection. Students with interests in development, environmental issues, poverty and hunger, agribusiness, international relations and public policy in general may all find material that is of interest in this book.

This book represents the culmination of many years of study, reflection, and teaching in the areas of agricultural and natural resource policy. Throughout this process, I have learned much from students and colleagues and wish to acknowledge their contribution to the present work. I owe many of my interests and much of my understanding of public policy to the professors who guided me through my graduate studies at Michigan State University, notably Professors Jim Shaffer and Al Schmid, both of whom have continued to challenge and inform my thinking in the years since I graduated. I have also greatly benefited over the years from discussions with professional colleagues and friends. Paul Thompson of the Philosophy Department at Purdue University is responsible for piquing my interest in ethics although he bears no responsibility for any mistakes on that subject that I may have made in this book. The general exchange of ideas at various professional meetings of agricultural economists has been of great importance in the development of this book. In particular, conversations with Dick Perrin, Lilyan Fulginiti, Dave Bessler, Jerry Skees, Samarendu Mohanty, Mechel Paggi, Dan Padburg, Fred Boadu, Fred Ruppel, and many others have been of great value.

In 1999–2000, I spent a year as a visiting professor in the Food and Resource Economics Department at the University of Delaware. It was during that year that most of this text was written and the support of the faculty and staff at the University of Delaware was invaluable in completing the manuscript. Outside the discipline of agricultural economics I have learned much from friends and scholars in anthropology, political science, and philosophy. Another important influence on my thinking has come from undergraduate and graduate students in France, Texas, Nebraska and Delaware who have challenged the ideas in this text and forced me to deal more realistically with public policy issues. I have also had the good fortune to be able to work with excellent staffs at both the University of Nebraska and the University of Delaware. In Nebraska, Diane Wasser has provided great general support and did most of the work on the figures appearing in the text.

Finally, I also owe a debt of gratitude to my brother, Eric Peterson, and my wife, Andrea Peterson. Eric has stimulated and challenged my thinking as no other friend or colleague has. Throughout the long and sometimes grueling process of composing this text, Andrea has put up with my bad moods and other foibles and helped me to keep on track. My gratitude to her for her support is boundless and to all the others who have been my mentors, guides and teachers, I express my admiration and deep gratitude.

List of Acronyms Used in This Book

AAA	Agricultural Adjustment Act
ADM	Archer Daniels Midland
AIDS	Almost Ideal Demand System
ARP	Acreage Reduction Program
ASCS	Agricultural Stabilization and Conservation Service
BMP	Best Management Practices
Bst	bovine somatotropin
CAP	Common Agricultural Policy
CCC	Commodity Credit Corporation
CEA	Council of Economic Advisers
CEEPES	Comprehensive Environmental Economic Policy Evaluation System
CGE	Computable General Equilibrium
CPI	consumer price index
CRP	Conservation Reserve Program
DAC	Development Aid Committee
DES	diethylstilbestrol
EAGGF	European Agricultural Guidance and Guarantee Fund
EC	European Community
ECSC	European Coal and Steel Community
EDF	Électricité de France
EEC	European Economic Community
EEP	Export Enhancement Program
e.g.	for example (from Latin *exempli gratia*)
EPA	Environmental Protection Agency
ERS	Economic Research Service, USDA
EU	European Union
FAIR	Federal Agriculture Improvement and Reform Act
FAO	Food and Agriculture Organization
FAPRI	Food and Agriculture Policy Research Institute
FDA	Food and Drug Administration
FOR	Farmer-Owned Reserve

FSIS	Food Safety Inspection Service
FTA	Free Trade Agreement
GAO	Government Accounting Office
GATT	General Agreement on Tariffs and Trade
GDP	gross domestic product
GMO	genetically modified organisms
GNP	gross national product
GSP	Generalized System of Preferences
HAACP	Hazard Analysis Critical Control Point
HFCS	high-fructose corn syrup
IBRD	International Bank for Reconstruction and Development
i.e.	that is (from Latin *id est*)
IMF	International Monetary Fund
IRR	internal rate of return
MERCOSUR	Southern common market (Argentina, Brazil, Uruguay and Paraguay)
MFN	Most-Favored Nation
MMPA	Marine Mammal Protection Act
MRS	marginal rate of substitution
MRT	marginal rate of transformation
MRTS	marginal rate of technical substitution
MTN	Multilateral Trade Negotiations
NAAEC	North American Agreement on Environmental Cooperation
NAALC	North American Agreement on Labor Cooperation
NAFTA	North American Free Trade Agreement
NAGPRA	Native American Grave Protection and Repatriation Act
NATO	North Atlantic Treaty Organization
NFA	National Food Authority (Philippines)
NIEO	New International Economic Order
NTR	Normal Trade Relations
OECD	Organization for Economic Cooperation and Development
PIK	Payment In Kind
P.L.480	Public law 480 (food aid)
PPF	production possibility frontier
PPM	Production and Processing Methods
PRI	Partido Revolucionario Institucional (Mexican political party)
rBGH	recombinant bovine growth hormone
RCWP	Rural Clean Water Program
SAM	Social Accounting Matrix
SAP	Structural Adjustment Program
SPS	sanitary and photosanitary
TRQ	Tariff-Rate Quota
TVA	Tennessee Valley Authority
UN	United Nations
UPF	Utility Possibilities Frontier

UR	Uruguay Round
USAID	U.S. Agency for International Development
USDA	United States Department of Agriculture
VER	voluntary export restraint
WTO	World Trade Organization
WTP	Willingness to Pay

Part I

Theoretical Foundations
for Policy Analysis

The purpose of public policy interventions is to correct problems that reduce well-being as defined by the collective judgment of the relevant population. The first question raised by this statement is the precise nature of the problems for which public policies may be the solution. There are many kinds of problems, including such theoretical problems as the origin of the universe, the reason for life, and the value of the integral of a mathematical function. There are also practical problems related to individual decisions on marriage, whether to accept a particular job offer, or what color car to buy. None of these problems is likely to require collective intervention by the public. The first chapter in this part is directed at the question of identifying and defining the collective action problems that may require some form of public policy intervention.

With an initial sense of the types of problems that are addressed by public policies, the next questions concern the concepts of well-being and the relevant population. If public policy aims to improve well-being, we need to know what is meant by that expression. If well-being is defined relative to the beliefs of people living in a particular jurisdiction, we need to know whether the relevant population is limited to that jurisdiction, whether it is some subset of that jurisdiction, or whether it extends beyond it to include human and non-human animals in other places or existing at other times. Most of the rest of the chapters in the first part of this book are connected to these questions. Chapter 2 is a brief overview of some of the economic concepts and terminology that will be used throughout the book. Students with strong backgrounds in economics can probably skip this chapter, although the discussion of basic economic assumptions may be of interest to these students as well. Chapter 3 describes two concepts—Pareto optimality and allocative efficiency—that form the basis for many economic accounts of well-being and improvements in well-being. The theory of the invisible hand is also presented in this chapter. This theory shows that under certain circumstances, markets will lead to the highest levels of well-being with little or no public interference. In Chapter 4, the particular conditions required for this serendipitous result are relaxed and an important role for the government emerges: correct the cases in which markets fail to deliver optimal results and leave everything else to individual choice in market settings.

In Chapter 5, the economic model of public policy is expanded to include some of the political realities of collective action. Decisions on the types of policies to implement are made in political settings where factors other than efficiency take on greater importance. In fact, it can often seem that political decisions fly in the face of economic rationality and that the interests of the few frequently outweigh broader social interests. Public choice economics, as described in Chapter 5, sheds light on the behavior of voters and politicians and the interaction of their interests with the economic concerns for efficiency and a fair distribution of resources. Chapter 6 completes the description of the theoretical foundations for policy analysis by introducing ethics as a guide to collective decision making. Decisions on public intervention are inherently ethical because they involve social choices on what might constitute the best course of action. Consequentialist ethics typical in economics and benefit-cost analysis are contrasted with alternative theories that emphasize rights and virtues. In this chapter as well as in Chapter 4, identification of the relevant population is addressed. The general framework for policy analysis developed in Part I emphasizes the importance of complementing the economic analysis of efficiency with economic, political, and ethical considerations related to income distribution, special interests, fairness, and justice.

1

Public Policy and the Problem of Collective Action

INTRODUCTION

Many of things we do have little or no effect on others. Whether I choose to sleep on my back or my side is nobody's business but my own. My wife does not eat mushrooms because she finds them revolting. This is not likely to affect mushroom producers or others involved in the mushroom business because the impact of her refusal to eat mushrooms is extremely small. Whether one chooses to eat or refrain from eating mushrooms, oysters on the half shell, or any other kind of food is generally considered a decision that should be left entirely to the free exercise of individual preferences. One reason for reserving such choices for individual rather than collective resolution is that the interests of the individual are much more directly implicated by the decision than the interests of anyone else.

Some kinds of individual behavior do have an impact on others. For example, suppose that you need to renew your driver's license and, along with several others, you are waiting in the reception area for your turn to take care of the renewal requirements. If someone in the room begins to smoke, you and the others in the waiting room will be affected in one way or another. In this case, the smoker's behavior turns out to be everybody's business because its impact on others is almost as significant as it is on the one who chooses to smoke. Smoking in private would not have this broader impact, and the decision to smoke or refrain from smoking in private can largely be left up to individuals. In the public reception area, on the other hand, we generally feel that the decision whether to allow smoking or not should be made by the larger group of affected people rather than by any single individual.

Today, a collective decision has been made by most local and state governments that smoking is not allowed in places such as the waiting room at the Department of Motor Vehicles. If someone attempts to smoke in such a place, he or she is breaking the law and will be subject to whatever legal action is set out in the relevant statutes. In the 1950s and 1960s, many people in the United States smoked, and laws regulating smoking in public places were rare. Even today, there are no regulations on smoking in public places in many countries, and in the United States, decisions about whether to segregate smokers and nonsmokers in such public places as restaurants and bars are often left up to the managers of these enterprises. Suppose you are a nonsmoker eating dinner in a restaurant somewhere that does not have a separate smoking section and a woman seated at the next table starts to smoke. This bothers you. What are your options?

The first, of course, is to suffer in silence by simply putting up with the secondhand smoke. A second option would be to ask the person to stop, claiming, perhaps, that you are allergic to tobacco smoke. Suppose you do so but the person refuses to put out her cigarette. In this case, among the options available are to leave, put up with the irritation, try to bribe the smoker to stop, threaten violence if she continues to smoke, organize the other people in the restaurant to put pressure on the smoker, or use force to make her stop. These responses solve the immediate problem in one way or another. A different kind of response, a less immediate one, would be to attack the underlying issue by, for example, organizing a lobbying group to pressure the government to adopt regulations on smoking in restaurants.

The case of smoking in public places is one example of a very common situation, a situation where the actions of some individuals affect the well-being of others. This interdependence between different people means that the individual is not the only one with an interest in what he chooses to do. Such situations can be characterized as cases where there is a collective interest in individual behavior. Frequently, such cases are dealt with through *collective action,* that is, through intervention by public authorities acting under the direction of government agencies. It is not always the case, however, that such interventions will be either appropriate or effective. Decisions about when and how the public should regulate or coordinate various kinds of human interdependence require an assessment of the likely effects of alternative forms of intervention. The purpose of this book is to describe the conceptual foundations for the evaluation of public policies on food, agriculture, the environment, and natural resources and to introduce some of the methods used in practical policy analysis. This initial chapter focuses on the general background for policy analysis and sets out some initial definitions and propositions concerning the relation of markets and governments in coordinating individual behavior.

A SIMPLE TAXONOMY OF COLLECTIVE ACTION PROBLEMS

The following taxonomy of *collective action problems* is offered to help clarify the nature of the interdependencies that are the subject of this book.

- Situations where the behavior of one or more individuals causes discomfort or harm to others.

The person smoking in the restaurant is engaging in a behavior that may be disagreeable to the others in the room. The individual who has chosen to smoke probably does not intend to cause harm or discomfort. In fact, her only motivation for smoking may be personal pleasure. The problem is that this pursuit of personal pleasure has consequences for others. It is easy to think of other examples of this type of situation. A person who drinks too much alcohol may cause great harm if he or she chooses to drive an automobile. A firm that generates air or water pollution as a by-product of its primary production activities is producing both goods that are valued by society and "bads" (pollution) that may cause harm to other people. Often, it is the behavior of groups of people that is detrimental to the interests of an individual or other groups. For example, large numbers of individuals attending a football game may create a great deal of inconvenience and congestion for those living near the stadium as they try to find places to park their cars.

• Situations where the behavior of one or more persons is beneficial to others.

Some people go to great expense to maintain the lawn in front of their houses or to plant attractive gardens. Their prime motivation for such efforts is the pleasure they feel at making their house, lawn, and garden look nice. However, others also benefit from the well-tended yard. Neighbors and passersby are able to enjoy the attractive landscape "free of charge." There are just as many examples of this type of situation as there are of the type discussed in the preceding section. People vote, not because they actually believe that their vote will make any difference in the outcome, but because they feel that it is the "right" thing to do. Because enough people share this belief, democracy is maintained and strengthened to the benefit of all, including those who choose not to vote. In these examples, the beneficial effects are the unintended consequences of actions undertaken on the basis of personal motivations. Sometimes the beneficial activity stems from altruistic motivations, as when people volunteer to work in charitable organizations that provide assistance for people who are less fortunate, doing so in part, perhaps, to obtain personal satisfaction, but mainly because their beliefs about morality and fairness require them to do so. Altruism can be taken to extreme levels, as when someone risks or even loses his or her life to save someone else's.

• Situations where cooperation by several individuals is necessary for the attainment of certain objectives.

Situations of this nature are closely related to the preceding examples in that individual behavior brings benefits to others. In these cases, however, it is the coordination of the behavior of several individuals that leads to a more desirable outcome. Working alone, an individual may be able to catch a total of five fish. Working as a team, a group of four people may be able to use different fishing methods and obtain a total of, say, thirty-two fish. Divided equally, this means that each individual ends up with eight fish rather than the five she could catch working alone. In this case, *cooperation* allows each person to have more fish than could be obtained through individual effort. Sometimes it is a lack of cooperation that causes the destruction of something of value. Because the natural supply of fish in the ocean is limited, some degree of restraint needs to be exercised to avoid overfishing. If too many fish are removed, there may not be enough to assure adequate reproduction, and the fishery can be destroyed. The problem is that there is no incentive for individuals to exercise restraint despite the obvious harm to all of them if the fishery is destroyed. Long-term protection of the fishery requires cooperation by all who exploit this resource. At a more complex level, groups of individuals—the citizens of different nations, for example—may need to cooperate in the solution of such problems as global warming, depletion of the ozone layer, terrorism, and the spread of infectious diseases.

• Situations where coordination of individual behavior is necessary for the realization of desirable outcomes.

In some cases, the solution to a collective action problem is a matter of uncontroversial *coordination of individual behavior*. For example, it makes little difference which side of the road people drive on as long as they all drive on the same side. In other cases, the necessary coordination of individual actions may be more complex. Belgium cannot build a national highway system or a railway system without taking into account the location of rail lines and highways in France or Germany to which the Belgian system must connect if it is to have much value at all. Some objectives can be accomplished only if actions are taken at the right time or in the right sequence. For example, raising agricultural productivity in developing countries may require the introduction of high-yielding varieties. If the improved seeds are not available when farmers need to plant them, the goal of increased agricultural output will not be met.

- Situations where organizations or firms are able to take advantage of others because of their economic or political power.

A common example of this type of problem arises when a firm has a monopoly and is able to charge consumers higher prices than would be the case if there were other competing firms offering similar products. The exercise of *market power* can lead to transfers of income or wealth that do not give rise to the mutual benefits normally associated with economic exchange. Many cattle producers in the United States, for example, believe that the meatpacking industry, which is made up of a small number of very large firms, is able to exploit its power over the market to lower the prices paid to producers while increasing the prices charged to consumers. Private enterprises with market power are not the only example of this type of situation. In many countries, government agencies that represent the interests of a small, influential elite or those of a relatively homogeneous majority violate the rights of disadvantaged groups. The taking of Native American lands by nineteenth-century European settlers in the United States was supported by U.S. military forces and encouraged by the government.

- Situations where individuals negotiate freely with others in order to obtain what they desire.

This final category is included to recognize that some collective action problems are problems of coordination, where individual decisions may well lead to desirable outcomes with a minimum of collective interference. The primary mechanism for assuring this coordination is known as the *market*. Markets are complex and powerful institutions that coordinate large numbers of individual choices. Under certain circumstances, they can lead to efficient use of scarce resources and increased wealth. The beauty of markets is that these outcomes can be realized through decentralized, individual decisions without extensive control by collective agencies. In a well-functioning market, no central authority is required to determine how many blue cars should be produced. In responding to market signals, automobile producers will end up producing the precise number of blue cars that consumers wish to buy. Such a system economizes on a wide range of costs associated with systems that require more centralized coordination mechanisms.

The preceding taxonomy is not meant to suggest that every collective action problem will fit neatly into one of the six categories described. In fact, particular collective action problems may fit into more than one category or require the creation of an entirely new category not previously identified. On the other hand, the preceding comments do draw attention to the range of collective action problems that are inherent in human interaction and can serve as background for the topics to be addressed in this book. Many collective action problems are best solved through the design and implementation of *public policies*, and it is the purpose of this book to explore procedures for analyzing and evaluating such policies as they apply to food, agriculture, natural resources, and the environment.

COLLECTIVE ACTION AND PUBLIC POLICY

Collective action can be thought of as the coordinated efforts of several individuals who cooperate with each other in order to accomplish some outcome. In some cases, collective action arises more or less spontaneously so that there is little need for special arrangements to bring it about. For example, in many neighborhoods, individuals maintain their property in a manner that assures an agreeable atmosphere with no coercion other than the social pressure imposed by their own desires that their neighbors not think badly of them.

In other cases, individuals may have to be forced to alter their natural inclinations if damage to the community is to be avoided. For reasons that will be developed more fully later, individuals often have an incentive to behave in ways that bring them short-run benefits even though such behavior will turn out to be unsustainable because it damages the broader community or because it leads to long-run harm to the individuals themselves. For example, farmers may apply fertilizers at rates that, over time, lead to polluted groundwater because the short-term effects on farm income are positive. Collective action to resolve problems of this nature usually does not arise spontaneously. In other words, some form of public policy may need to be implemented to avoid the longer-term, negative effects.

Public policy begins with judgments about which problems require the intervention of some centralized authority and which are best left to the spontaneous initiatives of individuals. Although most cities and rural communities regulate nuisances such as noxious weeds or unsightly garbage, the appearance of people's yards is usually left to individual creativity rather than community design. In a like manner, decisions about the amounts of chemical pesticides and fertilizers to be used in producing crops have traditionally been left mainly to individual producers. Over the past several decades, however, it has become clear that the use of agricultural chemicals has reached levels that are harmful to water quality and other environmental goods. The realization that agricultural production may have a negative impact on the environment forced communities to confront the issue of whether some form of centralized action would lead to a more desirable level of chemical use than could be achieved through continued deference to individual decisions. In general, the consequence has been a shift from individual decision making to more centralized control to solve this collective action problem.

Of course this shift has not been without controversy. Those whose behavior is constrained by new rules are unlikely to take kindly to such efforts to protect the broader community from their actions. In 1996, President Clinton set aside 1.7 million acres in southern Utah as a new national monument.[1] This action preserves the natural beauty of this landscape for future generations and met with widespread approval. However, it also prevents the development of coal-mining operations and other uses of the land that were being planned by some private firms and individuals. In the United States, publicly owned land can often be used for grazing cattle or mining in return for the payment of nominal fees. These uses are precluded by the decision to create a national monument in the area. The individuals who had planned to mine coal on the Utah lands expressed great outrage at what they perceived as the "theft" of their rights to exploit the natural riches of that region.

The decision to undertake collective action through some form of social or group initiative is a public policy decision. Public policy can be defined as the guiding principle or strategy for deciding on the best actions to take to resolve particular collective action problems. During the early decades of the twentieth century, the agricultural sector in the United States experienced increased instability culminating in the disastrous conditions of the 1930s, when a general economic depression was coupled with the environmental collapse of the dust bowl. The problems of the farm sector were judged to be beyond the control of farmers themselves and of sufficient national importance to warrant collective action. In 1933, the Agricultural Adjustment Act (AAA) was implemented as part of President Franklin Roosevelt's New Deal. At the core of this program was a system of price supports designed to stabilize agricultural prices and to assure that farmers would realize adequate incomes. The policy instruments put in place to achieve those goals were subsequently modified in light of such events as World War II. In the early 1970s, there was a major shift in agricultural policy with the introduction of direct payments to farmers. This shift occurred because the effort to use a single instrument, price supports, to achieve both price stability and increased farm incomes had failed. Direct payments (known as deficiency payments) represented a shift in farm policy in light of experience and changing circumstances. In the 1990s, another policy change was made with the adoption of the Federal Agriculture Improvement and Reform Act (FAIR). This act aimed to reduce the role of the government in managing agricultural markets in favor of "market-oriented" policies. Throughout the century, the broad objectives of agricultural policy were fairly constant, although the specific approaches taken evolved to respond to changing needs and conditions in agriculture.

Public policies are carried out by all levels of government and cover a wide range of activities. In Nebraska, policies to protect groundwater from contamination by agricultural chemicals are designed and implemented by local agencies known as Natural Resource Districts. As noted earlier, complex policies aimed at insuring reliable food supplies and raising the incomes of farmers have been in place in the United States since the 1930s. The policy instruments used to accomplish these objectives have included price supports, quantitative restrictions on the marketing of particular commodities, trade barriers, direct payments to farmers, agricultural research, and many more. Similar agricultural policy instruments have been deployed on a regional basis by the European Union (EU), which is made up of fifteen independent nations. In 1987, many countries adopted an international environmental convention, the Montreal Protocol on Substances that Deplete the Ozone Layer, which calls for the elimination of certain industrial chemicals that have been shown to destroy upper-atmospheric ozone that filters out harmful radiation from the sun. The

Montreal Protocol includes provisions to prevent trade in these substances as well as national commitments to eliminate their production and consumption.[2]

The collective action problems targeted by these policies include groundwater pollution, farm income, food security, and the hole in the ozone layer. The governmental entities involved in the design of the policies and the choice of policy tools range from local agencies to the international community. The effectiveness of these policies in solving the particular collective action problems targeted depends on numerous factors, including the initial conditions, the response of individuals to the incentives created by the policies, unforeseen events such as natural disasters, changes in important physical and social relationships, and many more. The purpose of policy analysis is to predict as accurately as possible the effects of alternative approaches to solving collective action problems in order to offer guidance to politicians and bureaucrats charged with choosing the best of the possible policies.

The range of alternative policies that may be considered in attempts to solve collective action problems is not infinite. In fact, many conceivable policy approaches may be ruled out by the cultural values of a society and its basic institutional framework. It is unthinkable that the government of the United States would attempt to solve the collective action problem of criminal activity by amputating the hands of convicted thieves as is done in some countries. Such a policy would violate American beliefs about appropriate punishments and might be judged to violate the Eighth Amendment of the U.S. Constitution concerning cruel and unusual punishment. Values, norms, customs, laws, rules, and regulations are known as *institutions*, the subject of the next part of this chapter.

INSTITUTIONS

Douglass North, who won the Nobel Prize in Economics in 1993, defines institutions as

> the humanly devised constraints that structure political, economic and social interaction. They consist of both informal constraints (sanctions, taboos, customs, traditions, and codes of conduct), and formal rules (constitutions, laws, property rights).[3]

In terms of the preceding discussions, institutions provide the matrix within which collective action problems are solved. The *informal institutions* listed by North are generally outside the control of policy makers, although it is not uncommon for political leaders to call on their constituents to return to some (often mythical) set of traditional values or to alter their conduct in a way that will help to solve important social problems. In the United States, a widespread perception has developed that an important cause of problems related to crime, drugs, and violence is the loss of traditional family values. Many allege that these values have been undermined by television, film, rock music, and other aspects of popular culture. Political leaders from both major political parties have spoken out extensively on the need to shore up various informal constraints on human interaction through voluntary changes in individual behavior. For example, voluntary rating systems for television, movies and rock music have been suggested, and antidrug campaigns have been based on encouraging people to "Just Say No."

In some cases, the desire to strengthen informal constraints leads to calls for formal mechanisms to force people to alter their behavior. For example, changing divorce laws to make it more difficult for parents to separate has been suggested as a way to promote traditional family values. Changes in *formal institutions* are generally more effective at correcting social and economic problems than exhortations or jawboning, and it is these institutions that will be the primary topic of this book. Most public policies eventually are codified into some type of legal institution, and in this sense, public policy analysis is the evaluation of the advantages and disadvantages of these institutional changes.

North's examples of formal institutions (constitutions, laws, and property rights) can serve as a taxonomy of these formal constraints. *Constitutions* set out the basic procedures, rules, and principles for making laws and governing a country. I will argue in a later chapter that national constitutions are social contracts that set out the rights and responsibilities of the citizenry and the government. Although various procedures are usually put in place for changing or amending them, the expectation is that constitutions are permanent documents that will rarely be altered. In fact, the procedures for changing constitutions are often quite cumbersome because of the desire that they be changed only after very serious consideration. In the United States, for example, amending the U.S. Constitution requires two-thirds majorities in both houses of Congress as well as ratification by three-quarters of the states. The difficulty of amending constitutions makes constitutional change a poor vehicle for carrying out public policy, which generally requires a more rapid response than can be achieved through such a process.[4]

On the other hand, North's second example of formal institutions, *laws*, can be changed more rapidly, and it is usually through changes in laws and statutes that public policies are implemented. For example, raising the minimum wage in the United States required that a law be passed by Congress and signed by the president. This law mandates the minimum level of wages that firms have to pay their workers. If a firm pays workers less than this amount, it is violating the law and will be subject to legal action (fines, imprisonment) meted out by the justice system (police, courts). Laws developed according to the rules set out in any constitution serve as constraints on individual behavior, constraints that are enforced by a legal system backed up by the power of the state.

Changes in the law often have the effect of changing *property rights*, the third type of formal institution listed by North. A change in laws on pesticide use will change the constellation of rights associated with a farmer's ownership of land and other types of property. In some cases, the authority for changes in property rights has been established by laws adopted in the past so that new arrangements can be put in place without the need to adopt new laws. Such is the case, for example, with the national monument in Utah. The authority to create national monuments had been established in previous law so that a presidential decree was all that was necessary to change the use of that section of the state of Utah. The creation of a national monument changes the property rights associated with the land included in the monument. While private firms and individuals could not own this land (it is owned by the federal government), they were legally entitled to use the land for various economic activities. The presidential decision takes away this "right" and establishes a new property right, the right of current and future generations to enjoy this scenic resource.

Property rights are an extremely important concept for the analysis of public policies. To determine whether the decision to create the new national monument in Utah is a good decision requires an evaluation of the costs and benefits associated with a property rights regime that accords rights to use public lands in certain ways compared with one that establishes a different set of rights to the land. Property rights are defined for resources and goods that have actual or potential value to human beings. They can be defined for individuals, for groups, or for society as a whole and are backed up by the coercive force of a country's legal system. The value of a property right depends on what the holder of the right is allowed to do with the property. For example, the rights to use public land are less valuable than would be full ownership rights to that land. Full ownership rights would allow an individual to sell the land and keep the income from the sale, something that cannot be done with public land even if one has use rights to that land.

When two individuals exchange goods in a market, they are actually exchanging bundles of property rights associated with the goods. For example, a grain producer has the right to transfer the ownership of grain produced on her land in return for some form of payment but does not have the right to dump the grain on a neighbor's house following a dispute over the relative virtues of rival football teams. Ownership of the grain confers certain rights (to store the grain, transport it, sell it, consume it, feed it to livestock, and so on) but not the right to do anything that one wishes with it. Well-defined and clear property rights are particularly important for the functioning of market economies. Without clear title to goods and services as well as laws and codes defining what can be done with one's property, every exchange would have to be negotiated directly and the parties to the exchange would also have to find some way of assuring that the terms of the bargain are met.

Olson makes the distinction between markets in which transactions are self-enforcing and those in which they are not.[5] Face-to-face exchange in traditional markets found in every country on earth is self-enforcing because one can inspect the goods being offered and there is little question of ownership or the right to exchange the goods. Many markets that are crucial for economic growth and prosperity, including distant (e.g., foreign) and financial markets, do not involve self-enforcing transactions. Market exchange in these settings is extremely costly, reducing the opportunities for trade that would otherwise be of benefit to both parties. Many of the economic problems in Russia that have followed the dissolution of the Soviet Union stem from the fact that there is often no clear title to property, making it difficult to know whether an individual offering something for sale actually has the right to sell it. Without legally enforced property rights, the distinction between "theft" and "exchange" becomes ambiguous. According to Runge, property rights reduce the risk and uncertainty associated with economic exchange and allow individuals to predict more accurately how others will behave.[6]

Ronald Coase, who won the Nobel Prize in economics in 1992, developed an important result concerning property rights. The *Coase theorem* shows that the way in which a property right is assigned has no impact on economic efficiency as long as the individuals can bargain freely and the costs associated with this bargaining are small.[7] The example Coase used to illustrate this result concerned a farmer with a field located next to a pasture grazed by a dairy producer's cows. If the dairy farmer increases the number of cows in the pasture, the chance that some will stray into the farmer's field and damage his crops is increased. If property rights are defined to protect the farmer's land, the dairy producer will have to compensate the farmer for any damage caused by his cows. The dairy farmer will not add cows to his herd beyond the point where the likely damage to the farmer's field will cost him more than the additional cows are worth. In a later chapter, we will see that such an adjustment of the herd size will result in the most efficient allocation of resources.

Suppose, however, that the effective property right is assigned to the dairy producer. In this case, it is the farmer who will have to pay to avoid damage to his field. The farmer will be willing to bribe his neighbor to reduce his herd size to the extent that the cost of the bribe is less than the value of the crop that would otherwise be damaged. In the end, the same number of cows will be put in the pasture whether the dairy producer has to compensate the farmer for damage to his crops or the farmer has to compensate his neighbor for reducing his herd. In terms of *economic efficiency*, it makes no difference who holds the property right. There is an optimum number of cows, and that optimum will be achieved no matter who holds the property right as long as the two individuals can bargain freely. Of course, it makes a lot of difference to the individuals because the property rights assignment determines who is compensated and who has to pay. Although the impact on efficiency is negligible, the impact on the incomes of the two individuals may be substantial.

We will develop this result in somewhat greater detail in Chapter 4. The important point at this stage of the discussion is that while well-defined property rights are essential for economic efficiency, the way in which property rights are defined will also influence the *distribution of wealth and income* among individual agents. As noted earlier, many public policies have the effect of changing the structure of property rights, and such changes have implications for both efficient resource use and individual well-being. In fact, the motivation for a particular public policy often resides in its expected effect on efficiency and/or the distribution of income or wealth. One rationale for farm policy in the United States has been the fact that, historically, farmers earned lower incomes than those in other sectors of the economy. Substantial amounts of income have been transferred to farmers over the years in an effort to reduce these disparities. These transfers are a direct cost to taxpayers but have also distorted incentives, leading to inefficiencies that reduce overall welfare. We will need to develop some further results in welfare economics to support the latter statement.

At the same time, another rationale for farm policy has been the potential inefficiencies in resource allocation that might result from the high degree of uncertainty in agricultural production introduced by variability in the weather. If the weather is particularly clement, harvests will be abundant and prices are likely to fall. Some farmers may be unable to make ends meet in these circumstances and may be forced to liquidate their operations. In subsequent years, weather that is either unfavorable or normal may result in lower production levels and high prices that provide an incentive for some individuals to enter farming only to find that an unpredictably good harvest a few years later forces them out again. This type of cyclical entry and exit could be very costly for the overall economy, and policies aimed at stabilizing the agricultural sector could generate widespread efficiency gains. Thus, policy interventions may be motivated by both efficiency and equity considerations. One of the things that makes the design of appropriate public policies so difficult is that it may be virtually impossible to insure internal consistency when there are multiple objectives. Most of the efficiency gains that could be attributed to U.S. farm policy, for example, have probably been more than offset by the distortions introduced by efforts to raise farm incomes.

INSTITUTIONS, MARKETS, AND POLITICS

Public policies condition and alter the structure of formal institutions such as laws and property rights in order to solve collective action problems. Many analysts distinguish approaches to solving collective action problems through public or governmental intervention from those that rely on the pursuit of private interests in market contexts. According to Thomas Dye,

> Both governments and markets provide mechanisms for social choice—for deciding "who gets what, when, and how." Governments organize people for collective choice, and once choices are made, rely upon coercion to enforce them. Markets organize people for individual choice, and once choices are made, rely upon voluntary exchange to implement them. Both politics and markets function to transform individual demands into goods and services, allocate costs, and distribute burdens and benefits.[8]

On this account *markets and governments* offer alternative ways to solve collective action problems. In both cases, institutions are the instruments that make governmental or market solutions effective.

While Dye's statement nicely captures conventional images of the contrast between markets and governments, it may overstate the distinction between them. In the first place, markets can only function in settings where well-defined property rights are enforced by the authority of the state. A functioning legal system defining fraud, establishing laws related to contracts, and providing for other rules and regulations is a prerequisite for a market system. For example, producers are free to use whatever methods they wish to produce sugar beets, subject to regulations to protect migrant workers who work in the fields from pesticides and other unsanitary or dangerous working conditions. There is always some degree of *government regulation* even when the economic system is organized primarily on the basis of markets and private property. Markets and governments should be seen as tightly related parts of a single system of social control.

In some cases, governments have attempted to completely regulate markets through *central planning* or some other type of authoritarian control. Prior to its dissolution, the Soviet Union determined the number and kind of various goods to be produced through government-run central planning. Many argued that such a system was unlikely to be effective at making the best possible use of available resources. As the Soviet Union began to come apart in the late 1980s, it became apparent that Soviet central planning had in fact resulted in massive inefficiencies. The failure of Soviet central planning does not mean, however, that completely unregulated markets will lead to a perfect society. One of the lessons of the post-Soviet experience is that a smoothly functioning market system will not arise spontaneously following the elimination of central planning. Laws, regulations, and enforcement mechanisms, as well as behavioral norms and other types of informal institutions, are required to create the setting within which markets can be established. If such institutions are absent, the result is likely to be chaos rather than an efficient market system.

Another problem with the dichotomy set out in Dye's comment is the notion that governments use coercion to enforce collective choices while markets do not. Lindblom points out that markets can also be coercive, as in the case where individuals can only choose between taking a job they find highly objectionable and starving.[9] The goods and services provided in a market system reflect the tastes and preferences of the majority. Those with different tastes are forced to adjust their lives to conform with the majority preferences. In a sense, the majority is able to "coerce" minorities to conform to certain patterns of consumption through its influence on the market. This type of coercion is not a product of government intervention.

Much of the public discourse, in the United States at least, seems to be based on oversimplified distinctions between the private and public spheres. For example, there has been much discussion over the past two decades of the need for deregulation of various economic sectors (airlines, cable television, trucking, etc.). Often these debates sound as if there are only two possibilities: regulations or no regulations. A more fruitful approach is to recognize that markets are always regulated in some way, so that the appropriate topic of debate is not whether to regulate or not but rather what type of regulations will best serve the broad interests of society.

A few comments on the nature of government are in order. In the popular press and in many other public discussions in the United States, government—particularly the federal government in Washington, D.C.—often seems to be represented as some sort of disconnected, autonomous agent imposed on us by unknown beings from some other place. Such a representation allows us to criticize and condemn government actions without confronting our own responsibilities for the things we are condemning. This view of government is not particularly conducive to constructive discussion of important public policy issues because it prevents any intelligent assessment of the advantages and disadvantages of government intervention to resolve particular problems. In Western democracies, the most appropriate way to think about government is to view it as the primary means for individuals to act collectively. In other words, governments are one of the instruments we use to solve collective action problems. This is the basic view of government that underlies this book.

It could be objected that such a view of government is unduly narrow. After all, there are many countries in the world where the government is the instrument used by a very small group of authoritarian political leaders to impose their will on the rest of the people. While we must never lose sight of this issue, it will not be central to the arguments advanced in this book. With some exceptions, I will assume that governments are the result of some type of political process and that they can be taken as "legitimate" for the purposes of the analyses in this book. At the same time, even if governments are presumed to be legitimate, it does not necessarily follow that they will always implement and enforce policies that advance broad societal interests. Governments are made up of individuals who may make mistakes, behave immorally, break existing laws, or otherwise fail to manage the public's business in a manner that is efficient or efficacious.[10] As a result, society may be better off simply ignoring certain collective action problems rather than attempting to deploy government agencies to correct them. An important challenge for policy analysts is to figure out whether government action in a particular case is preferable to simply accepting the status quo. It is important to emphasize that simplistic assertions that governments can never be effective are just as far off base as claims that governments are always the best way to resolve public policy issues.

It is also important to note that governments exist at many levels and in many forms. In the United States, for example, there are government entities at the city, county, state, and national levels. Beyond individual countries, there are many organizations and special-purpose agencies that have government-like functions. For example, assuring that the nation is defended against foreign attack is seen as one of the primary functions of every national government. In addition, however, individual nations may perceive a need to cooperate in regional defense and security organizations, such as the North Atlantic Treaty Organization (NATO). NATO allows its members to do a better job of defending their citizens than they could do if they did not cooperate. Other examples of international organizations that have purposes similar to those of national governments include the International Monetary Fund (IMF), the United Nations (UN) and its agencies, international environmental conventions, and many more.

Of course, the interaction of all these various governing organizations can give rise to complex and contradictory situations. When the federal government of the United States enters into an international agreement, there may well be implications for state and local governments. For example, an agreement to allow Mexican tomatoes to be imported into the United States without customs duties would be violated if the government of the state of Texas decided to place a duty on Mexican tomato imports. Likewise, state governments in the United States cannot establish laws on interstate commerce that contradict federal laws, and the federal government sometimes establishes mandates that require state and local governments to perform functions that have not been determined by discussion and debate at the local level. Some of these complex interactions will have to be taken into account in the discussions of the theoretical and applied bases for the economic analysis of public policy that are presented in subsequent chapters. As a prelude to those discussions, the next chapter is a review of some basic microeconomic principles.

SUMMARY

1. Individual behavior can often be left entirely to individual discretion. Sometimes, however, the things one chooses to do affect other people, and this can lead to collective action problems.

2. Public policy is the effort to solve collective action problems. The government is the instrument for taking collective action in most cases.

3. Institutions coordinate human behavior either informally through customs, traditions, and taboos or formally through laws, constitutions, and the definition of property rights. Institutions are necessary for the existence of markets in which transactions are not self-enforcing (face-to-face).

4. Public policy is usually carried out by creating or changing laws and property rights. The Coase theorem shows that economic efficiency is unaffected by the way in which property rights are assigned if transactions or bargaining costs are low. However, the assignment of property rights does affect the distribution of income and wealth.

5. Markets and governments are not really alternative mechanisms for social control. Rather, they are tightly related components of a single system of control and coordination.

6. The analysis of public policy involves determining which institutional arrangements are the most advantageous for society as a whole.

KEY CONCEPTS

collective action	formal institutions
collective action problem	constitutions
cooperation	laws
coordination of individual behavior	property rights
market power	Coase theorem
markets	economic efficiency
public policies	distribution of wealth and income
institutions	markets and governments
informal institutions	government regulations
	central planning

DISCUSSION QUESTIONS

1. Identify and describe examples of collective action problems from your own experiences. Can you think of examples that would fit into each of the categories described in the taxonomy of collective action problems presented in this chapter?

2. List some of the informal institutions that you believe to be important in the community in which you live. How do these informal institutions regulate individual behavior?

3. What are property rights, and why are they important for social and economic interaction?

4. Discuss the relationship between governments and markets. Is it possible to eliminate governments, leaving the coordination and control of human interaction entirely up to markets?

5. Is it true that governments rely on coercion while markets are based on voluntary decisions?

SUGGESTIONS FOR FURTHER READING

North, Douglass C. *Institutions, Institutional Change and Economic Performance*. New York: Cambridge University Press, 1990.

Olson, Mancur. *Power and Prosperity*. New York: Basic Books, 2000.

Olson, Mancur, and Satu Käkhkönen, eds. *A Not-So-Dismal Science: A Broader View of Economies and Societies.* New York: Oxford University Press, 2000.

Williamson, Oliver. *The Mechanisms of Governance.* New York: Oxford University Press, 1996.

NOTES

1. *Economist*, 24 September– October 1996.

2. Todd Sandler, *Global Challenges: An Approach to Environmental, Political and Economic Problems* (New York: Cambridge University Press, 1997), 109–15.

3. Douglass C. North, "Institutions," *Journal of Economic Perspectives* 5, no. 1 (winter 1991): 97.

4. In 1972, the Equal Rights Amendment guaranteeing equal rights for women was passed in both houses of Congress with far more than the two-thirds majorities required for constitutional amendments. A period of ten years was allowed for the necessary ratifications of thirty-eight state legislatures (three-quarters of the fifty states). On June 30, 1982, at the end of this period, only thirty-five states had ratified it and the amendment failed. At the time, polls indicated that a substantial majority of American citizens favored the addition of this amendment to the Constitution.

5. See Mancur Olson, *Power and Prosperity* (New York: Basic Books, 2000) for a full discussion of this point.

6. C. Ford Runge, "Strategic Interdependence in Models of Property Rights," *American Journal of Agricultural Economics* 66, no.5 (December 1984): 807–13.

7. R. H. Coase, "The Problem of Social Cost," *Journal of Law and Economics* 3 (October 1960), 1–44.

8. Thomas R. Dye, "Forward: The Political Legitimacy of Markets and Governments," Thomas R. Dye, ed., p.xi in *The Political Legitimacy of Markets and Governments* (Greenwich, Conn.: JAI Press, 1990).

9. Charles E. Lindblom, *Politics and Markets: The World's Political-Economics Systems* (New York: Basic Books, 1977).

10. Of course, the same observation could be made about the individuals working for private-sector firms. This fact is one of the reasons why legal institutions and polices are needed for the proper functioning of the market economy.

2

Markets and Collective Action:
The Economic Setting

INTRODUCTION

The purpose of this chapter is to review a few elementary economic concepts as a basis for subsequent discussions of the economics of policy analysis. Much of economic reasoning centers on the way in which individuals interact in markets. Markets are extremely useful mechanisms for coordinating large numbers of individual decisions to generate certain kinds of social outcomes. It is the market mechanism that insures that there will be an adequate number of apples available for sale in local supermarkets and that the number of red sports cars produced will be about equal to the number that individuals wish to buy. Not all collective action problems are as easily resolved as the examples of apples and sports cars. These more problematic cases are precisely the areas where public policies are likely to be necessary to solve the collective action problem. Before examining these cases in the following chapters, it will be helpful to discuss the way in which economists approach the analysis of markets and individual behavior.

ECONOMIC RATIONALITY AND OTHER PHILOSOPHICAL FOUNDATIONS

To figure out how a complex system such as the economy functions, it is often necessary to make simplifying assumptions. Along with philosophical perspectives, beliefs, and values, such assumptions define the approach to understanding and analysis associated with a particular scientific discipline. The purpose of this part of the chapter is to identify and describe some of the concepts and assumptions that provide the foundation for economic analysis. The ideas discussed in the following paragraphs will demonstrate how economists approach public policy issues and will highlight some of the important and contestable assumptions economists often make.

Economic Rationality

One of the fundamental assumptions that is commonly made by economists is that human beings have reasonably well-defined goals toward which they direct their efforts on the basis of reasoned assessments of the advantages and disadvantages of alternative courses of action. (See Box 2.1). This assumption does not mean that individuals always succeed in choosing the best course of action for achieving a desired end. People make mistakes, and the inevitable uncertainty of the future means that it will never be possible to predict the consequences of a course of action with complete accuracy. The problems of risk, uncertainty, and incomplete information are always present and may lead to decisions that are not in the best interests of the individual making them. The rationality assumption in economics simply means that individuals are assumed to strive to discover optimal solutions, not that they always find them.

Box 2.1. The calculus of economic rationality

Rational individuals as represented by economists strive to do the best for themselves that they can. This can be represented as a problem in optimization, that is, a problem in finding the course of action that will achieve the best result. The differential calculus offers a mathematical approach to solving this type of problem. Let,

1) $Y = f(X)$

represent a function such that given a value for X, we can determine the corresponding value for Y. For example, suppose that the actual function is:

2) $Y = 12 + 4X - X^2$

If values for X are inserted into this equation, the corresponding Y-values can be computed. If X-values are entered in very small increments (e.g. 1.0001, 1.0002, etc.), a continuous curve can be traced out. It will look something like this:

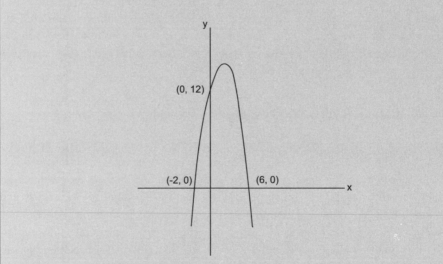

Figure 2.1: Graph of the function $Y = 12 + 4X - X^2$.

Consider only the interval from 0 to +6 on the x-axis. Over this range, the curve reaches a maximum at the point where $X = 2$. This result can be obtained through trial and error or with the use of some simple calculus. Recall that the derivative of a function provides information on the shape of that function. In particular, the derivative describes the slope of the function at any point. The derivative of equation 2) is:

3) $dY/dX = 4 - 2X$

Box 2.1. (Continued)

If $X = 1$, $dY/dX = 2$. In other words, at $X = 1$, the slope of the function shown in equation 2) is 2. If $X = 3$, $dY/dX = -2$. But if $X = 2$, the derivative is zero. In other words, equation 2 slopes up when $X = 1$, slopes down when $X = 3$, and is flat when $X = 2$. This is significant because if the slope is zero, the function is neither increasing nor decreasing. It must have reached either a maximum or a minimum. (There are other special cases where the derivative can be zero, but that need not concern us here).

This simplified excursion into the differential calculus shows that, under certain conditions, we can find out where the maximum on a function is by taking the derivative, setting it equal to zero, and solving for the critical values.

What does this have to do with economic rationality? As defined above, economically rational agents strive to do the best they can for themselves, and this can be represented as a problem in maximizing what one desires. In the case of business enterprises, this might mean maximizing profits. Profits are defined as the difference between total revenue and total costs.

Without introducing a specific functional form for the costs, let

P = the price of the product produced by the firm in a situation of perfect competition,
Q = the quantity of the product the firm produces, and
$C(Q)$ = a cost function showing that the costs of production increase as Q increases.

Then, $PROFIT$ = Total Revenue − Total Costs = $PQ - C(Q)$

To determine what quantity should be produced to maximize profits, the director of the enterprise can locate the value of Q that will cause the derivative of $PROFIT$ to be equal to zero. The derivative of $PROFIT$ is:

4) $dPROFIT/dQ = P - dC/dQ$

where dC/dQ is the derivative of the cost function. dC/dQ is called the marginal cost (MC). When 4) is set equal to zero, we obtain the standard result that profits are maximized where marginal cost is equal to the competitive market price: $P = MC$. In words, a firm maximizes profits by producing at the point where the additional cost of producing the next unit of output is exactly equal to the price that will be received for that unit of output.

The assumption of economic rationality means that individuals can be represented as optimizers. In much of economic theory, this translates into problems where individuals maximize something subject to constraints imposed by their circumstances. This latter way of expressing the optimization problem sounds very similar to the definition of economic rationality offered at the beginning of this discussion. The assumption of economic rationality allows economists to represent the problems of economic choices and decisions as problems of constrained maximization or minimization and use the differential calculus to derive economic principles concerning the behavior of economic agents.

The assumption of *economic rationality* has been challenged by other social scientists and some economists on the grounds that it seems to reflect an extremely narrow description of human motivation and behavior. These critics point out that individuals often sacrifice their personal interests and goals for the greater good. It seems difficult to account for these types of behavior if individuals are motivated exclusively by the pursuit of their own personal goals and interests. Economically rational individuals seem excessively cold and uncaring in direct contradiction to the observed behavior of real people. One response to these criticisms is to point out that even the most altruistic behavior can be thought of as self-interested. A mother risking her life to save her child is attempting to insure that her genes will be represented in future generations. Or she may be motivated by worry that people will think ill of her if she does not make an extraordinary effort to save her child.

But what about a mother risking her life to save the life of a child who is a total stranger to her? It seems a little extreme to claim that this mother is motivated by self-interest because she desires respect and recognition for her courage and willingness to make sacrifices. Adam Smith was an early proponent of the idea that one of the most important motivations driving human behavior is the desire to satisfy an "impartial spectator" who judges particular actions not from the point of view of the individual himself but rather from the point of view of broader societal interests.[1] But Smith would almost certainly not have claimed that an impartial spectator would require sacrificing one's life to save that of a perfect stranger. Such cases suggest that human motivation is much more complex than can be captured in a simple model of self-interested, rational individuals.

For our purposes, however, the assumption that individuals rationally pursue their self-interests may not be too far off base. In markets and in efforts to influence the design of public policy, individuals do seem preoccupied with advancing and protecting their personal well-being. Thus, while this assumption is inappropriate for the study of much human behavior, it serves a useful purpose in predicting how individuals will respond to public policy initiatives.

Methodological Individualism

A somewhat more controversial concept among economists is the notion that only individuals can have interests, tastes, or desires. This philosophical position is referred to as *methodological individualism*. Those who subscribe to this point of view argue that there is no such thing as group interests because only individuals can have interests. Some individuals may find that their personal interests are similar to those of others and that strategic alliances are useful in advancing their common interests. But these coalitions exist because the individuals have common interests, not because the group itself has interests separate from those of the members of the group. In contrast, some economists and other social scientists believe that groups are often more than just the sum of the interests of those who belong to them. Thus, for example, some might argue that there are "family farm" interests that are greater than the simple aggregate of the interests of family farmers. In general, the problem of policy analysis, as discussed in this book, will be approached from an individualist perspective. This does not mean that social groups are not seen as important actors in policy analysis. Rather, it means that groups are assumed to be made up of individuals whose interests coincide so that the pursuit of individual interests is best accomplished through group action. The groups themselves are not thought to have any independent interests beyond those of the individual group members.

Equilibrium

Another central economic concept is that of *equilibrium*. Economists define equilibrium as a state of affairs from which there is no tendency to depart. In the words of Thomas Schelling, "An equilibrium is a situation in which some motion or activity or adjustment or response has died away, leaving something stationary, at rest, 'in balance'. . ."[2] In general, economists assume that economic systems tend to return to equilibrium rather than to remain permanently out of balance.

Consider, for example, the market for tomatoes in New York City. Economically rational consumers will seek to satisfy their desire for tomatoes at the lowest cost possible. Producers, on the other hand, will search for outlets with the highest possible price. An equilibrium is established in this market where the quantity of tomatoes supplied to the market is equal to the quantity demanded. The price associated with this equilibrium is the lowest price consumers will be able to find as well as the best price producers can obtain. If for some reason tomatoes are available at a lower price in Queens, consumers who previously purchased tomatoes in other parts of the city will shift their purchases to stores in Queens if the costs of getting to Queens are not so great as to offset the price differential. Demand in these stores will expand, but there is no reason for supplies to increase because producers would still prefer to sell at the higher prices. Because demand is greater than supply at these lower prices, some consumers will not be able to purchase tomatoes. Some of these consumers will be willing to pay a little extra for tomatoes and will begin to bid up the price. This process will continue until an equilibrium price is reestablished across the entire city of New York.

This process can also be thought of in terms of *arbitrage*. Arbitrage is the action of taking advantage of local price differences by purchasing goods where prices are low and reselling them in locations where prices are high. Arbitragers in the New York tomato market will rapidly discover the low-priced tomatoes in Queens, buy up large quantities, and transport them to other parts of the city where they can be sold for higher prices. Again, the increase in demand will cause the prices of tomatoes in Queens to be bid up. In addition, the increased supplies in other parts of the city will put downward pressure on prices. The result is that the low prices in Queens will rise and the higher prices in other parts of the city will fall, and these processes will continue until an equilibrium price is established. The New York tomato market is an example of an economic system with a tendency toward equilibrium. Such systems are not expected to disintegrate into chaos when there is an external shock. Rather, the expectation is that external shocks will generate processes that return the system to an equilibrium state of affairs in which it will remain until disturbed again by another shock.

The assumption that economic systems generally return to an equilibrium seems to be borne out in the real world. Although markets are often characterized by erratic changes (bull and bear stock markets, for example), they usually do not explode into chaotic and destructive oscillations as might occur in a poorly designed suspension bridge during an earthquake. The suggestion that economic systems tend to return to equilibrium does not mean that there is necessarily something good or bad about the equilibrium. From a nutritional point of view, it is possible that the equilibrium in the New York tomato market is such that some people are unable to consume adequate amounts of vital nutrients. Economists would not claim that the New York tomato market is in a good state of affairs under these circumstances just because it is in equilibrium.

The market for tomatoes in New York is one example of an economic system. But economic systems can be defined at many levels and over many activities. Other examples of economic systems include the market for health care in the United States, the world market for potassium, and the entire U.S. economy. The particular economic system that provides the focus for policy analysis will depend on the problem under consideration. Many policy issues can be analyzed by studying a particular market. Such studies are often referred to as *partial equilibrium* analyses because they deal with departures from and returns to equilibrium in one part of the economy. Other policy issues require attention to the relationships between large numbers of markets, and studies that take account of these relationships are often referred to as *general equilibrium* analyses.

Opportunity Cost

No discussion of basic economic principles would be complete without mention of the concept of *opportunity cost*. Opportunity cost is the value of the best alternative to whatever course of action is under consideration. Suppose one is deciding whether it is a good idea to go to a movie. The economic approach to this decision is to recognize that going to a movie precludes a variety of activities, such as studying for final exams, watching television, reading a book, sleeping, and so on. Suppose that the most important or "valuable" of the alternatives to the movie is studying for final exams. Deciding which of these activities to pursue depends on the values of each to the individual making the decision. The opportunity cost of the movie includes the price of the ticket, the cost of transportation, and the cost of not being able to study for final exams. If the student decides to go to a movie, one can infer that the expected value of seeing the movie is greater than the full opportunity cost made up of out-of-pocket expenses and the value of studying for final exams.

The concept of opportunity cost is of fundamental importance in policy analysis because the decision to implement a particular policy precludes a variety of alternative actions, including the maintenance of the status quo. Effective policy analysis requires attention not only to direct, observable financial costs but also to the opportunity costs of the particular alternatives under consideration. Consider, for example, policies related to the supply of energy in the northwestern part of the United States. If the policy goal is to assure reliable supplies of relatively inexpensive energy, one option might be to construct a new dam on the Columbia River to increase the supply of hydroelectric power in the region. This dam would generate benefits in the form of inexpensive energy for the region. But there are also both direct costs for construction and maintenance and opportunity costs in the form of forgone opportunities to use the river for other purposes such as transportation, recreational fishing and boating, habitat protection for migrating salmon, or any other of the uses to which the river could be put if the dam were not built. A benefit-cost analysis of this project would seriously underestimate the total cost to society of building the dam if these forgone opportunities were not included in the analysis.

Likewise, any public policy will give rise to benefits, direct financial costs, and opportunity costs related to the options that are given up in choosing that particular public policy. Until 1996, the incomes of U.S. grain and cotton producers were supported by a system of direct payments. The payments were financed from the government budget on the basis of complex formulas related to historic patterns of crop production. The use of budget resources to raise farm incomes meant that these funds were not available for welfare programs, national defense, the arts, or other government programs that some might deem important. Moreover, the farm support programs had an additional opportunity cost as producers responded to the higher returns by increasing grain production. In increasing production, more inputs such as fertilizer and pesticides were used, to the detriment of the environment. In addition, the program made these crops so much more profitable than alternative crops that producers planted most of their land to the relatively narrow range of crops receiving support. A realistic assessment of these programs should take account of the opportunity costs associated with alternative uses of the budgetary resources, the impact on the environment, and the loss of crop diversity.

ELEMENTARY ECONOMIC CONCEPTS

The ideas discussed in the preceding section figure significantly in the philosophical and conceptual foundations of economic analysis. The brief outline of fundamental concepts is presented simply as an introduction to these themes and as a kind of glossary of terms that will be used throughout the book. A full economic epistemology would require a great deal more space than is available. In this section, some basic economic concepts derived from the ideas introduced earlier are reviewed. These concepts constitute important background for some of the more complex ideas that will be developed in the next chapter. This review is necessarily superficial and can be skipped by readers comfortable with basic microeconomic principles.

Markets

For most people, chocolate is one of life's great pleasures. It is widely available at prices that are not too high, at least for consumers in the high-income industrialized nations, and most people find that it has a taste and texture that are quite agreeable. Chocolate is made from cacao beans, large seeds encased in oblong pods that grow on the tropical cacao tree. After harvesting, the beans are roasted and the outer shell is removed to obtain what are called "nibs."[3] The nibs are then ground, and the result of this process is called cocoa liquor or chocolate liquor. Cocoa liquor can be used directly to make chocolate by adding milk, sugar, and other ingredients. Alternatively, the cocoa liquor can be pressed to squeeze out the fat, known as cocoa butter, leaving cocoa cake, which is ground into cocoa powder. To make chocolate from cocoa powder, the cocoa butter needs to be reintroduced. About 2.5 million metric tons of cacao beans are produced each year, with Africa accounting for about 57 percent of the total, the Americas for 26 percent, and Asia for 17 percent. The leading producers are Côte d'Ivoire, Brazil, Ghana, Malaysia, Nigeria, Indonesia, and Cameroon. In Africa, cacao beans are produced primarily on small farms, while in Asia and Latin America, they are usually produced on large plantations. In most African countries, a centralized government board purchases the beans from producers and sells them to brokers in Europe and North America. Commercialization in Brazil, Indonesia, Malaysia, and Nigeria is handled by private firms, although the governments of these countries intervene extensively in these markets.

The purchasers in Europe and North America are dealers and brokers who, in turn, sell the beans to processors for grinding into cocoa. Some processors are vertically integrated, producing everything from cocoa liquor to chocolate and chocolate products, while others are specialized only in the first stages of transformation. According to Dand, there are seventeen processors in Europe and North America specialized in the production of cocoa liquor and butter. These firms sell cocoa products to confectioneries and chocolatiers who produce the final chocolate products. Five firms (Mars, Nestlé, Hershey, Cadbury Schweppes, and Jacobs Suchard) account for 40 percent of the market for chocolate. About 30 percent of the beans are ground in the countries where they are produced, so the main cocoa-producing countries do export some cocoa products as well as the beans. In addition, some low-income countries, such as China and Korea, import beans for grinding so that altogether the developing countries account for about 43 percent of the initial grindings. Western Europe is the leader in grinding beans, with 43 percent of the total, while the United States and Canada account for 12 percent of the total. The Swiss are the leading consumers of chocolate, with per capita consumption of almost five kilograms (about eleven pounds) per year, followed by the Belgians, Austrians, Germans, Norwegians, British, French, Americans, Canadians, and Australians.

At each stage in the production and processing of chocolate, there are markets in which buyers and sellers are brought together to do business. In addition to markets directly related to cocoa and chocolate, there are markets in which other inputs, such as fertilizer or sugar, are traded. In all these markets, there are individuals who produce something and offer to sell it to other individuals who wish to purchase it either as a final consumer good (chocolate) or as an input into some producing, processing, or manufacturing activity. At each market level, the collective effort of producers generates the market supply while the needs and means of consumers or processors give rise to the market demand. The precise way in which supplies and demands are brought together in markets depends on many factors, including, for example, transportation or storage conditions. The most significant factor is the *market structure* or organization of the industries involved in the various markets. Some markets can be characterized as "competitive," which means that there are enough buyers and sellers in the market to assure that the actions of individual market participants will have no significant impact on the market and, most notably, on the market price.

In *competitive markets*, firms are forced to accept prices that are established through the interaction of supply and demand, as suggested by the example of the New York tomato market discussed earlier. If an individual store attempts to raise its price for tomatoes, consumers will shift their purchases to other stores and the higher-priced firm will be forced to reduce its prices to the level prevailing in the market. Moreover, in competitive markets, individual firms do not control a sufficient proportion of the market to be able to affect overall supplies by withholding some of their products in an effort to raise prices. There are parallels to these circumstances on the demand side. The upshot is that all participants in competitive markets are *price takers*, and market forces impose a discipline on individual agents that forces them to conform to market conditions.

On the other hand, if there is only one firm producing the good sold in the market, that firm can manipulate prices to its advantage. After all, if there is only one supplier, individuals cannot shift their purchases to other suppliers if they do not like the prices or quality of the product being offered. In fact, it turns out that a single supplier (a monopolist) will maximize his firm's profits by restricting output to raise the price. Recall from Box 2.1 that profits are maximized for firms in a competitive market by producing where the market price is equal to the marginal cost. The same is true for a *monopoly* except that the market price is no longer entirely outside the firm's control. In fact, the monopolist faces the market demand curve, which relates prices to the quantities consumers are willing to purchase. At higher prices, consumers purchase less. Using the symbols for prices, quantities, and costs defined in Box 2.1, this can be represented mathematically as

$$PROFIT = P(Q)Q - C(Q)$$

$P(Q)$ is the downward-sloping demand function faced by the monopolist, with price inversely related to the quantity demanded. $P(Q)Q$ is total revenue (price times quantity). As in the earlier example, profits are maximized at the point where the first derivative of *PROFIT* is equal to zero. The derivative for the first term is obtained by using the quotient rule and is referred to as marginal revenue: $MR = P(Q) + Q(dP/dQ)$. For the competitive firm, the expression dP/dQ is zero because firms are price takers, that is, the price they receive is not affected by the quantity they offer for sale. Under these conditions, marginal revenue is simply the market price and profits are maximized by setting marginal cost equal to this price.

But for the monopolist, marginal revenue depends on how much is offered for sale, so the expression dP/dQ is not equal to zero. Setting marginal revenue equal to marginal cost gives the following decision rule: $P(Q) + Q(dP/dQ) = dC/dQ$. Compared to the competitive equilibrium, the quantity offered for sale by the monopolist will be less and the market price will be higher. This is because the second expression on the left-hand side is negative, so that marginal revenue is less than demand represented by $P(Q)$. This is illustrated in Figure 2.2. Marginal revenue is equal to marginal cost when output is set at Q_o, with the price read off the demand schedule at P_o. In a competitive market, there would be an equilibrium where supply (represented in this case by the marginal cost schedule) equals demand. Equilibrium is established at quantity Q_1 and price P_1, where a greater quantity is exchanged at a lower price than is the case with a monopoly.

The preceding discussion has focused on the behavior of firms selling in either competitive markets or markets monopolized by one firm. Of course, markets include buyers as well as sellers, and the number of consumers is an important aspect of market structure. In competitive markets, consumers are price takers in the sense that the amount any individual elects to purchase has no impact on price. On the other hand, if there is only one consumer in a market, the situation is referred to as a *monopsony*, and it is expected that the single consumer will exercise market power to lower the prices it pays for its purchases. Markets with only one buyer and one seller are sometimes referred to as *bilateral monopolies*. In these cases, the outcome is less easy to predict because the two firms are both trying to exploit their market power, and the result will depend on their bargaining abilities. Indeterminate outcomes also occur in markets in which there are a few buyers or sellers.

Such markets are often referred to as *oligopolies*. These cases differ from perfect competition and perfect monopoly because the participants are neither pure price takers nor pure price setters. As in the case of bilateral monopolies, buyers and sellers will bargain to determine prices, and firms will generally employ methods, such as advertising, that differentiate their products from those of rival firms. They may also attempt to collude or form alliances (although this is often illegal under antitrust laws) or develop strategies that take account of the expected strategies of their rivals.

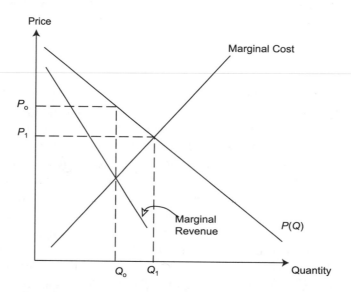

Figure 2.2: Monopoly.

Some of the markets for cocoa and chocolate are relatively competitive, while others are dominated by firms or government agencies that are large enough to exercise market power in one form or another. In some cases, price-taking producers face private firms or government marketing boards that are able to act as monopsonies. In several West African countries, government marketing boards were set up to prevent private monopsonies from exploiting peasant cacao bean producers. Sometimes these boards are able to fund research and develop infrastructure to the advantage of the producers. Often, however, they are used by the government to tax producers by paying prices for the raw products that are very low relative to the prices for which they can market the processed cocoa. Marketing boards for primary commodities were common in low-income countries prior to the policy reforms begun in the 1980s. (The use of a state-run marketing monopsony for cashew nuts in Mozambique is described in the conclusion to Chapter 10.) Much of the cocoa traded between the producing and consuming countries is handled by private traders in markets that are fairly competitive. At higher levels of processing, however, the industry is more concentrated, with small numbers of large firms dominating the market for the final products. The chocolate market is an oligopoly, and prices may be somewhat higher than would be found if the market were perfectly competitive.

Production and Supply

Markets exist to coordinate the exchange of goods and services. The goods and services traded in markets have to be produced and then supplied to the market. An extensive theoretical apparatus has been developed to explain production and supply. The following comments are intended to highlight a few important elements of this theory, recognizing that only a superficial treatment is possible here. The central notion that underlies theories of production is that goods and services are made by combining *inputs* such as labor, capital, and raw materials, all of which cost something. In competitive markets, the prices of these inputs are accurate reflections of their opportunity costs. If markets are not competitive or if there is some other type of distortion, input prices may not accurately reflect the true value of their use for some other purposes. Cases of this nature will be explored more fully in later chapters.

Assuming that markets are competitive, producers will only purchase inputs and produce goods if they can sell the goods for prices that will allow them to cover the costs of the inputs. This statement is strictly true only in the long run. In the short run, a firm may find that it is sensible to produce goods even if the costs during a particular production cycle cannot be covered. Such a situation could arise if the firm anticipated reduced costs or increased prices in the future. In the long run, however, firms will have to cover their costs if they are to stay in business. If the way in which inputs can be combined to produce a good is fixed and invariant, production decisions would be fairly easy. For example, suppose that the production of a ton of cacao beans requires one unit of land, a fixed number of cacao trees, and fixed amounts of fertilizer, labor, and farm equipment. Under these circumstances, producers do not need to decide how much of each input should be purchased and used to produce the cacao beans because the only way any cacao beans can be produced at all is to combine the fixed set of inputs in a particular way. The decision is simply one of determining whether it is reasonable to produce any cacao beans at all and, if it is, what quantity should be produced.

On the other hand, if it is possible to substitute some additional fertilizer for the land needed to produce cacao beans, for example, the producer will have to decide not only whether and how much to produce but how they should be produced. Should the producer use relatively more land or relatively more fertilizer? The decision will depend on the prices of the two inputs. If land is particularly scarce and expensive, it may make economic sense to use a lot of fertilizer on a limited amount of land to grow the cacao beans. Alternatively, if land is abundant and cheap, it may be more profitable to spread limited amounts of fertilizer over a greater land surface. Either combination of land and fertilizer will be efficient. This is a very important result because it suggests that efficiency depends on the prevailing prices for inputs. The efficient way to produce cacao beans in a country with abundant, inexpensive land may not be the efficient way to produce it in a country with scarce, expensive land. From an economic point of view, *input substitutability* means that efficiency depends on the prevailing economic conditions and the efficient way to do something will differ from place to place in line with differences in relative prices.

With the possibility of substituting more of a less-expensive input for one that is more costly, producers are faced with a problem of determining the optimum combination of productive inputs to produce a good or service. This is, in fact, another problem in constrained maximization. (See Box 2.2).

Box 2.2. Constrained maximization of production

The producer's problem is to maximize output subject to a budget constraint that determines how much can be spent on various productive inputs. Mathematically, this problem can be handled using the technique of Lagrange multipliers. Let

$$Q = f(L, F)$$

be a production function for cacao beans showing that the total quantity produced is some function of the amount of land (L) and fertilizer (F) used to produce it. The producer wishes to maximize Q but cannot purchase an unlimited amount of either input. In fact, the producer has a budget constraint represented by

$$M = P_L L + P_F F$$

where M is the budget and P_L and P_F are prices for land and fertilizer respectively. The total amount that can be spent on the two inputs is M. The Lagrange multiplier method requires creating a new, composite equation:

$$\mathscr{L} = f(L, F) + \mu(M - P_L L - P_F F)$$

The maximum for this equation can be found by taking the first partial derivatives with respect to land, fertilizer, and the Lagrange multiplier, μ, and setting them equal to zero. The resulting equations are referred to as first-order conditions:

$$\partial \mathscr{L}/\partial L = \partial Q/\partial L - \mu P_L = 0 \quad ==> \quad MP_L = \mu P_L$$
$$\partial \mathscr{L}/\partial F = \partial Q/\partial F - \mu P_F = 0 \quad ==> \quad MP_F = \mu P_F$$
$$\partial \mathscr{L}/\partial \mu = M - P_L L - P_F F = 0$$

$\partial Q/\partial L$ and $\partial Q/\partial F$ are known as the marginal products (MP) of land and fertilizer, respectively. The first-order conditions show that the optimum is achieved where the marginal products of the inputs are equal to a proportion (given by the Lagrange multiplier) of their prices. This is easier to understand if the first two equations are expressed as ratios:

$$MP_L/MP_F = P_L/P_F$$

The ratio of the marginal products is the marginal rate of substitution between the inputs and is equal to the ratio of the input prices at the optimum. The main lesson to be drawn from this is that the input prices determine the best way to combine inputs that can be used in variable proportions to produce an output.

Producers m.ust choose how much of each input to use to maximize output given that they do not have an unlimited budget for the purchase of inputs. The *optimum combination of inputs* will depend on their relative prices. Once the best input combination is found, it is possible to compare the costs of production with the price that can be obtained for the good or service being produced to determine whether it is profitable to produce at all and, if it is, how much to produce. The assumption that economic agents, including producers, are rational allows economists to represent this latter decision as a problem of profit maximization. The solution for the competitive firm, as shown in Box 2.1, is to produce where the *marginal cost* is equal to the market price.

For price-taking producers, the price is set by market forces. Under ordinary circumstances, marginal costs are an increasing function of output. That is, it is expected that the addition to costs will increase as the changes in production occur at higher levels of output. As long as the price is equal to or greater than the average cost of production (the per-unit cost), the firm will find it profitable to produce. Because higher marginal costs are associated with larger outputs and profit-maximizing firms produce where marginal costs are equal to prices, higher prices mean that more will be produced. Marginal costs can be presented as an upward sloping function relating costs to quantities produced. Such a marginal cost curve, in the zone where costs are greater than the per-unit costs, is the individual supply curve for a given producer. The market supply curve (Fig. 2.3) showing the relationship of quantities supplied to prices received is the horizontal summation of the individual supply relationships. The positive slope means that, at higher prices, more will be brought to the market for sale. Of course, to produce the additional goods and services in response to higher market prices, producers will have to purchase more inputs. Thus, cacao production generates demand for land, labor, fertilizers, and other production inputs as well as supplies of cacao beans, which are themselves inputs into the production of chocolate.

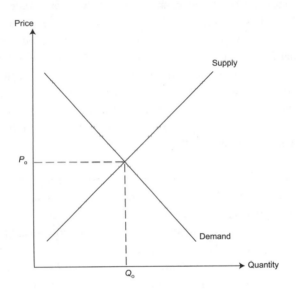

Figure 2.3: Supply and demand in a competitive market.

The *law of supply* shows that market supplies are an increasing function of prices. For empirical analyses, it is often useful to have a measure of the degree to which supplies respond to changes in prices. The slope of the supply curve is one indicator of producer responsiveness. The steeper the slope, the greater the price change required to obtain a given increase in supply. However, the slope of a supply curve is measured in terms of prices and quantities, and it is often more useful to have a measure that reflects changes without reference to price or quantity units. Supply elasticities are such measures. The *elasticity of supply* is defined as the percentage change in quantities supplied given a 1 percent change in the price. Percentage changes are calculated by dividing the change in a variable by its initial value. The percentage change in quantity supplied, thus, is $\Delta Q/Q$ and the percentage change in price is $\Delta P/P$. The ratio of these percentage changes can be rearranged to obtain: $(\Delta Q/\Delta P)(P/Q)$. Replacing the expression $\Delta Q/\Delta P$ with the derivative dQ/dP gives a point estimate of the supply elasticity.

Suppose the relationship between the supply (Q) of cocoa from Ghana to the world market and the world price (P) can be represented as a linear equation. Using data for Q and P between 1961 and 1992,[4] the following linear equation is estimated:

$$Q = 288.03 + 134*P$$

This can be used to calculate an elasticity of supply for Ghanaian cocoa. The derivative of the estimated supply equation, dQ/dP, is 134. The average value of P between 1961 and 1992 was 160 cents per pound, while the average quantity of cocoa supplied in Ghana was 320,000 metric tons. Using these figures, the average supply elasticity can be computed as $134*(160/320,000) = 0.067$. This shows that, on average, a 1 percent increase in the world price gives rise to an increase of Ghanaian cocoa supply of only 0.067 percent. Thus, if Ghana supplied 350,000 metric tons of cocoa last year and the price is expected to increase by 1 percent this year, the estimated elasticity could be used to predict that Ghana would supply an additional 234.5 metric tons ($0.00067*350,000$) for a total of 350,234.5 metric tons. A supply response of this nature is referred to as *inelastic*, which means that the change in supply in response to a price change is relatively small. An *elastic* response means that price variation leads to relatively large quantitative changes. Note that elasticities can be computed for any two variables that are functionally related. We will have occasion to use other kinds of elasticities later.

So far, the discussion has focused on producer decisions, given the prices of inputs and outputs. This type of analysis is referred to as static because it takes prevailing circumstances as given and compares them with hypothetical future situations without consideration of the way in which these circumstances might actually evolve over time. *Comparative static* analyses are useful simplifications that allow a better understanding of complex economic interactions. However, any discussion of production and supply would be incomplete if an important dynamic element, *technology*, were not addressed. Technology can be thought of as the particular ways in which inputs can be combined as well as the range of possible productive inputs available for consideration. In a static setting, the prevailing technologies are given and fixed. These technologies determine the possible input combinations from which producers choose on the basis of relative input prices. Over time, technologies can be expected to change, and it is these technological changes that expand the possibilities for increased productivity and economic growth. For example, FAO data indicate that average cacao bean yields for the world have grown about 15 percent, from 385 kilograms per hectare in 1979-81 to 444 kilograms per hectare in 1992-94. Part of this growth in yields is due to changes in input and output prices that have led to more intensive use of inputs, but technological changes (new varieties, new cultivation methods, better control of diseases and pests, etc.) have played an important role in these increases as well.

Consumption and Demand

In Figure 2.3, a downward-sloping line has been drawn intersecting the upward-sloping supply curve. This line is a demand curve, showing that consumers are willing to purchase greater quantities at lower prices. In Figure 2.3, market equilibrium is established at price P_0 and quantity Q_0 where supply and demand are equal. As in the case of production and supply, an elaborate theory has been developed to explain demand. The theory of demand begins with the idea that rational consumers can be expected to maximize their happiness or satisfaction given the resources at their disposal. Economists often refer to happiness and satisfaction as *utility*, in reference to early links between utilitarian philosophers such as Jeremy Bentham and various nineteenth-century political economists. It is frequently assumed that utility depends on the consumption of various material goods, although it is theoretically possible to incorporate nonmaterial sources of utility into the utility function. The problem faced by consumers is to maximize this utility function subject to the constraint imposed by the fact that the goods consumed are not free and the consumer has a finite budget to allocate across the various utility-producing items. This is another problem in constrained maximization and as such can be solved with the differential calculus.

It will not be necessary for our purposes to work through the mathematics of the consumer problem. Maximizing utility subject to a budget constraint leads to some basic principles concerning the best way to use scarce budgetary resources. The utility a consumer derives from the consumption of a particular good depends on that consumer's tastes and preferences. Some people dislike the flavor of such foods as broccoli or, as in the case of my wife, mushrooms. These individuals will not allocate any of their budgetary resources to the purchase of these items. For others with different tastes, broccoli and mushrooms may be on the menu if their prices are not too high, the prices of other desirable goods are not such that consumers have an incentive to curtail broccoli and mushroom expenditures in order to purchase these other items, and if consumer incomes are sufficient for them to buy these goods along with other necessities. The change in utility obtained from consuming an additional unit of some good is known as *marginal utility*. A consumer's utility is maximized where the marginal utilities of the various consumer goods are equal to a proportion of their respective prices. The marginal rate of substitution in consumption is the ratio of the marginal utilities of two goods and, at the point where utility is maximized, it will be equal to the price ratio for the two goods.

Thus, in the case of demand as well as in the case of supply, rational individuals determine how to allocate their scarce resources across either consumer goods or productive inputs by using the information contained in market prices. For individuals, the solution to the consumer demand problem generates a system of demand equations, with one equation for each available good or service. Each of these equations will show the quantity demanded of a specific good as a function of its own price, the prices of other goods that are *substitutes* (tea is a substitute for coffee) or *complements* (cream and sugar are complements for coffee), and the consumer's income. At the market level, demand for a good such as chocolate is the summation of all the individual chocolate demand equations. As in the case of supply, the responsiveness of consumers to changes in prices can be measured by elasticities. The *price elasticity of demand* shows the percentage change in quantity demanded of a good given a 1 percent change in its price. It is negative because demand is inversely related to price: a price increase will lead to a decline in the quantity demanded. Dand reports estimates of price elasticities of demand for chocolate in Europe that average about -0.2, suggesting that a 10 percent increase in the price of chocolate will lead to a decline in demand of 2 percent.[5]

The *income elasticity of demand* measures the percentage change in quantity demanded given a 1 percent change in income. Income elasticities can be either negative or positive. Income elasticities of demand for chocolate appear to be around 0.7. For most goods, a 1 percent increase in income will lead to an increase in the quantity demanded of somewhere between 0 and 1 percent, as in the case of chocolate, according to these estimates. In some cases, however, a larger response to income increases may be forthcoming. According to one estimate, the income elasticity of demand for housing in rural China is 2.17.[6] This suggests that a 1 percent increase in income will lead to a 2.17 percent increase in demand for housing. The income elasticity of demand for chocolate in the United Kingdom has been estimated to be 1.36. Goods characterized by this kind of response to income variations are referred to as *luxury goods*.

At the other extreme are goods with negative income elasticities, sometimes known as *inferior goods*. It has been estimated, for example, that rice in China is an inferior good with an income elasticity of demand in 1986 of -0.15.[7] This estimate means that income growth in China of 10 percent is likely to lead to a fall in per capita rice demand of 1.5 percent. Such a result is not surprising given the high levels of rice consumption in China (around 100 kilograms per person per year). As Chinese consumers become wealthier, they will allocate more of their budgets to other goods rather than adding to their already high consumption of rice. Individual rice consumption will decline as consumers seek to diversify their diets by consuming more livestock products, fruits, and vegetables.

World Markets and Trade

The decisions of rational producers and consumers give rise to the supply of and demand for particular goods and services. The confrontation of supply and demand in a market is expected to lead to an equilibrium at a price consistent with the quantity exchanged. However, as noted earlier, markets can be defined at many levels of aggregation. The local market for haircuts in Chicago and the U.S. market for soybeans are examples of markets at different levels of aggregation but still within a single country. The world cocoa market, on the other hand, involves a great many countries. Does this fact make any difference? Is there something about *international trade* that requires different analytical tools from those used to study domestic markets?

To many, it may seem obvious that there are important differences between international and domestic trade. But why exactly should trade in corn between the United States and Japan differ from trade in corn between Nebraska and Minnesota? One reason might be that Nebraska and Minnesota are, in some sense, more similar to each other than are Japan and the United States. Citizens of the same nation are often thought to have similar tastes and similar economic resources, and these national characteristics seem to differentiate them from citizens of other nations. Moreover, distances between locations within a nation are usually smaller than the distances between locations in different countries so that transportation barriers are greater with international trade than with domestic exchange.

The problem with these observations is that they are not entirely true. Large sections of eastern Canada are in closer geographic proximity to New York than is the entire state of California. The variation in tastes, preferences, and economic resources within China may be just as great as similar variations across the countries of Western Europe. While it is true that there are identifiable differences in culture and national characteristics across nations, there are similar differences within nations, and these internal differences are not thought to be of great significance in the functioning of domestic markets.

There is one other distinguishing characteristic of nations that may be of greater importance in thinking about trade. The governments of different countries operate diverse policy regimes, and it is this variation in the institutional settings that may justify treating nations as relatively homogeneous units interacting in the international arena. The exchange of goods and services within a nation is driven by differences in local economic conditions. Dairy producers in New England import feed grains from the Midwest because the New England climate and soils are not suitable for corn production. The same is true, however, for international trade. Dairy producers in eastern Canada also import feed grains from the U.S. Midwest and for the same reasons that the New England farmers do. The only difference is that in this latter case, the grain crosses a national border and is subject to different government regulations concerning its handling, transportation, storage, taxation, and exchange. If the government regulations of all countries were the same or if an invasion from outer space forced planet Earth to set up a single world government with unified policies in order to present a united front to the invaders, there would be no need to distinguish between international and domestic trade.

In fact, there are many national governments with different policy regimes, and there has been a long history of economic enquiry concerning the reasons why nations trade and whether international trade is good or bad for them. Because differences in policy regimes are important distinguishing features of trading nations, it is not surprising that much of the study of international trade has focused on policy, notably the advantages and disadvantages of implementing policies to protect domestic producers from foreign competition as opposed to allowing goods to flow freely between countries. In attempting to understand why nations trade, the early political economists developed an extremely important concept that can actually be applied to both domestic and international exchange. This is the concept of *comparative advantage*, which is usually attributed to the nineteenth-century banker and economist David Ricardo, although others had articulated similar concepts at about the same time. According to this concept, goods will be exchanged between regions of a country or between countries if there are differences in relative prices that allow individuals to purchase a good more cheaply from another region or country than it can be found locally.

Suppose there are two countries (the United States and the Rest-of-the-World) producing and consuming two goods (food and clothing). Let P_{UF} and P_{UC} be the prices of food and clothing, respectively, in the United States, while P_{RF} and P_{RC} are food and clothing prices in the Rest-of-the-World. Comparative advantage is determined by examining the relative prices for food and clothing when there is no trade between the two countries, a situation referred to as *autarky*. The United States has a comparative advantage in the production and trade of food if, in autarky, $P_{UF}/P_{UC} < P_{RF}/P_{RC}$. In other words, the United States has a comparative advantage in the good with a lower relative price as compared to the Rest-of-the-World. The Rest-of-the-World has a comparative advantage in the other good, clothing.

This may seem somewhat obscure, but it is really a matter of common sense. Consider the example of New England and Canadian dairy farmers. In autarky, these farmers would have to use some of their resources to produce feed grains. The opportunity cost of producing these feed grains is the forgone dairy production resulting from the diversion of land, labor, and capital from the dairy activities to growing feed grain crops. If the farmers can obtain feed grains at a lower cost from the U.S. Midwest, they can put their resources to better use producing dairy products as opposed to producing feed grains. Trade is beneficial because producers are able to use their resources in more productive activities. Cacao trees do not grow naturally in temperate climates. Without trade, it would be possible to produce cacao beans in the United States, but the trees would have to be grown in greenhouses. This would be very costly because resources would have to be withdrawn from more productive activities and used to grow cacao trees. The result would be very expensive chocolate as well as fewer and therefore more expensive goods that could have been produced with the resources diverted to cocoa production. The well-being of U.S. citizens is greater if the goods produced domestically are those in which the United States has a comparative advantage while the goods that are imported are those in which the United States has a comparative disadvantage.

Comparative advantage explains specialization and trade on the basis of differences in relative autarky prices. The question of why relative autarky prices differ remains, however. Much of international trade theory is devoted to explaining this difference. One of the most prominent explanations of comparative advantage is the *factor proportions* theory first developed by two Swedish economists, Eli Heckscher and Bertil Ohlin. The Heckscher-Ohlin model explains differences in relative autarky prices with reference to differences in national factor endowments. The expression "factors of production" is roughly equivalent to the word input used in previous discussions. The classic factors of production are land, labor and capital. Heckscher and Ohlin noticed that nations generally have different amounts of these factors. The United States has more land than Belgium, while China has more labor than either Belgium or the United States. In addition, the production of different goods requires different amounts of these factors. For example, the ability to produce agricultural products is particularly sensitive to the amount of arable land in a country. Likewise, the production of textiles requires relatively more unskilled labor than the production of, say, computer chips. In the language of international trade, goods differ in their *factor intensities* so that agriculture, for example, is land-intensive while textiles are labor-intensive.

Based on these observations, Heckscher and Ohlin showed that countries that have large endowments of a particular factor will have a comparative advantage in the production and trade of goods, the production of which requires the intensive use of that factor. The reason for this is that the abundant factor will be relatively inexpensive, so the costs of production for goods produced through intensive use of that factor will be lower than in countries where the factor is scarce and expensive. Thus, the United States has a comparative advantage in agricultural goods because these goods are land-intensive and the United States is well endowed with land. Likewise, China has a comparative advantage in textiles because production of these goods is labor-intensive and China has abundant unskilled labor.

When taken to its logical extreme, the Heckscher-Ohlin model suggests that international trade in goods is really a method for evening out the distribution of productive factors around the world. One of the assumptions underlying the Heckscher-Ohlin model is that factors are not internationally mobile. If land, labor, and capital cannot move internationally, countries that have small supplies of capital, for example, will import capital-intensive goods, and these imports will compensate for the effects of scarce domestic capital. In this sense, international trade is really the exchange of the productive factors that are embodied in traded goods.

There are several other explanations of comparative advantage, but a full description of these explanations is beyond the scope of this chapter. The important result of the preceding analysis is that basic economic principles, such as opportunity cost, can be used to understand both international and domestic exchange. It is interesting to note that it is the differences between countries or regions within countries (differences in factor endowments, climate, tastes, and so on) that give rise to the possibility of beneficial exchange. Because countries differ in certain fundamental ways, there is scope for specializing to take advantage of particular characteristics while trading with other regions to obtain the goods and services lost through this specialization. One implication of international trade is that national economic systems do not exist in isolation from the rest of the world. Consider, for example, the market for chocolate candy in the United States. U.S. firms (e.g., Hershey) and the U.S. subsidiaries of foreign-owned firms (e.g., Cadbury Schweppes) import cocoa and other ingredients, combine them with some locally produced ingredients (e.g., milk), and market the chocolate candy in the United States and in other countries, such as Canada and Mexico. The market price for chocolate is determined by the economic conditions in all the countries involved in the production and processing of this product.

Producer and Consumer Surplus

Much of the preceding discussion can be brought together by introducing the concepts of producer and consumer surplus. In Figure 2.4, a simple model of the U.S. market for chocolate candy is illustrated. The supply schedule shows the quantities of chocolate that will be offered for sale at different prices. This schedule is generated by domestic chocolate producers who purchase domestic and foreign raw materials, labor, capital, and other inputs and combine them to produce chocolate. The demand schedule is made up of individual demands for chocolate and shows the quantities that will be purchased at different prices. In autarky, an equilibrium is established with the quantity Q_a being exchanged at price P_a. For some producers, this price may seem too low, while others find that it is high enough to make their activities profitable. A producer's attitude toward the price will depend on her costs of production. If for some reason the producer price were lowered to P_1 in Figure 2.4, some producers would still be willing to supply chocolate. Less would be offered for sale, but there would still be some chocolate supplied at this lower price. Thus, chocolate production is still profitable for some producers at a price of P_1.

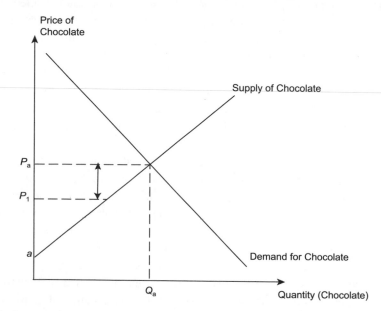

Figure 2.4: Producer surplus.

However, the equilibrium price in the market is not P_1 but P_a. Those who would be willing to sell at the lower price benefit from being able to market their chocolate at the higher equilibrium price. The difference between P_1 and P_a is a measure of the per-unit benefit these producers obtain. If this exercise is carried out for every possible price between P_a and the intercept of the supply schedule on the P-axis at a, it will give a measure of the total benefits to all producers of being able to sell chocolate at P_a rather than at lower prices. This measure is known as *producer surplus*, which is defined as the area above the supply schedule and below the price producers receive. For the chocolate industry as a whole, total revenue is equal to the price received per unit times the number of units sold. At equilibrium in autarky, this is P_aQ_a. Recall that the supply schedule is made up of individual marginal cost schedules. In a sense, the supply schedule measures the industry's marginal costs. The marginal cost schedule is the derivative of total costs, so the integral of the marginal costs gives back the total cost function (with the constant of integration representing fixed costs). Since the integral of a function is the area under that function, the area under the supply schedule is a measure of total variable costs. Subtracting these costs from the total revenues leaves the area identified as producer surplus. Thus, producer surplus is a measure of total industry profits.

A similar analysis can be conducted for consumers. At prices above P_a, some consumers would still wish to purchase chocolate. These consumers benefit from being able to buy at P_a rather than at higher prices. The total benefits to consumers of being able to buy at the equilibrium price as opposed to higher prices is called *consumer surplus*, which is measured by the area below the demand schedule and above the price consumers pay. These two measures are jointly referred to as "economic surplus." They can be thought of as measures of economic welfare reflecting the benefits of being able to buy and sell chocolate in the market. In Figure 2.4, P_a is the autarky price. If this market is opened to international trade, chocolate will become available at the world price. Note that in all of the market analyses in this section, transportation costs are ignored. In fact, prices will always differ between locations by the costs of transporting the goods from the place where they are produced to the place where they are consumed. It is a convenient fiction, however, to ignore these costs for the present. The basic principles derived under the assumption that there are no transportation costs are not changed when such costs are reintroduced in more realistic models. With this assumption, the movement from autarky to trade means that the relevant price in the U.S. chocolate market is the world price rather than the autarky price.

Suppose that the United States does not have a comparative advantage in chocolate, so the world price is below the autarky price at, say, P_w in Figure 2.5. At this price, consumers will wish to purchase a larger quantity of chocolate, while U.S. producers will find that the profitable amount of chocolate to produce is less than when the price was higher. In Figure 2.5, consumers purchase the quantity Q_D while producers supply quantity Q_S. The difference between Q_D and Q_S is imported from chocolate firms in other countries. In this case, opening the market to foreign suppliers will increase consumer surplus. The relevant price in this market is P_w and consumer surplus, the area below the demand schedule and above this price, is now larger by the areas marked A and B. Conversely, producer surplus, the area above the supply schedule and below this new price, has been reduced by area A. From the point of view of society as a whole, the reduction in producer surplus has been transferred to consumers. If the welfare of producers and consumers is weighted equally so that a dollar for consumers has the same value as a dollar for producers, the transfer of area A from producers to consumers has no effect on total social welfare. However, the addition of area B to consumer surplus represents a gain to society. In other words, opening the market to trade has resulted in increased overall welfare even though producers are worse off.

Suppose that lobbyists for the U.S. chocolate industry successfully lobby the government for protection from foreign competition. The government could apply an import tariff to imported chocolate that would raise the price received by producers. Let T represent the value of the tariff. Then the price in the U.S. chocolate market will be P_w+T, as shown in Figure 2.6. The tariff raises the price for both producers and consumers. At this higher price, consumers purchase less chocolate than they would with free trade, while producers have an incentive to offer more for sale. In terms of social welfare, consumers lose the areas marked A, B, C, and D. Area A is transferred from consumers to producers while area C is the tariff revenue collected by the government (the value of the tariff, T, times the quantity of chocolate imported, $Q_D' - Q_S'$). For society as a whole, areas A and C are transfers from consumers to producers or the government while areas B and D, known as deadweight losses, simply disappear. Deadweight losses are efficiency losses due to the market distortion of the tariff and reflect inefficiencies resulting from the use of resources at levels that are not consistent with their full opportunity costs.

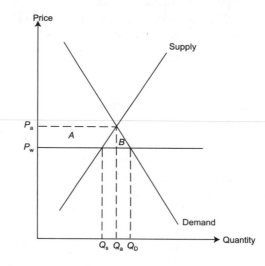

Figure 2.5: Gains from trade.

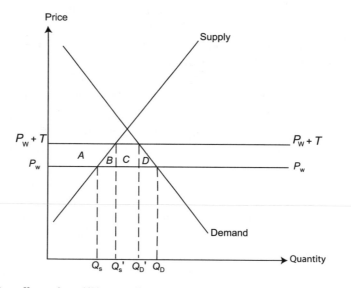

Figure 2.6: Welfare effects of a tariff in a small country.

This final example brings us back to the central focus of this book, the evaluation of public policy. The introduction of an import tariff is a distortion of the market that masks the true opportunity costs of resources used to buy and sell goods such as chocolate. The efficiency losses stemming from this distortion reduce total social welfare even though producers have been made better off. The costs of the tariff for consumers are much greater than the benefits to producers. If it is assumed that the tariff revenues (area C) are returned to consumers in the form of reductions in other taxes, consumers still lose the part of consumer surplus transferred to producers as well as the efficiency losses. These are the bare facts of the import tariff illustrated in Figure 2.6. They would seem to show that imposition of such a tariff is not a good idea. However, there may be other considerations that should be brought to bear on the decision. For example, if the industry being protected is of great strategic significance for a country (e.g., producers of military equipment in the United States), some might argue that it is more important to insure the health of this industry than it is protect the welfare of consumers (in this case, taxpayers who fund the Pentagon). These are the kinds of questions we will need to explore in the rest of this book.

SUMMARY

1. Four concepts that are basic to the economic approach are economic rationality, methodological individualism, equilibrium, and opportunity cost.

2. Market structure ranges from perfect competition, in which all agents are price takers, to perfect monopoly/monopsony, in which some agents are price setters. Between these ideal types are various oligopolistic structures in which outcomes depend on relative bargaining power.

3. Production usually involves deciding on the optimal combination of inputs as well as the optimal output. These are problems in constrained maximization.

4. Supply is derived from production costs. Technological change can shift supply relationships.

5. Market demand is the horizontal summation of individual demand curves that depend on prices and income.

6. Elasticities are measures of responsiveness. They show the magnitude and direction of changes in one variable that are likely to follow changes in a related variable.

7. International trade differs from domestic exchange because of national differences in policy and institutions. Comparative advantage based on different factor endowments and different factor intensities is one explanation for trade.

8. Consumer and producer surplus are welfare measures that show the advantages for producers and consumers of being able to trade at a market equilibrium price.

KEY CONCEPTS

economic rationality
methodological individualism
equilibrium
arbitrage
partial/general equilibrium
opportunity cost

market structure
competitive markets
price taker
monopoly/monopsony
bilateral monopoly
oligopoly

inputs/input substitutability	substitutes/complements
marginal product	price/income demand elasticity
optimal input combination	luxury/inferior goods
marginal cost	international trade
law of supply/demand	comparative advantage
elasticity of supply	autarky
elastic/inelastic	factor proportions/intensities
comparative statics	producer/consumer surplus
technology	economic surplus
utility/marginal utility	

DISCUSSION QUESTIONS

1. Explain the concept of economic rationality and show how this concept can be represented mathematically. Do you feel that Mother Theresa was economically rational? Why or why not?

2. Do you think that groups can have interests that are independent of the people making up the group, or are group interests simply the sum of the interests of the individuals in the group?

3. Explain the concept of comparative advantage and show how it is related to the concept of opportunity cost.

4. Technological change can shift out the supply schedule and, if demand is unchanged, will normally lead to a fall in the equilibrium price. Explain whether producers are better off or worse off as a result of this change. Can technological change make anybody worse off?

5. Is it better to be a price taker or a price setter? Explain.

SUGGESTIONS FOR FURTHER READING

Case, Karl E., and Ray C. Fair. *Principles of Microeconomics*. New York: Prentice Hall, 1998.

Minkiw, N. Gregory. *Principles of Microeconomics*. New York: Harcourt Brace, 1998.

Rhoads, Steven E. *The Economist's View of the World*. New York: Cambridge University Press, 1985.

Schelling, Thomas C. *Micromotives and Macrobehavior*. New York: W. W. Norton & Co., 1978.

NOTES

1. Adam Smith, *The Theory of Moral Sentiments* (1759; reprint, ed. D. D. Raphael and A. L. MacFie, Indianapolis: Liberty Fund, 1982).

2. Thomas C. Schelling, *Micromotives and Macrobehavior* (New York: W. W. Norton & Co., 1978), 25.

3. The information on cocoa and chocolate used in this chapter is taken from Robin Dand, *The International Cocoa Trade* (Cambridge, Eng.: Woodhead, 1993).

4. Production data are from Food and Agriculture Organization of the United Nations, *Production Yearbook* (Rome: FAO, 1995). Price data are from International Monetary Fund, *International Financial Statistics Yearbook* (Washington, D.C.: IMF, 1995).

5. Dand, *International Cocoa Trade*, p. 173.

6. Shenggen Fan, Eric J. Wailes, and Gail L. Cramer, "Household Demand in Rural China: A Two-Stage LES-AIDS Model," *American Journal of Agricultural Economics* 77, no. 1 (February 1995): 54–62.

7. E. Wesley F. Peterson, Lan Jin, and Shoichi Ito, "An Econometric Analysis of Rice Consumption in the People's Republic of China," *Agricultural Economics* 6 (1991): 67–78.

3

Economic Foundations for Public Policy

INTRODUCTION

Welfare economics provides the theoretical foundation for the economic analysis of public policies. Most public policies will have repercussions for economic performance, although in some cases such effects are not their primary purpose, and their economic impacts may be limited. Foreign policy, for example, may be thought of primarily as the public effort to maintain peaceful relations with a country's neighbors. Its primary aim is not to increase economic growth, but because it requires the use of resources and may influence such economic variables as foreign trade or currency exchange rates, it is not unrelated to the performance of a nation's economic system. Likewise, public policies that are focused mainly on the economic system cannot be thought of as entirely independent of other social goals such as racial harmony or world peace. This book deals primarily with the set of public policies that target economic performance, although much of the theory discussed in this chapter as well as the analytical methods described in the next part of the book provide important insights in thinking about public efforts to achieve social goals less directly connected to the economic system.

As noted in Chapter 1, public policies are often given substance through legal instruments that redefine property rights, alter individual incentives, and constrain the use of resources. In pursuit of environmental policy objectives, for example, laws may be passed that tax, regulate, or prevent certain activities. Thus, to reduce nitrate contamination of groundwater (see Chapter 9), laws regulating the use of nitrogen fertilizer may be adopted. In some instances, the effects of such policies are not very widespread, and it is reasonable to limit the analysis to one market, that is, to do a partial equilibrium analysis, as defined in the last chapter. It is often the case, however, that public policies will affect several interrelated markets or broader economic systems. Thus, regulations on the use of nitrogen fertilizer have an impact on the entire agricultural system, including transportation, food processing, the markets for different crops, the markets for labor and other inputs, and the market for fertilizer. This is a problem in general equilibrium analysis because changes in one market will have repercussions for many other parts of the economy. General equilibrium analysis deals with the way in which a broad economic system responds to changes, including those stemming from public policy initiatives, that cause departures from the initial equilibrium of the system as a whole.

The full economic theory of general equilibrium has a long history and, in its modern version, is one of the most abstract and complex theoretical structures within economics.

Because welfare is best assessed in the context of the economy as a whole, it is customary to begin its study with a presentation of general equilibrium theory. Fortunately, some of the most useful insights of welfare economics can be derived without working through the full general equilibrium apparatus. In fact, our purposes will be well served by beginning with one of the most famous and readily understood propositions in economics, Adam Smith's theory of the *invisible hand*. This theory shows that, under certain circumstances, the pursuit of individual self-interest will lead to the best of all possible worlds with a minimum of collective intervention. The key phrase, however, is "under certain circumstances," and we will discover that the particular circumstances that produce this serendipitous result often are not realized in the real world. The analysis of these departures from ideal circumstances provides a rich foundation for understanding the collective action problems identified in the introduction to this book.

THE INVISIBLE HAND

As noted in Chapter 2, one of the basic assumptions of modern economics is that individuals are rational in the sense that they are expected to do their best to advance their self-interests. This notion has actually been controversial for a long time, and modern critics of the economic approach to social analysis often reject the entire apparatus of economic theory and analysis because they find this assumption unappealing. The origins of this controversy are lost in the mists of time, but it is helpful in thinking about the question of human motivations to begin with Thomas Hobbes, the great seventeenth-century English philosopher. In *The Leviathan*, Hobbes describes the behavior of human beings in their relations with each other as a continual search for the means to satisfy their desires.

> And therefore the voluntary actions, and inclinations of all men, tend, not only to the procuring, but also to the assuring of a contented life; and differ onley in the way; which ariseth partly from the diversity of passions, in divers men; and partly from the difference of the knowledge, or opinion each one has of the causes, which produce the effect desired.[1]

Such an account of human motivation and behavior is not too far removed from the definition of economic rationality presented earlier. For Hobbes, the consequences of these inclinations are that human beings continually seek to assure the satisfaction of their desires by gaining more and more power over those who may affect their lives, and this quest for power will lead inevitably to conflict unless behavior is constrained by some other force.

To explore the implications of this characterization, Hobbes examines what would happen in the absence of some sort of civil constraint on human behavior. He refers to this situation as the "state of nature" in which there is no authority or law to control the passions and the fears of these self-interested human beings. For Hobbes, life in the state of nature is full of strife as people attempt to gain their particular ends, often at the expense of others: "Hereby it is manifest, that during the time men live without a common Power to keep them all in awe, they are in that condition which is called Warre; and such a warre, as is of every man against every man."[2] This war of all against all precludes any possibility of human flourishing, as every individual is engaged exclusively in gaining and protecting whatever he can. In one of the most colorful lines in the English language, Hobbes concludes that life in this state of nature is "solitary, poore, nasty, brutish, and short."[3] For Hobbes, the way out of these horrible circumstances is for human beings, who are endowed with greater intelligence than their equally brutish cousins of the animal kingdom, to use their reason to devise and enter into a social contract with a sovereign power (the monarch) that can serve as the needed force to constrain their destructive behavior.

Hobbes's view of human motivation seemed excessively harsh to many people, and it was not long before alternative descriptions of human nature began to be offered. Albert Hirschman provides an interesting description of how various seventeenth- and eighteenth-century thinkers developed the notion of countervailing "passions" as an alternative to the subjugation of humans to the absolute authority of the sovereign.[4] According to these accounts, certain passions might serve the purpose of constraining the exercise of other passions which if left unchecked could give rise to a Hobbesian state of nature. Ultimately, Hirschman suggests, the passion that ended up playing the role of behavioral controller was greed. In the pursuit of material wealth, individuals would find that their personal ends are best served by controlling the passions that if left unchecked would lead to conflict. In this way, the pursuit of self-interest becomes a way to insure social harmony without recourse to an authoritarian monarch. The culmination of this line of reasoning is Adam Smith's famous statement in *The Wealth of Nations*:

> As every individual, therefore, endeavours as much as he can both to employ his capital in the support of domestic industry, and so to direct that industry that its produce may be of the greatest value; every individual necessarily labours to render the annual revenue of society as great as he can. He generally, indeed, neither intends to promote the public interest, nor knows how much he is promoting it. By preferring the support of domestic to that of foreign industry, he intends only his own security; and by directing that industry in such a manner as its produce may be of the greatest value, he intends only his own gain, and he is in this, as in many other cases, led by an invisible hand to promote an end that was no part of his intention. Nor is it always the worse for society that it was no part of it. By pursuing his own interest he frequently promotes that of society more effectually than when he really intends to promote it.[5]

The implications of this conclusion are of great interest. The pursuit of self-interest turns out to be beneficial, according to Smith, the exact opposite of Hobbes's belief that self-interested individuals would inevitably end up in strife and conflict. In an early passage in his book, Smith suggests, "It is not from the benevolence of the butcher, the brewer, or the baker, that we expect our dinner, but from their regard to their own interest."[6] These observations give rise to very different ideas about social organization and control than is found in Hobbes. Instead of relying on a social contract with an absolute monarch to assure social harmony, the invisible hand operating through the market mechanism can accomplish the same end without the sacrifice of individual freedom inherent in the Hobbesian solution. In other words, according to Smith, not only is it possible for an individual to behave in an economically rational manner, it is desirable that he do so, because the result of such behavior will be to advance the interests of society as a whole even though such an outcome is "no part of his intention."[7]

EQUILIBRIUM IN A COMPETITIVE ECONOMY

The concept of the invisible hand can be stated more formally and more precisely using some of the economic tools outlined in the preceding chapter. The purpose of this exercise is to specify clearly the particular conditions that must hold if the invisible hand of the market is to perform as expected. The implication of Smith's insight is that collective or governmental control of individual behavior may not be needed if the economic system satisfies these necessary conditions. Human interaction can be left up to rational individuals pursuing their own interests with a maximum of freedom but under the gentle guidance of the invisible hand. If the conditions are violated in real economic systems, however, the invisible hand may lead to undesirable outcomes, and, as we shall see, that may open the door for collective intervention to correct the failures of the market mechanism. It should be clear that an understanding of the nature of the ideal economy in which reliance on the invisible hand is possible is of great interest in identifying the situations in real economies where these processes can be expected to break down. It will turn out that these situations will bear a striking resemblance to the brief taxonomy of collective action problems presented in Chapter 1. But there is much work to do before we can develop that result.

It will be useful for this discussion to imagine that we have a very simple economic system to analyze. Very little is lost in making these simplifying assumptions, which allow the basic theory to be presented with diagrams rather than mathematical formulas. So let us assume that the economy to be analyzed is made up of two people (Sam and Mary) and that two goods are produced (food and shelter) through a mysterious process that uses two inputs or factors of production (capital and labor). This imaginary society faces three problems:

1. how much of the two factors to use in the production of each good
2. how much of the two goods to allocate to each individual
3. how much of the two goods to produce

All economic systems have to solve these problems. Because resources are not infinite, it is not possible for this economy to produce enough food and shelter to completely satisfy the wildest desires of Sam and Mary, assumed to be unbounded at least over the relevant range. As is often the case with abstract economic reasoning, many important elements, such as money or the ownership and supply of the labor and capital, have been left aside (they will be brought back into play in Chapter 13). Leaving them out makes the discussion seem unrealistic but has the advantage of allowing us to focus on the most important issues in the present context. These issues are the allocative and distributional questions we have enumerated. There are many feasible answers to these questions. Recall, however, that we wish to understand how the invisible hand of the market might lead to the best solution to these problems. In other words, how can spontaneous, voluntary market transactions lead not only to a feasible outcome but, more importantly, to the best one from among the many that are feasible?

Clearly, further consideration of this problem requires some definition of what we mean by "best." Note that the range of possible outcomes has already been circumscribed somewhat by the use of the word "feasible" in the previous paragraph. I take this to mean that the best outcome cannot be some kind of utopia because such states of the world are not included in the set of feasible outcomes. Those who believe that utopias are possible will find this assumption unnecessarily limiting. However, it is consistent with the main elements of the intellectual tradition in economics and the social sciences and will be maintained in what follows. What is sought, then, is not utopian perfection but a social optimum. This social optimum might be the solution that maximizes some set of goals subject to the fact that resources are finite. This is a constrained maximization problem very similar to the producer and consumer problems outlined in the preceding chapter. Economists are comfortable with this kind of analytical approach, and one might expect that they would define the social optimum in this way. Thus, for example, the best outcome might be achieved where the nation's gross national product (GNP) is maximized subject to the amount of land, labor, and capital available in that nation. Another possibility would be the solution that maximizes the total amount of consumer and producer surplus generated by the economy, subject to the constraints imposed by existing supply and demand schedules.

These and other approaches to the definition of a social optimum are undermined by one of the great conundrums in the history of economic thought. Given two states of the economy, it will almost always be the case that some individuals will be better off in one state than the other. Suppose that the maximum for the sum of producer and consumer surpluses or the size of the GNP in a particular economy is achieved by eliminating small farms and using those resources to produce other goods. If the owners of the small farms suffer a catastrophic loss in happiness or utility, can it be concluded that this is a social optimum even if everyone else in society is much happier? What if everyone is only a little bit happier than before but, because there are so many of them and so few owners of small farms, total utility (the sum of the small increases in happiness of a large number of people minus the sum of the large decreases in utility of a small number of people) has increased? The problem with these cases is that they require comparing the pleasures or pains of some individuals with the pleasures or pains felt by others. Many consider it impossible to find a way to measure the pain of my toothache so that this pain can be compared with the discomfort of your runny nose.

The way around the problem of interpersonal utility comparisons most commonly adopted by economists was suggested by an Italian economist, Vilfredo Pareto (1848–1923). *Pareto optimality* is the notion that a social optimum is achieved if no individual can be made better off without making someone else worse off. If through some rearrangement in the distribution or the use of resources and goods, someone will end up at a higher level of satisfaction while no other person will end up at a lower level, such a rearrangement is an unambiguous improvement (as long as we rule out envy and animosity). It is unambiguous because it does not require that changes in the utility of one individual be weighed against changes in the utility of others. Consider two states of the economy, one in which no further changes that will improve the well-being of one individual without harming some other individual can be made, and one in which there are still possibilities for such changes. Pareto optimality suggests that the former state is superior to the latter and that the best policy in the second state would be to make the change that will result in a Pareto optimum.

Pareto optimality is applied to individuals who can experience increases or decreases in utility. A similar concept, known as *allocative efficiency*, can be applied to the use of productive resources to produce goods and services. Efficiency in the allocation of resources is achieved when it is impossible to change the employment of these resources so that the output of one good increases while production of other goods remains unchanged. The economic system is operating inefficiently if resources can be reallocated so as to increase the amount of one good that is produced and there is no fall in the output of any other good. One way to think about this is to consider unemployed labor. If labor is unemployed, society is not getting as much output as it could. Likewise, if some workers are employed but are not contributing anything to total production, they can be withdrawn from that activity and put to work to increase output in some other industry without causing a fall in production in the industry in which they were previously employed.

These two concepts, Pareto optimality and allocative efficiency, can be used to define the social optimum used in welfare economics. The best situation for the economic system is the one that satisfies both criteria. In other words, a social optimum in our simplified economy would be achieved when no change in the use of capital and labor can be made that would not reduce the output of either food or shelter, when no change in the amount of food and shelter that Sam and Mary have can be made without reducing the well-being of one of these individuals, and when the amounts of food and shelter produced cannot be changed without making someone worse off. Under these circumstances, the economy is turning out as much food and shelter as it can and these goods are shared between Sam and Mary in a way that cannot be improved upon. This, then, is the economic state that is to be realized by the invisible hand of the market.

But the invisible hand can accomplish this feat only under very specific conditions. In particular, the market will lead to Pareto optimality and allocative efficiency if it is perfectly competitive, there is perfect information and foresight, and the costs of reallocating resources are negligible. If these conditions are not met, the invisible hand may go astray. These problems will be addressed after we see how the invisible hand works in a market setting that satisfies these basic conditions.

A Pareto Optimal Distribution of Food and Shelter

Sam and Mary each derive utility from the consumption of food and shelter. Their tastes may differ, but it is assumed that marginal utilities for both individuals and both goods are positive. We must assume that the *utility functions* of the two individuals satisfy certain technical characteristics so that we can be sure that the consumer maximization problem described in the previous chapter is feasible. This assumption is not particularly restrictive and can be expected to be met in most real-world situations. For each individual, an indifference mapping showing the combinations of the two goods that generate various levels of utility can be drawn. *Indifference curves* for Sam and Mary are shown in Figures 3.1a and 3.1b.

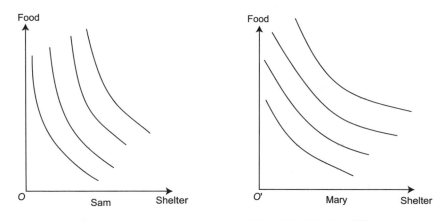

Figure 3.1a: Sam's indifference curves. Figure 3.1b: Mary's indifference curves.

The vertical axes in both graphs register quantities of food that may be consumed, while the horizontal axes measure quantities of shelter. At the origins, 0 and 0', these quantities are zero. Because units of the same goods are measured on these vertical and horizontal axes, it is possible to use a trick to combine Figures 3.1a and 3.1b. Imagine that Mary's utility mapping is rotated to the left until it is upside down and placed on top of Sam's utility surface. The result, referred to as an *Edgeworth-Bowley box* after two economists who devised this technique, is shown as Figure 3.2. Note that the origin for Mary's utility map (Fig. 3.1b) is now in the upper right-hand corner. Mary's food consumption is measured from the top to the bottom while Sam's food consumption is measured from bottom to top. Likewise, Sam's shelter consumption is measured from left to right on the horizontal axis while Mary's is measured from right to left.

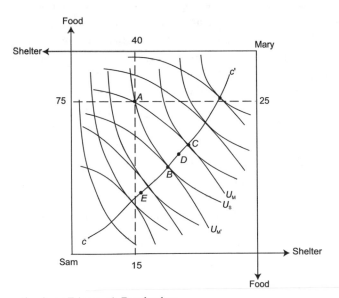

Figure 3.2: Consumption in an Edgeworth-Bowley box.

Any point in the Edgeworth-Bowley box shows the amounts of food and shelter consumed by each person. Consider the point labeled A. At that point, Mary is consuming 40 units of shelter and 25 units of food, while Sam is consuming 15 units of shelter and 75 units of food. At these consumption levels, Sam realizes utility of U_S while Mary has utility of U_M. Note that the indifference curves in Figure 3.2 have been drawn in such a way as to trace out the points where they are tangent. These points have been connected to obtain the *contract curve*, labeled cc'. If food and shelter were distributed as at point A, the economy would not be at a Pareto optimum because it would be possible to make one person better off without lowering the utility of the other. To see this, consider the distribution at point B. At this point, Sam consumes less food but more shelter and is still obtaining the same level of utility, U_S. Mary, however, is better off because at point B, her utility has increased to U_M'. Mary's shelter consumption has fallen somewhat, but her food consumption has increased. According to the utility mapping in these figures, the new consumption bundle is more highly valued by Mary, and this is reflected in the move to a higher level of utility.

It should be clear that any points that do not lie on the contract curve, cc', are not Pareto optimal. What is less clear, but of great significance, as will be seen later, is that Pareto optimality is incapable of picking out any particular point on the contract curve. If the economy is initially at point A, a move to point B or point C or any point in between will be a Pareto better change. A move to point D, for example, represents greater economic well-being for both people. However, if the economy is at point B, a move to point C is not a Pareto better change because although Sam's utility increases, Mary is made worse off by that change, and this violates the definition of Pareto optimality. The upshot is that all points on the contract curve are Pareto optimal, and a move to the contract curve from a point such as A to any point on the contract curve between B and C is a move to Pareto optimality. Such moves can be considered improvements because one person is made better off while the other is left no worse off than in the initial situation, and it is even possible in some special cases for both individuals to realize a higher level of utility. Moving from A to E, however, is not Pareto optimal because while E is on the contract curve, Sam is on a lower indifference curve at E than at A.

So far, no mechanism for causing movements within the Edgeworth-Bowley box has been described. What might cause Sam and Mary to reallocate their consumption from a point such as A to one such as B, C, or D? More interestingly, what does this have to do with invisible hands? To answer this, we need to recall the consumer decision problem described in the preceding chapter. It was shown there that consumers maximize their utility by choosing the consumption bundle that equates the slope of the indifference curve with the slope of the price ratio. The slope of the indifference curve is the *marginal rate of substitution* (MRS). Because arbitrage prevents the existence of multiple prices for the same goods at the same time, there is one common price ratio in this economy. In other words, both Sam and Mary face the same ratio of shelter and food prices. If both are rational, they will each choose to consume where their MRS is equal to the price ratio. But since this price ratio is the same for both, the result will be that Sam's MRS will end up being equal to Mary's (because both are equal to the price ratio). Note that the points on cc' are defined by the tangencies of the two sets of indifference curves. But the slope of a curve is equal to the slope of another curve at the point where these two curves are tangent. Thus, by choosing to consume where individual MRSs are equal to the common price ratio, Sam and Mary will be led "as if by an invisible hand" to choose consumption bundles that lie on the contract curve.

Efficient Allocation of Productive Inputs

The same logic applies to the production side of this economy. The problem to be solved in this context is the efficient allocation of the two factors of production, labor and capital, to the production of food and shelter. The prevailing technologies are reflected in the shape of the *production functions* relating factor inputs to outputs of the two goods. These production functions can be represented by sets of *isoquants* as in Figures 3.3a and 3.3b. The main difference between this problem and the consumer problem described in the preceding section is that output can be measured in physical units while utility can only be measured ordinally. This means that it is reasonable to label the isoquants with hypothetical quantities of output rather than the less precise utility levels U_S and U_M identified in the previous discussion. As might be anticipated, the same trick can be applied to the two production surfaces to obtain the Edgeworth-Bowley shown in Figure 3.4.

The allocation of labor and capital to the production of the two goods is considered efficient if it is impossible to obtain more output of one of the goods through a different allocation without reducing the output of the other good. The contract curve defined by the tangencies of the isoquants shows the efficient combinations of inputs and outputs available to this economy. Suppose that the economy is operating at a point such as the one labeled N. At that point, 10 units of capital and 60 units of labor are being used to produce 50 units of food (as shown by the isoquant from the food production function passing through N) while 50 units of capital are being used with 20 units of labor to produce 80 units of shelter. Note that this implies that this economy is endowed with a total of 80 units of labor and 60 units of capital. If these inputs are reallocated so that 25 units of capital and 50 units of labor are used to produce food, with the rest of the inputs (30 units of labor and 35 units of capital) used to produce shelter (point R), it will turn out that the same amount of shelter is produced while food output has risen to 70 units. At N, resources are not being used efficiently, because it is possible to change their use and obtain more food without giving up any shelter. As in the consumer problem, any point on the contract curve reflects an efficient allocation of the two inputs, and these efficient allocations are better (in the sense that more of at least one of the goods is available) than allocations that are not on the contract curve. The mechanism that causes producers to move to the contract curve is the same as in the consumer problem. Recall that producers maximize profits by producing where the *marginal rate of technical substitution* (MRTS) is equal to the ratio of the input prices (wages for labor and interest for capital). Markets for these inputs are assumed to be competitive so that producers of both goods face the same input prices. As a result, equating the MRTS for food production to the common price ratio while the MRTS for shelter production is equated to the same ratio means that the invisible hand of the market leads producers to a point where the slopes of the isoquants (the MRTS) are equal. In other words, rational profit maximizers are led by the invisible hand to a point on the contract curve which includes all possible resource combinations that are allocatively efficient.

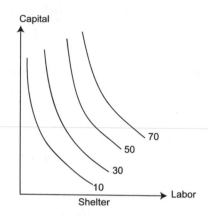

Figure 3.3a: Isoquants for the production of shelter.

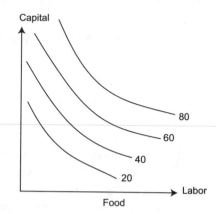

Figure 3.3b: Isoquants for the production of food.

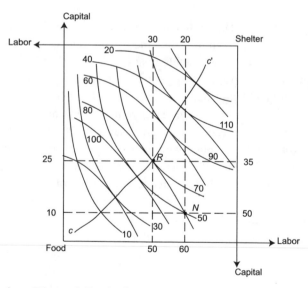

Figure 3.4: Production in an Edgeworth-Bowley box.

The Optimum Level of Output

Rational consumers and producers responding to market prices will automatically attain a Pareto optimal distribution of the two goods as well as an efficient resource allocation. However, it might be possible in this economy for Sam and Mary to attain higher levels of utility if there were a different combination of food and shelter available. Stated differently, even though food and shelter are produced efficiently and the resulting products are distributed optimally, it may be possible to change the quantities of food and shelter that are produced and make at least one person better off. Reaching the optimum combination of outputs is the final problem that needs to be solved. As in the previous cases, this problem will be automatically resolved by competitive market mechanisms.

To show how this will come about, it is necessary to define the concept of a *production possibility frontier* (PPF). The PPF is the locus of feasible combinations of food and shelter that can be produced. Recall that the Edgeworth-Bowley box presented in the preceding section contained a contract curve showing the output combinations that could be produced efficiently. In addition, the isoquants from the two production surfaces reflect the actual levels of output of the two goods. Thus, if the total amounts of labor and capital (80 units and 60 units, respectively) are allocated as at point *R*, then 70 units of food and 80 units of shelter are produced. Suppose we define a graph with the quantity of food produced measured on the vertical axis and the quantity of shelter produced shown on the horizontal axis. Point R could be located in this graph at the coordinates (70, 80). Likewise, every other point on the contract curve could be located in this graph by recording the combinations of food and shelter produced as shown by the isoquants. The result of such a procedure is the production possibility frontier shown in Figure 3.5.

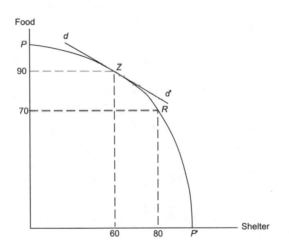

Figure 3.5: Production possibility frontier.

There are some important observations to make about the production possibility frontier. First, all the points on the curve labeled *PP'* in Figure 3.5 satisfy the definition of allocative efficiency. Points lying inside the curve are not on the contract curve and represent inefficient allocations in the sense that more output of one good could be obtained without reducing the amount of the other good that is produced. The portion of the graph to the northeast of *PP'* is beyond the reach of this economy. The resource endowment (80 units of labor and 50 units of capital) is too small to reach any of the output combinations in this zone. If all available resources are used to produce food, total output will be at point *P* on the food axis. If all resources are used to produce shelter, output will be P' on the shelter axis. In between these points, the production possibility frontier bows out. The reason for this shape is that, inevitably, some resources are better adapted to the production of one of the goods. If, for example, the main resource needed for food production is labor, allocating all labor and capital to food production means that some of the capital, which is better adapted to shelter production, is likely to be contributing very little to total output. If that capital is withdrawn along with a small amount of labor and used to produce shelter, the result will be only a small decline in food production, although shelter output may be substantial. If the process of removing resources from the food sector and transferring them to shelter production continues, it will eventually be the case that labor, which is better adapted to food production, is forced into the shelter sector so that a fairly large decrease in food output leads to only a small increase in shelter production.

These considerations give rise to the expectation that production possibility curves will have the shape shown in Figure 3.5. The slope of PP', the *marginal rate of transformation* (MRT), shows the rate at which the economy can transform food into shelter by changing the use made of its endowment of labor and capital. Note that this slope changes along the production possibility frontier. Suppose that the economy is operating at point Z on the PPF. The MRT is shown by the line dd' that is tangent to the PPF at Z. The slope of dd' shows how much food will have to be given up for a marginal increase in shelter production. In other words, the MRT reflects the marginal opportunity cost of changes in the combination of food and shelter produced by the economy. If the economy is operating at Z, we know that a total of 90 units of food and 60 units of shelter are being produced. Knowing the amounts of the two goods that are available means that we know the dimensions of the Edgeworth-Bowley box containing the utility mappings of Sam and Mary. In Figure 3.6, this box is drawn inside the production possibility frontier. Because the MRT shows the relative opportunity costs of the two goods, it is equivalent to the price ratio that would arise in a perfectly competitive economy. The optimum distribution of the two goods between Sam and Mary will be found on the contract curve within the Edgeworth-Bowley box at the point where the two marginal rates of substitution are equal to the marginal rate of transformation.

To see why this is so, recall that the MRS measures the rate at which one good can be substituted for another while leaving the individual at the same level of utility. Suppose that Sam is consuming at a point where his MRS is equal to 2. This means that he is indifferent between an additional two units of food and one additional unit of shelter. If the economy can transform a unit of food into a unit of shelter (the MRT is equal to 1), an additional unit of shelter will only "cost" one unit of food. But for Sam, a unit of shelter is worth two units of food. What about Mary? In a competitive economy, both individuals will choose to consume where their MRS are equal. Thus, Mary's MRS will also be 2, and both individuals value additional shelter more than the food they would have to give up to produce it. A change in the combination of goods produced will increase their welfare. The optimum combination will be found where the MRT is equal to the common MRS of the two consumers. At that point, the rate at which the economy can transform one good into the other will be the same as the value Sam and Mary attach to such a transformation.

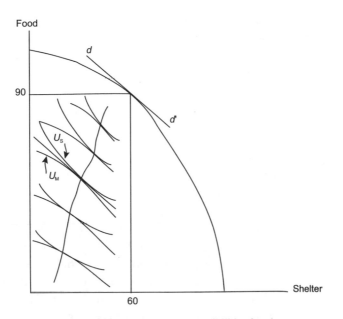

Figure 3.6: Consumption combination within the production possibilities frontier.

The Competitive Equilibrium

The general equilibrium in this economy that is attained through competitive forces will satisfy the conditions of Pareto optimality and allocative efficiency. This outcome will have the following characteristics:

1. The two goods will be distributed between Sam and Mary such that no other feasible distribution can be found that will increase the utility of one without reducing the utility of the other (Pareto optimality). This distribution is found where the MRS of the two consumers are equal. It is reached automatically by rational consumers equating their individual MRS to the price ratio.

2. Labor and capital are allocated efficiently in the sense that no other feasible allocation is possible that would increase production of one good without reducing production of the other. This allocation is realized automatically by rational producers equating the MRTS for each industry to the common price ratio.

3. The combination of goods produced is Pareto optimal in the sense that no other feasible combination can be reached that would raise the utility of one person without lowering the utility of the other. This optimum is found where the equal MRS of Sam and Mary are also equal to the MRT. Note that the MRT is also equal to the common MRTS for food and shelter because its slope is measured at a point on the PPF defined by the contract curve from the production side of the economy.

Pareto optimality and allocative efficiency are attained in this economy by rational consumers and producers facing prices that are fixed and the same for all, with everything that is produced being sold and in the absence of barriers that prevent the entry of new firms or consumers. An economy with these characteristics is known as perfectly competitive. The theory of the invisible hand shows that rational producers and consumers making choices in the context of perfect competition will automatically end up in a situation that is Pareto optimal and allocatively efficient.

THE FUNDAMENTAL THEOREMS OF WELFARE ECONOMICS

The preceding conclusion is also known as the *first fundamental theorem of welfare economics*. Its corollary, the *second fundamental theorem of welfare economics*, states that any of the infinite number of potential Pareto optimal and efficient arrangements of the economy can be attained by adjusting the initial endowments held by individuals and allowing competitive markets to work their magic. This second theorem is of particular importance in thinking about a problem that has been ignored in the preceding discussion. Pareto optimality and allocative efficiency have been taken to define an economic optimum. The problem is that neither of these conditions precludes the possibility that one individual, say Mary, is able to consume enormous amounts of food and shelter while Sam is left with virtually nothing. Suppose that the initial distribution of food and shelter between Sam and Mary is given by point E in Figure 3.2. This point is Pareto optimal (it is on the contract curve) but leaves Sam with very small amounts of both goods compared to Mary's rather extravagant consumption. Any move that would increase Sam's utility from the level realized at point E would reduce Mary's utility, thereby violating the Pareto criterion. To many people, it seems that there is something wrong with an "optimum" that would allow the possibility for such an unfair outcome.

The second fundamental theorem of welfare economics circumvents this problem by noting that changing the initial endowments of each individual will allow this economy to achieve any distribution of well-being that is desired without overriding the operations of competitive markets. In Figure 3.7, two Edgeworth-Bowley boxes have been drawn inside the production possibilities frontier. In each case, resources are allocated efficiently and a Pareto optimal distribution of the two goods can be found where the marginal rates of substitution (MRS_1 and MRS_2) are equal to the marginal rates of transformation (MRT_1 and MRT_2). Note that these points of equality are on the contract curves c_1c_1' and c_2c_2'. The optimal quantities produced are different in these two cases, and the combinations of utilities of the two individuals are different.

If we continue to derive resource and utility combinations for each of the points on PP', we can trace out all of the feasible combinations of utility that can be realized in this economy. The result is referred to as a *utility possibility frontier* (UPF), UU' in Figure 3.8. At any point on UU', Sam's and Mary's MRS are equal to each other and to the MRT at some point on PP'. In other words, all the points on UU' are Pareto optimal and allocatively efficient. Because UU' is presumed to slope downward, an increase in Sam's utility is associated with a decrease in Mary's. UU' has not been drawn as a smooth curve convex to the origin at all points to reflect the uncertain nature of this relationship. While it may make sense to believe that UU' slopes downward, the problem of aggregating utilities is such that little more can be inferred about its shape.

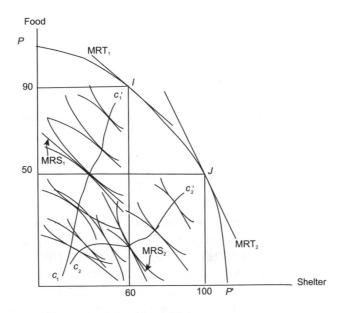

Figure 3.7: Alternative production and consumption equilibria.

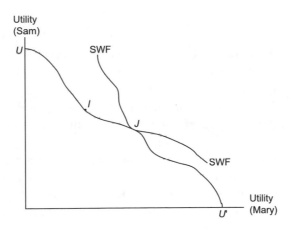

Figure 3.8: Utility possibility frontier.

Clearly, a change from point J to point I in Figure 3.8 violates Pareto optimality. However, if we imagine that Sam and Mary are attempting to reach an agreement over the distribution of utility prior to actually setting the economy in motion, we can think of them as choosing a point on UU' that reflects their collective beliefs about what constitutes a fair distribution. How might they arrive at a decision that J is preferable to I? One very controversial approach to this question is to imagine that there is a *social welfare function* (SWF) defined as

$$SWF = g(U_M, U_S)$$

It is usually assumed that both partial derivatives of this function are positive and that its shape is similar to that of individual utility functions, although, as in the case of UU', this relationship has been shown as a wavy line to reflect the uncertainty about its precise shape.

The SWF is controversial, in part, because it suggests that groups (e.g., "society as a whole") can have utility, and this would violate economists' cherished belief in methodological individualism. Assuming that the social welfare function is a meaningful concept, however, it can be used to explain how point J might be chosen over point I. And once such a point has been selected, the corresponding utility levels, MRS, MRT, and resource allocation from Figure 3.7 can be located. If society can decide on a fair distribution of well-being, it can choose the initial endowments for each individual, and competitive markets can be relied upon to generate the best resource allocation, the best distribution of goods between the consumers, and the best level of overall output. The two fundamental theorems assure us that individuals pursuing their self-interests in a competitive economy will make decisions that allow the entire society to attain a social optimum with no guidance from collective authorities. The implications of these results and some very important qualifications are the subject of the next chapter.

SUMMARY

1. The theory of the invisible hand shows how individual pursuit of self-interest in the context of competitive markets can actually lead to the best possible outcome for society as a whole with a minimum of government intervention.

2. The formal representation of this theory requires a definition of "best possible outcome for society." To avoid problems of interpersonal utility comparisons, economists use the concepts of Pareto optimality and allocative efficiency to define this state.

3. Rational consumers choose to consume where their marginal rate of substitution (MRS) is equal to the common price ratio, and this means that they end up on the contract curve where all the points are Pareto optimal. Rational producers equate their marginal rate of technical substitution (MRTS) to the common input price ratio and are led to the contract curve and allocative efficiency.

4. The optimum combination of outputs is achieved where the marginal rate of transformation (MRT) is equal to the common MRS of the two consumers. Pareto optimality and allocative efficiency are automatically reached by rational producers and consumers facing competitive prices.

5. The first fundamental theorem of welfare economics states that a competitive equilibrium is Pareto optimal and allocatively efficient. The second fundamental theorem of welfare economics shows that a competitive economy can attain any of the potential Pareto optimal and efficient points through adjustment of the initial endowments.

6. If a social welfare function exists, it would be possible to determine an initial fair dis-

tribution of resources from which a competitive economy could reach a point that is not only efficient and Pareto optimal but consistent with social preferences about equitable distribution as well.

KEY CONCEPTS

welfare economics
invisible hand
Pareto optimality
allocative efficiency
utility functions
indifference curves
Edgeworth-Bowley box
contract curve

marginal rate of substitution
production function/isoquants
marginal rate of technical substitution
production possibility frontier
marginal rate of transformation
first and second fundamental theorems of
    welfare economics
utility possibility frontier
social welfare function

DISCUSSION QUESTIONS

1. From similar assumptions about human behavior, Hobbes and Smith derived very different predictions about social coordination. What did they assume, and who do you think is right?

2. The metaphor of the invisible hand can be used for any process that leads to unintended consequences. Can you think of other examples where individuals pursuing some objective end up accomplishing something that was "no part of their intentions"?

3. Explain how rational producers and consumers choose points on the contract curves. What is the role of prices in this process?

4. Suppose that Sam and Mary actually face different relative prices. Would their rational pursuit of self-interest lead them to a Pareto optimal position?

5. Could it ever be possible for some individuals to starve to death in an economy that is at a Pareto optimal point?

SUGGESTIONS FOR FURTHER READING

Baumol, W. J. *Economic Theory and Operations Analysis*. Englewood Cliffs, N.J.: Prentice Hall, 1977.

Eaton, B. C., and D. F. Eaton. *Microeconomics*. New York: Prentice-Hall, 1994.

Kogiku, K. C. *Microeconomic Models*. New York: Harper and Row, 1971.

Schotter, A. R. *Microeconomics: A Modern Approach*. New York: HarperCollins, 1993.

NOTES

1. Thomas Hobbes, *The Leviathan* (1651; reprint, Buffalo, N.Y.: Prometheus Books, 1988), 49.

2. Ibid., 64.

3. Ibid., 65.

4. Albert O. Hirschman, *The Passions and the Interests* (Princeton, N.J.: Princeton University Press, 1977).

5. Adam Smith, *An Inquiry into the Nature and Causes of the Wealth of Nations* (1776; reprint, Chicago: University of Chicago Press, 1976), 477–78.

6. Ibid., 18.

7. Ibid., p. 478.

4

Welfare Economics
and the Role of the State

INTRODUCTION

Although the first and second fundamental theorems of welfare economics presented in the previous chapter are more elaborate than Adam Smith's original statement on the invisible hand of the market, they essentially make the same point. Under conditions of perfect competition, with adequate foresight and information, individuals left to their own devices will act in ways that turn out to be in the best interests of society even though such an outcome is no part of their intentions. If the world were actually characterized by these conditions, there would be little need for government intervention in the economic system. Real economic systems do not fit this idealized model, however, and that fact has implications for the role of government and public policy in improving social well-being. The purpose of this chapter is to explore these implications.

IMPERFECT COMPETITION

Cases where the invisible hand may not lead to the best of all possible worlds are often referred to as *market failures*. A classic example of market failure is an imperfectly competitive market such as a monopoly. In Chapter 2, it was shown that the monopolist maximizes profits by equating marginal revenue to marginal costs where marginal revenue is not a fixed price but a variable that depends on the quantity the monopolist sells. The result is that the monopolist restricts output to raise the market price and earn economic profits (also referred to as monopoly profits or rents). In perfect competition, economic profits are driven to zero as firms enter the market in response to returns above the opportunity cost of capital. The problem for the invisible hand posed by a monopoly can be grasped intuitively by recalling that the social optimum is found where competitive firms hire inputs according to the relation between the marginal rate of technical substitution (MRTS) and the relative factor prices. A monopolist produces less output with fewer inputs than an equivalent set of competitive firms. This misallocation of resources is inefficient because the mix of outputs from the competitive and monopolized industries does not reflect true opportunity costs. Reallocating resources from the overproducing competitive industries to the underproducing monopoly would result in enough additional output from the monopolist to more than compensate for the lowered output from the competitive industries, leaving the society better off.

An interesting example of the monopoly problem is provided by the case of an industry with *increasing returns to scale*. To this point, it has been assumed implicitly that production is characterized by constant returns to scale, that is, an increase of 10 percent in the resources used to produce a good will lead to precisely 10 percent more output. If an industry is characterized by increasing returns to scale, increased resource use will lead to more than a proportionate increase in output, and this means that the per-unit costs of production decline. Such industries will tend to become highly concentrated as individual firms become ever larger to take advantage of the economies of scale. This tendency toward increasing concentration gives these industries their common name, *natural monopolies*. Examples of natural monopolies include such industries as electric utilities or telephone services. It is difficult to imagine a large number of competing firms stringing separate electricity lines across a city. In the United States, technological changes have made it possible to break up the natural telephone monopoly, although competition in this sector is limited almost exclusively to long-distance telephone service while local telephone service is provided by a local monopoly. In the United States, natural monopolies such as electric utilities are often owned by private firms but regulated by the government. In other countries, it is often the case that such industries are nationalized and run by the government. For example, Électricité de France (EDF) is a government-run monopoly that provides electricity in France.

Consider the case of an industry that produces a good using labor as an input.[1] Assume that the production technology, $Y = f(L)$, can be represented by the following equation:

$$Y = \begin{bmatrix} 0 & \text{if } L \leq c \\ a(L - c) & \text{if } L > c \end{bmatrix}$$

where, L is labor, a is a positive constant, and c is fixed costs measured in labor units. This relationship can be used to derive a cost function relating costs of production to output. If output is zero, nothing is produced and there will be no costs. If output is greater than zero, costs are

$$C(Y) = c + 1/a(Y)$$

If $L > c$, $Y = a(L - c) = aL - ac$. Note that ac is a constant equal to the fixed costs of production. Solving this expression for L gives: $L = Y/a + ac/a = 1/a(Y) + c = C(Y)$. This is the cost in labor units shown as a function of output. Note that cost increases as output (Y) rises. Average cost, however, declines as output increases. Average cost is equal to total costs divided by output, $C(Y)/Y$ or $c/Y + 1/a$. Marginal cost, the derivative of $C(Y)$ with respect to Y, is equal to $1/a$. Clearly, average cost is above marginal cost and declines as Y increases, approaching the marginal cost asymptotically. These relationships are shown in Figure 4.1.

Figure 4.1 also includes a negatively sloped demand curve which intersects the constant marginal cost at an output of Y_1. This is the competitive optimum (price equals marginal cost). The problem is that the firm is not covering its average costs selling Y_1 at price $1/a$ and would go out of business. Note that this is a case of increasing returns to scale. The larger the firm (the more output it produces), the lower its average per-unit costs of production. Because average costs are always above marginal costs, marginal-cost pricing would make the firm nonviable. Figure 4.1 illustrates two alternatives. One would be to allow the firm to behave as a monopolist, restricting output to Y_3 where marginal revenue is equal to marginal cost with price at P_3. Another alternative would be to regulate the firm, forcing it to sell at P_2 where average cost intersects the demand schedule. This would allow the firm to cover its costs without earning monopoly profits. A final possibility would be to subsidize the firm's costs so that it could sell its product at a price equal to marginal cost at output Y_1.

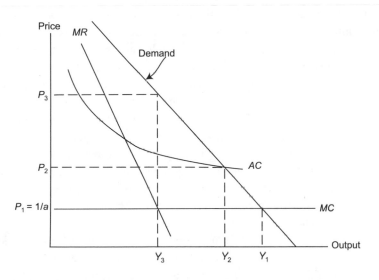

Figure 4.1: Natural monopoly.

Which of these alternatives would be the best for society is difficult to determine. In general, marginal-cost pricing is the best option, but in this case it would be necessary to subsidize the firm if the price is to be set equal to marginal cost and raising tax revenues to fund the subsidy introduces other distortions in the economy that reduce efficiency and social welfare. In any case, rational economic agents pursuing their self-interests in a market would not automatically arrive at whatever alternative is ultimately chosen as desirable. That choice has to be made by society as a whole, and the only way to make and enforce such choices is through government intervention. Thus, the existence of a natural monopoly provides one rationale for government regulation. The invisible hand of the market will not lead to optimal outcomes if markets are not perfectly competitive, and this means that government intervention may lead to greater efficiency. It turns out that imperfectly competitive markets are the rule rather than the exception, and this creates a potentially extensive role for public regulation of the private sector. Competition policy (antitrust laws), regulation, and nationalization of the industry are examples of the types of intervention used to correct these types of market failures.

RISK AND UNCERTAINTY

The conditions required for the effective operation of the invisible hand include not only perfect competition but also perfect information and foresight and the absence of impediments to the reallocation of resources. If it is expensive to shift resources from one use to another, allocative efficiency will be difficult to achieve. Costly resource reallocation may arise in markets characterized by imperfect competition or other structural rigidities. For example, it is generally quite costly for workers to move from one job to another because it takes time to locate a new position, an individual's training may not match the needs of other firms located in the area where the person is searching for work, employer practices may discriminate against certain individuals, and so on. The high cost of searching for jobs means that labor resources cannot be reallocated easily as the economy adjusts to changing circumstances, giving rise to allocative inefficiencies.

Other resources such as land or capital may also be subject to similar problems. Land is tied to particular locations, making it impossible for it to be freely reallocated in response to economic signals. Physical capital such as agricultural machines and buildings cannot be instantaneously transformed into manufacturing plants for computers, textiles, or automobiles. As in the case of imperfect competition, there may be a role for the government to facilitate resource reallocation by regulating discrimination and other practices that increase reallocation costs; job retraining, and other measures to lower reallocation costs; furnishing physical infrastructure; and providing *information* that makes it easier for spatially separated buyers and sellers to find each other.

Market information is needed for producers and consumers to make efficient allocative decisions. But there are other kinds of information that are essential for the smooth operation of a competitive economy. If information on the true characteristics of goods is hidden from consumers, they may not actually be able to achieve a Pareto optimal outcome through the equation of marginal rates of substitution to market price ratios. For example, the harmful bacteria (salmonella, *E. coli*) that may contaminate fresh meat are not visible, and consumers may not be able to rationally reject meat that is unsafe. Some information may be provided by the private sector in the form of advertising, for example, but advertising is aimed at convincing consumers to buy particular products and may not provide full information on the characteristics of the advertised product. The information needed for producers and consumers to make efficient decisions is a type of public good and will be discussed further subsequently.

Even if economic agents have perfect information concerning the characteristics of goods and services produced in a perfectly competitive economy, there is still the problem that decisions can only be made on the basis of expectations about what will happen in the future. The unavoidable nature of the world is that the future can never be known with certainty. The existence of *risk and uncertainty* means that producers and consumers will inevitably make mistakes in their economic decisions. The question that must be addressed here is whether or not these facts generate a need for collective or governmental action. After all, private-sector insurance companies were created for the precise purpose of providing individuals with a means to deal with the inevitable risks and uncertainties of life.

Many government programs are best described as "social insurance" programs. In the United States, for example, the federal government subsidizes disaster insurance for farmers. One explanation for such programs is that the market for disaster insurance or other social insurance programs is "incomplete." Stiglitz defines incomplete markets as markets that fail to provide goods or services even though individuals are willing to pay prices that are greater than the costs of providing the good or service.[2] There are several reasons why a market such as that for disaster insurance might be incomplete. Insurers use probabilities to determine the rates that must be charged to cover the amounts that will have to be paid out to their clients. For example, an insurance company could use information on the frequency of droughts in an area to determine the probability that farmers would lose substantial portions of their crops, and this information could then be used to determine the rate. On the other hand, in some cases, there may be no actuarial data upon which to base the insurance rates. It is conventional to distinguish between situations of "risk," in which the probabilities of alternative outcomes are known, and situations of "uncertainty," in which the probabilities are unknown. In some cases, people may be willing to pay for insurance at rates that would turn out to cover the costs of providing the insurance, but such insurance is not available because of uncertainty surrounding the likelihood of the events against which people wish to insure themselves.

There are other problems with insurance markets that may justify various kinds of social insurance programs. In the case of many disaster insurance programs, the probability that one individual will be afflicted is not independent of the probability that his neighbor will also be harmed. Such is the case for floods, droughts, earthquakes, and other natural disasters in which insurance companies are exposed to huge liabilities because large numbers of people are all affected at the same time. Because such events are rare and somewhat unpredictable and their impacts are widespread, insurance companies often charge high premiums that include additional charges for the uncertainty surrounding them. Many people are unwilling to purchase such expensive insurance. Of course, this reluctance is often compounded by the knowledge that the government is likely to come to their rescue in any case. The question we are exploring here, however, is not whether the government often does rescue people from natural and other types of disasters but whether there are reasons why the invisible hand of the market cannot take care of these problems. Government intervention is justified if markets fail to provide insurance at rates that reflect the true actuarial risks involved, and this may, in fact, be the case for disasters that occur rarely but affect large numbers of people when they do.

Two other aspects of insurance markets that merit discussion in this context concern individual behaviors associated with the purchase of insurance. The first of these is referred to as *moral hazard*. Moral hazard arises when an insured individual is less careful than he or she would be if not insured. Suppose that insurance companies know the probability, with some degree of confidence, that a particular type of automobile will be stolen. This knowledge would allow them to determine an actuarially fair premium for insuring that automobile against theft. However, if an individual is insured against the theft of her new BMW, for example, she may be less careful about locking the car and taking other precautions to reduce the likelihood that it will be stolen. If all BMW owners behave this way, the insurer will have to either raise rates or suffer losses as more cars are stolen than anticipated.

The other behavioral response to insurance is known as *adverse selection*. It occurs when insurers are unable to distinguish between different levels of risk associated with individual consumers. Suppose that a health insurer cannot identify which individuals smoke and which do not. The fair rate for the group as a whole will be attractive to smokers, who face a higher probability of disease, but unattractive to nonsmokers. If the pool of insured begins to include more smokers and fewer nonsmokers, the premiums will have to rise and fewer nonsmokers will find the insurance attractive. If the health insurer could identify smokers and nonsmokers, it would be able to offer different actuarially fair rates to the two groups. Because the insurer cannot make this distinction, there is a market failure. Moral hazard and adverse selection can lead to inefficiencies in the markets for insurance, and this may mean that there is a role for the government in correcting these inefficiencies.

It turns out that a very substantial portion of public expenditures in the United States and other welfare states is best thought of as social insurance for such vicissitudes of life as unemployment, natural disasters, old age, disabilities, and so on. The economic justification for such programs is the failure of private insurance companies to provide such protection at actuarially fair rates. It should be noted, however, that many of these programs actually result in a redistribution of income among different groups. Whether such redistribution is justified or not is independent of the market failures associated with risk. We will have more to say about redistribution later.

EXTERNALITIES

It should be clear by now that prices play an extremely important role in guiding the invisible hand of the market. If prices are "wrong" in some way, they may lead consumers and producers to make decisions that give rise to inefficiencies or socially undesirable outcomes. Imperfect competition, risk, and uncertainty may all give rise to inaccurate price signals. Such inaccuracies also occur when markets are either missing or incomplete so that certain costs and benefits are not taken into consideration in the determination of market prices. These effects are known as *externalities*, because there are benefits or costs that are external to the set of information upon which the firm or household is basing its decisions. Externalities exist when there is a divergence between private and social costs and benefits, that is, when the costs and benefits of an activity as perceived by a private individual differ from the costs and benefits of that activity as seen by society. Externalities also arise in cases of interdependency between two private-sector agents, but even here, the crux of the problem is social because such interdependencies result in a misallocation of resources, depriving society of benefits that could be achieved with a more efficient use of the resource base.

Consider a profit-maximizing firm that produces fertilizer using labor, raw materials, and capital. Assume that all of the input and output markets are competitive. The managers of this firm will hire inputs according to their relative prices and produce the amount of fertilizer that is consistent with equating marginal costs to the competitive market price. As shown in the preceding chapter, this situation is socially optimal. Suppose, however, that in the process of producing the fertilizer, the firm also generates air pollution that harms nearby orange trees. There is a cost to orange growers and orange consumers because the quantity of oranges available is reduced, the price consumers pay is higher, and the returns to orange production lower than would otherwise be the case. For the fertilizer producer, the only costs that matter are those associated with labor, capital, raw materials, and other inputs, but from the point of view of society as a whole, there is an additional cost—air pollution—that is part of the cost of producing fertilizer. The fertilizer producer is not obligated to include this cost in making the profit-maximizing production decision, and this gives rise to a difference between private and social costs.

Economists often use the example of the relationship between apple and honey production to illustrate the concept of an externality (see Box 4.1). Honey is produced by bees that obtain nectar from flowers, such as those of apple trees. The nectar is needed for honey production, but there is no market in which nectar is bought and sold. Rather, it is simply collected by the bees from whatever flowers are found near the hive. Apple producers generate a positive externality in the form of free nectar available to the honey producer's bees. The result is less apple production than would be socially optimal because apple producers receive no compensation for their contribution to honey production. In the case of the fertilizer plant, a negative externality (air pollution) leads to a misallocation of resources such that more fertilizer is produced than is socially optimal. In a sense, society is subsidizing the fertilizer producer by absorbing a part of the firm's costs of production. These inefficiencies reduce social welfare relative to situations where there are no externalities.

Box 4.1. Apples, honey, and externalities

Suppose that there is one factor of production, labor, used to produce apples and honey. Total labor, L, is allocated between apples (L_A) and honey (L_H) so that $L = L_A + L_H$. Honey production (H) depends not only on the labor allocated to its production but also on apple production (A) and, hence, on the labor allocated to apples:

$$A = f_1 (L_A)$$
$$H = f_2 [L_H, f_1 (L_A)]$$

To determine the socially optimal allocation of labor between the two industries, maximize the total returns ($P_A A + P_H H$, where P_A and P_H are the relevant prices) subject to the constraint that apple and honey labor have to be equal to L. This can be set up as the following Lagrangean (see Chapter 2):

$$F = P_A f_1 (L_A) + P_H f_2 [L_H, f_1 (L_A)] + w (L - L_A - L_H)$$

Setting the partial derivatives (F_{LA} and F_{LH}) of this expression with respect to L_A and L_H equal to zero gives:

Box 4.1 (continued)

$$F_{LA} = P_A(dA/dL_A) + P_H(\partial H/\partial A)(dA/dL_A) - w = (P_A + P_H\partial H/\partial A)dA/dL_A - w = 0$$
and $F_{LH} = P_H(\partial H/\partial L_H) - w = 0$

Note that in the case of honey production, the optimum requires setting the value of the marginal contribution of labor equal to w, which can be taken to represent wages. This is the precise outcome that is expected in a perfectly competitive economy. In the case of apple production, however, there are some extra terms related to the interdependence between apple and honey production. This externality shows up in the expression $\partial H/\partial A$, which measures the change in honey production if there is an increase in apple production. If there were no externality, this term would be zero (variation in apple production has no impact on honey production) and the expression would reduce to the same result as for honey production (the value of the marginal product is equal to the wage).

In a competitive economy, rational producers will maximize profits by equating the perceived value of the marginal product of labor to the wage rate. Because the wage rate is the same for both, the value of the marginal products to both firms will be brought into equality:

$$P_H\partial H/\partial L_H = P_A dA/dL_A \quad or \quad (\partial H/\partial L_H)/(dA/dL_A) = P_A/P_H$$

The ratio of the two marginal products is the marginal rate of transformation, so this is the same result as shown in figure 3.5 where the MRT is equal to the price ratio. Note that these marginal products are those perceived by the individual producers. They do not include the term reflecting the interdependence between apple and honey production. This poses a problem, because the correct expressions for locating the social optimum are

$$(\partial H/\partial L_H)/(dA/dL_A) = [P_A + P_H (\partial H/\partial A)]/P_H = P_A/P_H + \partial H/\partial A$$

This differs from what would be produced in competitive equilibrium by the partial derivative showing the impact of apple production on honey output. If there were no externality, that expression would drop out and the result would be the same as the competitive equilibrium. In this case, however, the market is unable to reach the social optimum. In fact, what is happening is that the apple producers are not being compensated for their contribution to honey production, so they do not produce the socially optimal amount of apples.

It is important to emphasize that externalities of one sort or another are extremely common. At the beginning of this book, a number of situations characterized as collective action problems were described. We can now see that many of these problems are actually externalities as defined in this chapter. The smoker in the waiting room does not take account of the cost to others of her smoking and as a result smokes more than would be the case if she had to bear the full cost of smoking by bribing the others to let her smoke or paying them compensation for having to put up with the secondhand smoke. The good neighbors who take care of their yards are generating a positive externality, providing benefits to others for which no compensation is forthcoming. Because such interdependencies are so widespread, it would seem that the invisible hand must fail in leading rational producers and consumers to a social optimum.

While it is true that externalities pose a serious problem for the invisible hand, it turns out that things are not as bad as they seem. To begin with, it is often the case that private citizens are able to work out solutions to the inefficiencies associated with externalities. It is well known, for example, that honey producers and orchard growers have reached agreements covering the external effects of honey and fruit production. It turns out that there are actually two interdependencies in this example, one involving the use of nectar to produce honey and the other involving pollination as the bees collect nectar. Cheung found that where the main effect pertains to the impact of orchard production on honey output, bee-keepers pay fees to be able to locate their hives near orchards containing particularly desirable blossoms. In cases where the fruit trees are highly dependent on bees for pollination, orchard owners frequently pay beekeepers to place their hives near the trees needing pollination.[3] In these cases, economic agents have managed to bring the external effects into their internal economic calculus by creating markets in nectar and pollination.

It is not always the case that individual economic agents will be able to resolve externality problems in this neat and spontaneous way. In the case of the fertilizer producer, it may be necessary to call on the government to introduce regulations that force the firm to internalize the externality. It is in this context that the *Coase theorem* described in Chapter 1 takes on particular importance. Recall that the problem addressed in Coase's study concerned the impact of cows straying from a pasture onto the field of a neighboring farmer. Suppose that there is no legal instrument to force the owner of the cows to take into account the damage they cause to the farmer's field. This situation is virtually identical to the case of the fertilizer producer who makes production decisions without considering the costs of fertilizer production borne by others. If the owner of the cow herd is considering the purchase of an additional animal, he will not factor in the costs that animal may impose on the farmer in the form of damage to his field. If it is possible for the neighbors to bargain, the farmer will be willing to pay the cow owner not to buy the additional cow if the damage that cow will cause is greater than what it would take to bribe the owner not to make the purchase.

An important result of this analysis is that the absence of laws to regulate this interaction reflects the way in which *property rights* have been defined. In terms of efficiency, the particular definition of property rights that is in effect makes no difference as long as it is possible for the parties to bargain. If the farmer's property is protected, the cow owner will have to compensate the farmer for any damage, and the result will be the same number of cows as in the case where the farmer has to compensate the cow owner for not buying an additional cow. If transactions costs are high, however, the parties may be unable to reach agreement, and the inefficiency may persist as the cow owner maintains a herd that is larger than the socially optimal size. Another important aspect of this problem is to note that there can never be a situation in which property rights are undefined. If there is no law concerning stray cows, the cow owner has a de facto property right in the neighbor's field. A law on stray cows gives the farmer the exclusive property right to his field, with the cow owner excluded from its use. This is extremely important, because it leads to the conclusion that society can never avoid making a choice concerning whose property rights are to be sanctioned. Accepting the status quo means recognizing the current structure of property rights. If that structure seems to be generating unsatisfactory outcomes, it can be changed, although such changes generally require some form of collective action when transactions costs are high enough to prevent spontaneous, bargained solutions.

Prior to the 1970s, there were few national environmental regulations in the United States. Air was generally seen as a free good available to all for whatever use individuals wished to make of it. Some individuals wished to use it as a repository for various kinds of industrial by-products, such as the smoke and gases created by burning fossil fuels. With the passage of the 1970 amendments to the Clean Air Act of 1967, U.S. society changed the structure of property rights to air, making it illegal to use air as a dump for smoke and industrial gases. This change in property rights forced firms to take account of the costs of the air pollution they had been generating. As noted in Chapter 1, Coase's result that the way in which the property right is assigned is immaterial for questions of efficiency does not mean that decisions on the way to define property rights are without importance. In particular, ownership of a property right can lead to the generation of income streams, and whether the farmer has to compensate the neighbor with the cows or the other way around will clearly make a difference to the two individuals involved. Redefining property rights to the air led to changes in the income and well-being of large numbers of people, and the decision to make that change required collective action. It could not have arisen from the self-interested behavior of individuals responding to market signals.

COMMON POOL RESOURCES

Externalities are generated by the way in which property rights are defined. If property rights are inappropriately assigned, individual behavior will not lead to a social optimum. The link between externalities and property rights is nicely illustrated by the case of *common pool* or *common property* resources. In an article entitled "The Tragedy of the Commons" published in 1968, Garrett Hardin described the destruction of communally owned grazing land in England.[4] The commons was an area set aside for use by everyone in a village community. Shepherds in these communities could decide how to satisfy the grazing needs of their animals using either their own resources or the commons to which everyone had access.

Although everyone knew that putting too many animals on the commons would result in its destruction through overgrazing, no individual had an incentive to limit his own personal use of the communal resource. The costs of additional animals grazing on the commons are shared among all the owners—that is, the entire village—while the benefits are fully captured by the owner of the particular animals that are grazing the commons. Because clear title to the property is not defined, there is a divergence between the private and social costs of using the common grazing resources. Under these circumstances, each individual has an incentive to use the commons as much as possible. Conserving the resource requires the participation of all who have access to it. If there is no assurance that others will choose to limit their use, there is no reason for anyone to exercise restraint. The shepherd who unilaterally acts to conserve the resource loses twice: first, because he is not benefiting from the use of the commons as much as he could, and second, because the commons ends up being overgrazed despite the efforts to conserve it.

The case of ocean fisheries described in Chapter 1 is an example of a common pool externality. An individual fisher who restricts the catch in order to maintain the viability of the fishery will end up with lower returns, and her actions will not save the fishery unless they are part of a coordinated effort that involves all the other fishers as well. As in the case of ordinary externalities, individuals in these circumstances are not taking into account the full cost of their activities. Part of the cost of fishing is borne by the entire group, which sees the stock of fish decline. It is interesting to note that it is the assumption of economic rationality that leads to these results despite the apparent irrationality of destroying a useful resource. Although it is not in the collective interests of the community to destroy the resource, the strategy that will most advance the interests of rational individuals pursuing their self-interest is to exploit the resource as much as possible while it lasts. To see how this kind of counterintuitive result might arise, it is useful to consider a famous example from game theory known as the *prisoners' dilemma.*

Imagine that two men have been arrested on suspicion of committing armed robbery. The police discover that the men are carrying concealed weapons (which is illegal in this story), but they have no eyewitnesses to verify that these two individuals committed the crime. To obtain a conviction, the police will need to have a confession from at least one of the accused. They offer them the following deal:

Prisoner A⇒ Prisoner B⇓	CONFESS	REFUSE TO CONFESS
CONFESS	(5,5)	(10,0)
REFUSE TO CONFESS	(0,10)	(1,1)

Figure 4.2: The prisoners' dilemma.

The figures in parentheses show the amount of time in jail, first for Prisoner A, and then for Prisoner B. Thus, for example, if Prisoner A refuses to confess while Prisoner B confesses, A will be sent to prison for ten years and B will be set free. The logic behind this arrangement is that if one prisoner confesses and testifies against the uncooperative prisoner, there will be a reward for this helpful behavior and a severe penalty for the one who did not cooperate. On the other hand, if both confess, there is no need to be lenient for either of them and so they both get sent to prison for five years. Finally, if neither prisoner confesses, the police will have to settle for conviction on the concealed weapons charge that carries a penalty of one year.

The best outcome for the two prisoners would be to refuse to confess so that each would receive a one-year prison sentence. For each individual, this is better than the other possibilities unless one prisoner can be sure that the other will refuse to confess, in which case confession seems preferable. If one considers these two as members of a group, it is clear that the collective interest is advanced through refusing to confess: all the other options lead to a total of ten years of prison time as opposed to only two years total if both refuse to confess.

The problem is that, in game theoretic terms, the dominant strategy for each individual is to confess regardless of what the other does. To see this, imagine that you are Prisoner A and that along with Prisoner B you come before the judge, who asks how you wish to plead. If you refuse to confess, Prisoner B can simply confess and he will go free. If, on the other hand, you confess, he will certainly not refuse to confess because that would result in a long prison term. If the other prisoner refuses to confess, your best option is to confess. If the other prisoner confesses, your best option is still to confess. The same is true for the other prisoner. Thus, regardless of what the other prisoner does, the dominant strategy in this game is always to confess, and the result will be that both prisoners confess and receive five-year sentences.

The prisoners' dilemma explains how rational individuals pursuing their self-interest can bring about collective harm. Many of the fisheries off the coast of New England have been so extensively exploited that it has become necessary to close them in an effort to allow the stock of fish to be replenished. The New England fishers are in a prisoners' dilemma in that the dominant strategy is to catch as many fish as possible as long as there is no guarantee that everyone is actively cooperating in protecting the fishery. There are many other examples of this type of problem. Overgrazing, deforestation, and global warming all involve situations in which individuals have open access to collectively held resources. Because the property right is held collectively, the costs of overuse of the resource are shared. Privately held property rights, on the other hand, force each individual to take full account of the costs of his or her activities.

One way to handle the problems of common pool resources, then, would be to assign private property rights to the overused resource. This may be possible in some cases but certainly not in all. It would not be easy, for example, to figure out how to assign ownership of specific schools of fish to particular individuals. In cases of this nature, it will be necessary to introduce some other type of institutional arrangement to prevent the destruction of the resource. There are two underlying causes of these problems. The first is the fact that there is *open access* to the resource, that is, there are no institutions to regulate the way in which individuals use it. The second is that rational individuals cannot develop a system of self-regulation because they are caught in a prisoners' dilemma. Solving the problem of common pool resources requires either that the open-access condition be eliminated or that something be done about the prisoners' dilemma, or both.

One way to eliminate open access is to define private property rights to the resource. If that cannot be done, there are many formal and informal institutional arrangements that can be introduced to limit access. For example, Rappaport describes a complex system of religious ritual, taboos, and stylized warfare among the Tsembaga, a group of hunter-gatherers in New Guinea, that regulated the rate at which common pool environmental resources, such as wildlife, were exploited.[5]

The problem of self-regulation in a prisoners' dilemma involves an inability to cooperate. Cooperation would lead to a preferred outcome, while defecting from the cooperative solution leads to a state that is inferior. Note that in this case the result violates Pareto optimality: both individuals could be made better off if they could only cooperate. It turns out that there are ways to promote cooperation. If the prisoners can trust each other, they might be able to make a pact swearing not to confess. If each is absolutely certain that the other will not defect, each can refuse to confess without fear that the other will. But how can this certainty be provided? In some circumstances, the bonds between the individuals involved may include sufficient trust to support cooperation. This might be the case if the prisoners are brothers, for example.

If such ties are lacking, it may be necessary to develop institutions that allow the participants to enter into contracts that will be enforced by some external authority. In the case of the two prisoners, cooperation would be the dominant strategy if they belong to an organized crime syndicate in which confession is punishable by death and it is clear that such punishment will be meted out to those who defect. More generally, *enforceable contracts* backed up by the coercive power of the state can help overcome a wide variety of problems in the ordinary conduct of business that have the characteristics of a prisoners' dilemma. For example, if I wish to purchase fertilizer to produce corn and can only pay after the corn has been produced and sold, I can enter into a legal contract with a bank which advances the money to purchase the fertilizer. The obligation to repay the loan is enforced by the authority of the state through laws on contracts. Without this legal infrastructure backed up by the power of the government, most business transactions would become problematic because the parties would be caught in a prisoners' dilemma.

Research on the prisoners' dilemma has shown that defection can often be overcome if the game is repeated. In repeated games, players may develop the trust that allows them to cooperate. For example, the arms race between the Soviet Union and the United States during the cold war can be seen as a repeated prisoners' dilemma. Each country expanded its arsenal even though both would have been better off using their scarce resources for more useful purposes. Over time, repeated contacts and negotiations began to lead to a degree of trust between the leaders of the two powers, and agreements on arms reduction were reached. Verification of compliance was always a central element in these agreements.

In this as well as the other examples, the prisoners' dilemma is overcome when the participants can feel assured that enough cooperation will be forthcoming to solve the underlying problem. Various kinds of institutions may be effective in providing such assurance. In the case of common pool resources, assigning property rights to the resource, creating enforceable contracts, strengthening informal institutions to limit access to the resource, increasing transparency so that defections are made public, facilitating repeated interaction to develop trust, and many other approaches have been followed to overcome the tragedy of the commons. Note, however, that all these approaches require some form of collective action, and this means that there may be a role for the state. There is strong evidence that the New England fisheries will be destroyed if their future is left entirely to decentralized market forces.

PUBLIC GOODS

Public goods were mentioned in the foregoing discussion of information. Public goods involve consumption interdependencies that can lead to the same kinds of barriers to cooperation as found in the case of common pool resources. Pure public goods have two character-istics. The first of these, referred to as *jointness*, *nonrivalry*, or *indivisibility*, is that the good in question is consumed by all at a level that is the same for all. For example, no matter how many people wish to tune into a given television broadcast, the "amount" of that broadcast will be the same for all. Moreover, the cost of absorbing an additional viewer of the broadcast is zero. Whether the number of people watching the Super Bowl is 100 million or six, the cost of broadcasting the game is the same. This would not be the case for a private good such as an apple. Increasing the number of apple consumers by 25 percent would lead to price increases that would induce producers to allocate additional resources to apple production. Nonrivalry means that the good in question can be supplied only as an indivisible lump and that the only choice consumers have is whether or not to consume the entire thing. Sometimes, as in the case of such services as national defense, police protection, or public health servic-es, they do not even have that choice, as the goods in question are furnished by the state regardless of whether an individual truly wishes them to be provided.

The second characteristic of public goods, *difficult exclusion*, concerns the ability of suppliers to exclude people from consumption of the good. In the case of private goods, ownership is defined and protected by law so that the owner can refuse to transfer the good to another individual unless payment is made. With public goods, however, the supplier is unable to force people to pay, because once the good is made available to one person, it becomes available to everyone else. Consider television broadcasts prior to the introduc-tion of cable. Once the broadcast was sent out over the airwaves, it became available to everyone with a television receiver. The difficulties in obtaining payment for television broadcasts were overcome in several ways. In the United States, the early broadcasts were provided by private firms that earned income from advertising. Viewers were forced to pay by being subjected to advertisements financed through increases in the prices consumers pay for the advertised goods. In France, the broadcasting companies were public firms financed through an annual tax levied on television receivers. With the advent of cable, exclusion was made much easier, although private broadcasters in the United States con-tinue to exact payments from consumers through frequent interruptions for advertisements.

The classic example of a pure public good is national defense, which has both of these characteristics. All individuals "consume" the same amount of national defense, and it is not possible to exclude people from the protection provided by this level of defense. Market information and information pertaining to the characteristics of goods or services are other examples of public goods. In these cases, nonrivalry means that it is difficult to determine the socially optimal price and quantity to be supplied (see Box 4.2). The prob-lem of excludability also undermines the ability of private markets to regulate the supply of and demand for public goods. Individuals have an incentive to hide their true prefer-ences, because once such a good is supplied, it is available to all whether payment has been made or not. This is known as the *free-rider* problem.

Many listeners of public radio broadcasts are free-riders who do not contribute because they know that they cannot be excluded from the audience. The free-rider problem can mean that the good is either undersupplied or not supplied at all even though members of the community actually would like to have it supplied. Public goods are similar to common pool resources in that the optimum social arrangement cannot be achieved because self-interested individuals have incentives not to cooperate. Free riding is the same as defection in the prisoners' dilemma. As with common pool resources, the problem of defection may mean that there is a role for governments in determining the amount of a public good to produce and the price to charge for it, and in using the coercive power of the state to exact payment.

Box 4.2. The mathematics of public goods

The market failure associated with public goods can be illustrated with some simple mathematics. Assume that there are two individuals who consume two goods, one of which is a public good. In the case of the private good, the total amount consumed is equal to the sum of the amounts consumed by the two individuals. Let G represent the total amount of the private good, with G_1 and G_2 indicating the consumption of G by individuals 1 and 2: $G = G_1 + G_2$

For the public good (P), however, the amount consumed by each individual is the same and equal to the amount supplied. Thus, $P_1 = P_2 = P$

Each individual receives utility from consuming both goods:

$U_1 = U_1(G_1, P)$ and $U_2 = U_2(G_2, P)$

The problem is to maximize the utility of one individual subject to the constraint that the utility of the other individual not fall below some constant amount. Note that this is a Pareto optimal representation because one individual's utility, say individual 1, is increased as much as possible without reducing the utility of individual 2. This constraint is represented by: $U_2(G_2, P) = U_2'$.

In addition to the constraint of maintaining individual 2's utility at some constant, it is necessary to include a constraint reflecting the economy's production possibilities:

$F(G_1 + G_2, P) = 0.$

These relationships can be written as the following Lagrangean function:

$\mathcal{L} = U_1(G_1, P) - \lambda_U(U_2' - U_2(G_2, P)) - \lambda_F F(G_1 + G_2, P)$

Let MU represent marginal utility and MF the derivatives of the implicit production function with subscript 1 and 2 for the two individuals and G and P for the two goods. The first-order conditions from maximizing the Lagrangean are:

1. $MU_{G1} - \lambda_F MF_G = 0 \quad ==> MU_{G1} = \lambda_F MF_G$
2. $\lambda_U MU_{G2} - \lambda_F MF_G = 0 \quad ==> \lambda_U MU_{G2} = \lambda_F MF_G$
3. $MU_{1P} + \lambda_U MU_{2P} - \lambda_F MF_P = 0 ==> MU_{1P} + \lambda_U MU_{2P} = \lambda_F MF_P$

Divide the third expression by the second to obtain:

$MU_{1P}/\lambda_U MU_{G2} + MU_{2P}/MU_{G2} = MF_P/MF_G$

From the first two expressions, $MU_{G1} = \lambda_U MU_{G2}$. This can be used to simplify the preceding expression to:

$MU_{1P}/MU_{G1} + MU_{2P}/MU_{G2} = MF_P/MF_G$ or $MRS_1 + MRS_2 = MRT$

This result shows that the optimum is achieved where the sum of the two individuals' marginal rates of substitution (MRS) between the two goods are equal to the marginal rate of transformation (MRT). Recall that in competitive equilibrium, the invisible hand leads each individual to equate her or his MRS to the ratio of the competitive prices for the goods and that competitive producers will be led to produce where the MRT is also equal to the price ratio. In other words, the invisible hand will lead to:

Box 4.2. (Continued)

$MRS_1 = MRS_2 = MRT$

which differs from the result shown above. The consequence is that the free market will not lead to the social optimum, which is found where the sum of the marginal rates of substitution is equal to the marginal rate of transformation.

Source: K. C. Kogiku, *Microeconomic Models* (New York: Harper and Row, 1971), 115-19.

There are many varieties of public goods defined by the degree to which the characteristics of nonrivalry and difficult exclusion are present. Cable television broadcasts are still nonrival, but technology has given suppliers a means to insure that consumers pay for the services delivered. In many respects, local fire or police protection is rival in that responding to one emergency may mean that it is not possible to respond to another (the cost of an additional user of the services is not zero) but nonexcludable in the sense that there is an expectation that emergency personnel will respond, if possible, to any incident that occurs, including those that involve indigents who pay no taxes to support the service. The designation "public" is somewhat unfortunate in that the various kinds of services with some measure of these two characteristics are not all supplied by the public sector. Television broadcasts are one example. In some areas, fire protection is provided by private firms hired by local governments and paid for from tax revenues. In general, however, socially optimal amounts of public goods will not be supplied without some form of government or collective action. The amount to be supplied can be politically controversial, as shown by the debates in the United States concerning the budget for the Defense Department. But the alternative of simply leaving these questions up to the market will not resolve the difficulty.

THE PROBLEM OF INCOME DISTRIBUTION

In Book 5 of the *Wealth of Nations*, Adam Smith discusses the role of the state by first describing the "expenses of the sovereign or commonwealth" and then listing the sources of revenue for funding these expenditures.[6] For Smith, there are three primary functions to be carried out by the sovereign: national defense, the administration of justice, and certain public works (roads, education, and so on). In the language of modern welfare economics, the activities that Smith saw as falling under the responsibility of the state cover several of the most prominent examples of market failure, as described in the preceding sections. National defense is a public good that will not be provided at socially optimal levels by private markets. Private individuals will tend to underinvest in education because it generates a beneficial externality that cannot be fully captured by the individual. This educational externality arises because all members of society are better off if the general level of education is higher. For example, citizens familiar with the history of their country and the way the government operates are likely to contribute to better public decisions than are those who are not. For individuals, this external effect may be left out of the decision on education, the amount pursued being determined by marginal private benefits and costs.

Justice is also a public good which has both characteristics of nonrivalry and difficult exclusion. We will have more to say about justice in later chapters when we confront the ethical aspects of public policy analysis. Here, it is instructive to anticipate those later discussions by introducing the notion of justice defended most notably by John Rawls.[7] Rather than seeing justice as retribution for misbehavior, Rawls views it as fair and equal treatment. One part of this kind of justice, sometimes referred to as *distributive justice*, has to do with the access people have to the basic goods and services needed to live reasonably satisfactory lives in a given society. There are two parts to this issue. First, justice or fairness may be thought to require that everyone be guaranteed some minimum entitlement that allows satisfaction of basic survival needs. Second, even if everyone possesses at least this minimum, it could be considered that fairness is violated if income and wealth are distributed across the population in a highly unequal manner.

With respect to the first issue, there are many who believe that people are entitled to only whatever they are able to earn through their own efforts. Such a position might be defended in the same way that the problem of *income distribution* was handled in Chapter 3. Recall that a movement from a point on the contract curve reflecting a very unequal distribution of the two goods (a point such as E in Fig. 3.2) to any other point on the contract curve with a more equal distribution violates Pareto optimality. If a more egalitarian initial distribution can be established, however, the second fundamental theorem of welfare economics can be invoked to achieve a fairer final distribution. In Chapter 3, we posited a social welfare function that allowed Sam and Mary to agree on an initial distribution from which the invisible hand could take them to an efficient and Pareto optimal final distribution that they would presumably consider fair.

This hypothetical procedure is generally precluded in existing economic systems, however. For all practical purposes, we are already on the contract curve, and any distributional adjustment will violate Pareto optimality. In these circumstances, the argument that people deserve only what they can obtain through their own efforts would have to be defended in some other way—for example, by arguing that initial distributions of endowments are fair. We will explore such defenses and arguments against them later. For now, let us assume that as a society, we do believe that justice requires that individuals be guaranteed some minimum entitlement to meet their basic needs. This leaves the second aspect of this issue, whether the overall distribution of income and wealth is more or less egalitarian. It is conceivable that a society might have a highly unequal distribution of income but still pass the test of ensuring that all citizens are able to meet their basic needs. For example, suppose that 90 percent of the individuals in a given society have entitlements that allow them to satisfy their basic needs, while the remaining 10 percent have entitlements that are larger than these minima by a factor of one thousand. While this distribution would be highly unequal, it is not clear that there is any injustice that would require some form of collective action.

In many societies, however, such inequalities are seen as unfair even if everyone is able to avoid falling into abject poverty. This point of view may be bolstered by some practical considerations. Many analysts believe that poverty is a relative rather than an absolute concept. Such beliefs amount to the assertion that the minimum entitlement needed to satisfy basic needs is a socially determined datum that varies from society to society. Part of what might be included in the basic needs of individuals is knowledge that they are not so far behind others in society that they cannot aspire to move up in the social hierarchy. This is the concept of *relative deprivation*.[8] In the foregoing example, the 90 percent with the bare minimum could be thought to be so deprived relative to the elite that the general welfare of society would be improved by some effort to equalize the distribution.

Another argument for more equal distributions of income and wealth could be based on the observation that wealth confers social and political power on those who have it, so that a highly concentrated distribution undermines and corrupts democratic processes. In the United States, for example, it would not be too difficult to argue that the great wealth held by certain individuals allows them to have more influence on public policy than the rest of the citizenry and that these individuals, being rational pursuers of their own self-interest, often succeed in convincing legislators to adopt policies favorable to them even though such policies are not in the broader public interest.

The justice involved in a fair distribution of society's resources, however such a distribution is defined in a particular society, is a public good that is likely to be undersupplied by competitive markets. There is nothing in the Edgeworth-Bowley boxes of Chapter 3 that will smooth out an initial set of endowments that are highly unequal. In this sense, unequal distributions of income and wealth can be seen as an important type of market failure. In fact, many countries actively pursue public policies that have a redistributional objective, that is, policies that aim to reduce income inequality. For example, in Europe and the United States, it was long thought that farmers were relatively less well-off than urban workers. Historically, many agricultural policies were aimed at supporting farm prices and transferring income to farmers in an effort to correct what was seen as an inequitable distribution of income. Likewise, most industrial countries have some type of public policy directed at poverty and the provision of a social safety net. These social programs are the basis for the name *welfare state* often attributed to these societies. It should be noted that the extent of *redistribution* and poverty eradication varies from one welfare state to another. Thus, the U.S. welfare system is considerably less extensive than those found in Western Europe. This state of affairs may reflect social preferences as expressed through democratic decision making or it may reflect the power of the wealthy in the United States to divert public resources toward themselves and away from poverty programs.

It is important to note that many of the social welfare programs in the United States are more appropriately seen as social insurance programs than as efforts to reduce poverty. In the 1998 U.S. federal budget, expenditures for means-tested income security amounted to about $233 billion, which represented 14.1 percent of total expenditures.[9] Defense spending was 16.2 percent of the total, while the main social insurance programs, social security and medicare, accounted for 34.6 percent of expenditures. The remaining 35.1 percent of the budget was used for all other programs, including agriculture, health, education, transportation, housing, and so on. It should also be pointed out that there was over $547 billion in tax expenditures in 1998. Tax expenditures are the amounts of government revenue that are not collected because of various loopholes such as the deductions for home mortgage interest, charitable contributions, and local taxes. The main beneficiaries of these loopholes are middle- and upper-income taxpayers, who are also the primary beneficiaries of the other social insurance programs. While many of these social insurance programs can be justified as ways to correct certain failures in private markets for insurance, their main effect is to redistribute income among middle-income citizens rather than to redistribute income from the rich to the poor.

In the preceding discussion, various expressions commonly found in discussions of the distribution of income or wealth were introduced, and it is important to make their meaning clear. First, the economic resources that are distributed across the population include both income and wealth. *Income*, of course, is the amount of economic resources, usually measured in monetary units, received by an individual or a household during some specified period of time, usually a year. *Wealth* is the value of accumulated or inherited savings retained from income received in the past. The wealth of an individual or household is a stock of economic resources that can be employed to increase future income or drawn against when current income is lower than usual.

The initial *endowments* that set the parameters within which the invisible hand determines a Pareto optimal distribution of goods include both income and wealth. *Entitlements*, as understood in this book, are somewhat broader than endowments in that they can be thought to include individual opportunities to earn income and wealth. Amartya Sen, who won the Nobel Prize in economics in 1998 for his work on social choice and is best known for his studies of poverty and famines, defines entitlements as an individual's owned resources along with his opportunities for exchanging owned resources in a given society.[10] Opportunities for exchange are established by laws and customs, while owned resources include those that are inherited or accumulated from past transactions. One could also broaden the sense of an entitlement to include such things as individual capacities, both those that are inherent (Michael Jordan has certain physical characteristics that were of particular use in becoming a great basketball player) and earned (Michael Jordan would still have been a great basketball player even if he were less than six feet tall because of his own efforts at developing his talents).

Wealth and income are the results of economic processes that start with the initial endowments or entitlements. As noted earlier, an unequal distribution of initial endowments is not likely to give rise to a more equal distribution of wealth and income in the absence of redistributional policies. But what does it mean to say that a given distribution is more equal than another? Consider the problem of dividing a cake among six people. Most would agree that an equal division would mean that each person receives exactly one-sixth of the cake. Anything else would be less equal, and it would be completely unequal if one person ate the entire cake.

Similarly, perfect income equality would mean that everyone earns identical incomes, while perfect income inequality would obtain when one person commands the total income of the group. It is rare to find individuals who would argue for complete income equality, a state that has never been realized anywhere on earth. Likewise, even the fortunate individual who would have everything in the case of perfect inequality would probably prefer to share at least part of her hoard with others. Actual distributions of income and wealth fall somewhere between these two extremes, and egalitarian objectives usually aim at reducing inequality rather than achieving perfect equality. In this context, it would be useful to have some measure of income or wealth inequality to be able to make comparisons between societies and across time.

The most commonly used measures of income inequality begin by ranking individuals or households from the ones with the lowest incomes to those with the highest. These rankings are then expressed in percentages indicating the percentage of total income received by each individual. Because it would be fairly difficult to record such percentages for the millions of people living in a given society, analysts usually divide the population into income groups. Thus, one might consider the percentage of total income received by the poorest 10 percent (decile) or 20 percent (quintile) of the population. Such an array for the United States in 1994 is shown in Table 4.1.

Table 4.1. Distribution of income in the United States, 1994

Percentage share of population	Percentage share of income
Lowest 20%	4.8%
Second 20%	10.5%
Third 20%	16.0%
Fourth 20%	23.5%
Highest 20%	45.2%

Source: *World Development Report*, World Bank, New York: Oxford Press, 1998/1999.

These data can be cumulated to develop a visual measure of income distribution known as a *Lorenz curve*. Thus, for example, the poorest quintile receives 4.8 percent of income, while the cumulative percentage of income received by the poorest 40 percent is 15.3 (4.8 plus 10.5). The endpoints of the Lorenz curve are found at zero and 100 percent for both population and income.

In Figure 4.3, the Lorenz curve is shown below a diagonal running from the two endpoints. This diagonal represents perfect equality in the distribution of income. Perfect inequality would be shown by a Lorenz curve lying on the lower horizontal axis and the right-hand vertical axis. Clearly, the closer the Lorenz curve is to the diagonal, the more equal is the distribution. A numerical measure of this distance, the *Gini coefficient*, can be found by comparing the area between the Lorenz curve and the diagonal (area A) to the area below the diagonal (area $A + B$). With perfect equality, the Lorenz curve is congruent with the diagonal, area A is equal to zero, and the Gini coefficient is also zero. With perfect inequality, area A expands to fill the area under the diagonal and the Gini coefficient is equal to 1 ($B = 0$ and A divided by $A = 1$).

According to the data reported in the World Bank's 1998–1999 annual *World Development Report*, the lowest Gini coefficient in the world is found in Belarus (0.216), while the highest is found in Sierra Leone (0.629). Most of the industrialized nations of Western Europe have Gini coefficients in the range of 0.25 to 0.33, while many low-income countries in Africa and Latin America are above 0.50. The Gini coefficient for the United States, based on the data used to construct Figure 4.3, is 0.401, which is higher (more unequal) than most other high-income countries. Even some low-income countries such as India (0.297), Indonesia (0.342), Pakistan (0.312), Ghana (0.339), and Tanzania (0.381) appear to have more equal distributions than the United States.

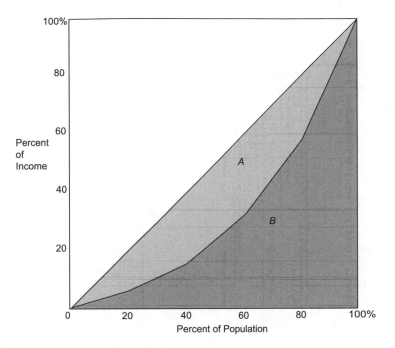

Figure 4.3: The Lorenz curve and Gini coefficient.

   The data used in computing these Gini coefficients may not be as reliable as data on other economic variables, and caution is needed in interpreting them. Other studies have placed U.S. Gini coefficients much closer to those found in Europe. As a rule, it does appear that income is less equally distributed in lower- and middle-income countries than in high-income countries. Simon Kuznets, who won the Nobel Prize in economics in 1971, observed that income inequality appears to increase in very poor societies initially characterized by fairly equal income distributions as they begin to experience greater economic growth.[11]

   At higher levels of income, countries are able to implement welfare programs that reduce income inequality. Some of these programs (education or public health, for example) help people increase their ability to earn income. Others, such as the U.S. programs for food stamps or direct aid to families with dependent children, transfer cash or in-kind income using general tax revenues. It is important to recognize that measures of income distribution for a given country will differ depending on whether they are computed before or after government taxes and transfers. For the United States, Ruggles estimates that the poorest 20 percent in 1988 received 3.1 percent of total income before taxes and transfers, compared with 3.6 percent after.[12] At the upper end, the richest 20 percent saw their income reduced from 51.8 percent of the total to 49.9 percent by the U.S. system of taxation and transfers. It is not clear whether the World Bank estimates reported earlier pertain to pre-tax/transfer or posttax/transfer income distributions, another reason to exercise caution in interpreting them.

   The welfare programs commonly found in high-income countries are collective responses to concerns about the fairness or justice of the prevailing distribution of income or wealth. These distributions can be measured and compared with income inequality in other countries or within the same country over time. Such comparisons can show that income is more equally distributed in one country than another or that the income distribution in a given country has become more equal over time. However, because there is no unique value for Gini coefficients that can be taken to represent absolute fairness, collective decisions to redistribute income can be based only on political and ethical beliefs about the degree of equality required for the situation to be considered fair or equitable. Some may consider that the degree of inequality reflected in the estimated Gini coefficient for the United States is inequitable for such a wealthy country. Others may feel that the relative inequality measured by Lorenz curves and Gini coefficients is much less important than the fact that some individuals in the United States are homeless and unable to satisfy basic survival needs. Either way, self-interested individuals interacting in impersonal markets are not likely to behave in ways that reduce or eliminate either of these types of inequality even though a substantial number may prefer that everyone be able to meet basic needs or that the overall distribution be more equal. Equity is an important part of social justice that will be undersupplied without some form of collective action.

COMPENSATION

Suppose that the members of a given society judge that a more equal distribution of income is required to attain its goals of social justice. One way to achieve a more equal distribution would be to tax the income of wealthier people and use these financial resources to subsidize the consumption of low-income households. This type of transfer violates Pareto optimality because the wealthy are made worse off to increase the utility of the poor. Any policy aimed at redistributing income will have this effect, and that is one reason why there is frequently substantial opposition to such policies. In the United States, it is not uncommon to encounter arguments, often made by wealthy individuals, that taxes are a form of theft, particularly when the revenues are used to support the welfare of the poor. Some level of taxation may be justified by the need to supply public goods such as national defense, but for many people, insuring equity and social justice is not seen as an appropriate function of the state. In support of this position, it might be argued that too much taxation and redistribution removes the incentives individuals ordinarily have to work hard and create new products and forms of wealth. Aid to the poor can make them permanently dependent on the government by allowing them to survive without having to take responsibility for their own welfare.

There are some policies, however, that are consistent with Pareto optimality. Recall, for example, that an antitrust law that breaks a monopoly into a set of competitive firms leads to improved resource allocation. This process involves a reduction of output from the competitive sector and an increase in output from the previously monopolized industry, which appears to violate the efficiency criterion defined in Chapter 3. The way around this is to note that the fall in output from the competitive sector is compensated by an increase in output from the industry subjected to the antitrust laws. In other words, the value of the increased output is greater than the loss of output from the competitive sector so that there is a net increase in the value of total output. The concept of *compensation* can be extended to cover other types of polices. In Chapter 2 (Fig. 2.6), the welfare effects of an import tariff were described. The tariff raises domestic prices, benefiting producers and harming consumers. It also generates government income, but the sum of the tariff revenue and the increase in producer surplus is less than the decrease in consumer surplus. The result is a net loss to society due to the efficiency losses associated with the tariff.

Suppose that a country decides to eliminate an existing tariff. This will clearly lower the welfare of producers and reduce government revenue. However, consumer surplus expands by an amount that is greater than these losses, with the difference again being represented by the efficiency triangles. Because consumers and taxpayers are essentially the same, the effect on government revenue can be ignored for the moment. But suppose that this society does not wish to see producer welfare lowered. It would be possible to tax away part of the increase in consumer surplus to exactly compensate producers for their loss while still leaving consumers better off. Such a policy change thus is consistent with Pareto optimality if compensation is made, because producers end up no worse off than before the change while consumers see their welfare increase.

In the United States, trade liberalization is politically controversial because it frequently means the elimination of tariffs that protect certain industries and their employees. Many have argued that such trade agreements as the North American Free Trade Agreement (NAFTA; see Chapter 14) or the Uruguay Round trade agreement that established the World Trade Organization (WTO; see Chapter 12) would lead to job losses. Empirically, the number of jobs lost to these trade agreements has been quite small relative to the total number of jobs in the United States or the number of jobs that disappear naturally as a result of technological change or other factors. Moreover, the number of jobs created by the trade agreements is greater than the number of jobs lost. For those who see a job disappear, such changes can be the source of hardship. Should society give up the jobs created by trade in order to preserve those that may be lost?

Compensation offers a way out of this dilemma. The gains from these agreements generate government revenue (through the normal taxation of the growing income streams resulting from increased trade) which can be used to compensate those who lose jobs as a result of import competition. In fact, the U.S. government administers a program to compensate individuals who lose their jobs as a result of trade. The compensation usually takes the form of job training or educational support rather than direct income support. Such programs can be seen as part of the general social insurance provided by welfare states. Whether a loss of welfare is brought about by government policy, technological change, or random events such as severe weather, it is not always possible for individuals to purchase insurance against the possibility of such losses. In these cases, social insurance can compensate for a wide range of changes that lower welfare for certain individuals.

There are very few government policies that will satisfy the Pareto criterion without compensation. This is another way of saying that government policies aimed at correcting a perceived social problem will often reduce the welfare of some individuals. Consider the case of the fertilizer producer who also produces air pollution. The benefits of reduced air

pollution almost certainly outweigh the costs of pollution controls imposed on the fertilizer producer and her clients. Compensation for these costs could be taken from these
increased benefits, leaving the producer and her clients no worse off while the general society would be better off. In the case of the U.S. Clean Air Act, however, it does not appear
that government resources were used to compensate those who had been polluting the air.
Determining when the government should compensate people who are harmed by a public policy is a matter of judgment. The Fifth Amendment to the U.S. Constitution requires
that the government provide compensation when it takes property belonging to private
individuals. While this may appear to be a clear prescription for compensation, determining which governmental actions actually constitute a *taking* is not that easy.

Most legal analysts would probably agree that the exercise of "eminent domain"—that
is, the government condemnation of private property in order to accomplish some public
purpose, such as building improved transportation infrastructure—is a taking that requires
compensation. There are many examples of public policies that provide compensation for
those who may lose out from the exercise of eminent domain. Public works such as dams
or highways, for example, often include provisions for compensating individuals who lose
property as a result of the project. In the United States, the government had to purchase
land at market values for use in constructing the interstate highway system.

Interestingly, in this case, compensation was paid for property losses, but the property
rights created by the highways were given away for free. Individuals owning land adjacent
to the highway can rent out billboard space, and those near interchanges can establish
restaurants and service stations. The location of these highways generated windfall gains
to those situated in the right places. Meanwhile, some rural towns that saw traffic diverted
from the old highway to the new interstate have declined substantially due to business losses. While compensation was paid for land used to build the interstate highways, no compensation was paid for the losses generated by the location of the highways, and no effort
was made to capture some of the windfall gains generated by this system for the benefit of
the general public.

These considerations show that the determination that a taking has occurred in particular cases is not always a simple matter. In recent years, there has been great interest in the
United States in the concept of a *regulatory taking*. Regulatory takings are thought to occur
when the government implements a policy that reduces the value of an individual's property. In *Lucas v. South Carolina Coastal*, the U.S. Supreme Court ruled that an individual
was owed full compensation for the loss in property value that resulted when the state of
South Carolina adopted a law preventing development of beachfront property that may be
vulnerable to storm damage and is environmentally fragile.[13] The owner of the beachfront
land had purchased it to build a housing development, and the change in the law reduced
the value of his property. The law was not intended to cause harm to Mr. Lucas but rather
to achieve a broader societal goal related to coastal protection. In this case, a judgment was
made that the realization of these goals, regardless of their merit, placed an unfair burden
on Mr. Lucas, entitling him to full compensation for his lost property value.

Decisions about compensation are generally made on the basis of judgments about the
fair treatment of those who are affected by public policies. From an economic perspective,

it might be thought that the Pareto criterion would provide an adequate basis for making such judgments. After all, any policy that reduced the well-being of one person would violate the Pareto criterion, and unless that individual could be compensated for his losses, the policy would be rejected on those grounds no matter how great the benefits to the rest of the society. Economists are divided over whether the compensation principle requires compensation to be paid or whether hypothetical compensation is enough to insure that the policy change is socially beneficial. Kaldor argued that a policy change would satisfy the Pareto criterion if it produced enough benefits to compensate the losers while still leaving the winners better off, even if no compensation was actually made.[14] In contrast, Baumol believes that the compensation actually has to be paid if there is to be a clear-cut welfare improvement.[15] In many cases, there may be no reason to provide either actual or hypothetical compensation to those who lose from a policy change or a redefinition of property rights, because they should not be entitled to the income stream they are losing in the first place.

Policy changes almost always cause both increases and decreases in welfare. If these changes are to be aggregated to determine whether the benefits outweigh the costs, it may be necessary to make interpersonal utility comparisons. Faced with the apparent subjectivity of such an endeavor, economists are drawn to the Pareto criterion, which essentially gives each individual a veto on a policy change. The problem with this criterion is that few policy changes could be made if everyone had to agree to them. This excessive conservatism has seemed unacceptable to many analysts. The compensation principle appears to salvage Pareto optimality, making it possible to implement beneficial policy changes even though they harm some individuals.

Compensation is not available for policies aimed at redistributing income or wealth, however, and such policies would be ruled out altogether if decisions are made on the basis of the Pareto criterion with or without actual or hypothetical compensation. In other cases, one might question whether compensation should be made even if it is possible to do so and still realize net social benefits. Why, after all, should society compensate a firm that finds its costs of production increased by a law that prevents it from polluting scenic rivers, something many would argue it should not have been doing in the first place? In this context, it may make more sense to reconsider the moral force of Pareto optimality than to continue searching for ways to make it fit our intuitions about justifiable public policies.

Most economic analysis is implicitly utilitarian in that its aim is to discover policies that will increase the general welfare.[16] The Pareto criterion seems to make sense primarily in the context of utilitarian objectives, and it is important to note that the decision to pursue such objectives presupposes a prior value judgment concerning what is important. In addition, it is clear that Pareto optimality takes the initial distribution of entitlements, including the ways in which property rights are defined, as given and inviolable. Such a position, which would preclude such policies as environmental laws designed to change the property rights in air or water, for example, also reflects a prior value judgment. It could be argued that value judgments related to public policy, including the Pareto criterion, can be dealt with only through political and ethical discourse. If true, this suggests that the effort to use the Pareto concept to keep policy analysis objective, value free, and apolitical fails before it even gets off the ground.

In terms of practical policy analysis, it seems that limiting public policies to those that satisfy the Pareto criterion, either directly or through compensation, would rule out a great many collective efforts generally seen as beneficial. Public policy choices may be consistent with the Pareto criterion with or without compensation, but such decisions, including the decision whether to compensate, reflect political and ethical judgments about the appropriate collective action and whether individuals disturbed by the effects of a policy shift deserve to have their interests protected. We may believe that individuals should be compensated in the case of property losses occasioned by public infrastructure projects but not in the case of eliminating a de facto property right to pollute air or water. Public policy analysis has to incorporate political and ethical considerations along with the economic analysis described to this point. These issues are addressed in the next chapters.

SUMMARY

1. The invisible hand leads to a social optimum when individuals respond to competitive price signals. Market failures arise when prices are distorted through imperfect competition, risk and uncertainty, the existence of externalities, common pool resources, and public goods. The distorted price signals cause individuals to misallocate resources, leading to inefficiencies.

2. Natural monopolies occur when industries are characterized by increasing returns to scale or decreasing average costs. Some form of collective regulation is often required for such industries.

3. Risk and uncertainty can lead to market failures, and it may not be possible for individuals to purchase private insurance to cover risky or uncertain situations. Many government policies amount to public insurance when private insurance markets fail.

4. Externalities arise when property rights are misspecified. The Coase theorem shows that the assignment of property rights is immaterial in terms of efficiency if transaction costs are low enough to allow the parties to bargain. Property rights also influence income, however, so the way in which they are assigned can have an important impact on equity.

5. Open-access common pool resources are often destroyed by overuse because people are caught in a prisoners' dilemma in which the optimum strategy is always to defect. Government intervention is often necessary to prevent their destruction.

6. Public goods have two characteristics: nonrivalry and difficult exclusion. Nonrivalry means that the marginal cost of an additional consumer is zero. Difficult exclusion means that once the good is supplied, it becomes available to all whether they have paid for it or not. Difficult exclusion gives rise to the free-rider problem.

7. Income distribution and distributive justice are public goods that are likely to be undersupplied by free markets. Income distribution is measured by Lorenz curves and Gini coefficients.

8. Compensation is often suggested as a way to maintain Pareto optimality. If a policy leads to benefits that are greater than the costs, the winners can compensate the losers and still be better off. Compensation cannot be used for pure redistribution

of income, which may be justified on political or ethical grounds that overrule the Pareto criterion.

KEY CONCEPTS

market failures
increasing returns to scale
natural monopolies
information
risk and uncertainty
moral hazard
adverse selection
externalities
Coase theorem
property rights
common pool or common property
  resources
prisoners' dilemma
open access
enforceable contracts
public goods

nonrivalry, jointness, indivisibility
difficult exclusion
free-rider problem
distributive justice
income distribution
relative deprivation
welfare state
redistribution
income
wealth
endowments
entitlements
Lorenz curve
Gini coefficient
compensation
taking
regulatory takings

DISCUSSION QUESTIONS

1. Give five examples of externalities that you have encountered and explain whether each is a positive or a negative externality.

2. Why does the U.S. government need to subsidize crop insurance?

3. One solution to the problem of common pool resources is to assign private property rights to the resource. What are some other solutions?

4. Pure public goods are characterized by nonrivalry and difficult exclusion. What are some examples of goods that have only one of these characteristics? Are there collective action problems associated with any of these goods?

5. Many government programs aim to improve efficiency by correcting market failures. Others redistribute income either intentionally or inadvertently. Do you believe that government interventions should be limited to improving efficiency, or is there a place for policies designed to reduce income inequality?

6. What alternatives are there for correcting the market failure associated with natural monopolies?

7. Many government programs benefit one group at the expense of another, thereby violating the Pareto criterion. In some cases, compensation provides a way to satisfy the Pareto criterion. Explain how compensation can make government interventions consistent with Pareto optimality, and give examples of programs in which compensation can and cannot be carried out.

SUGGESTIONS FOR FURTHER READING

Barzal, Yoram. *Economic Analysis of Property Rights*. New York: Cambridge University Press, 1989.

Cornes, R., and T. Sandler. *The Theory of Externalities, Public Goods and Club Goods*. New York: Cambridge University Press, 1996.

Ostrom, Elinor. *Governing the Commons*. New York: Cambridge University Press, 1990.

Sandler, T. *Global Challenges*. New York: Cambridge University Press, 1997.

Stiglitz, Joseph E. *Economics of the Public Sector* Third Edition. New York: W. W. Norton and Co., 2000.

NOTES

1. The mathematical formulation for this example is from K. C. Kogiku, *Microeconomic Models* (New York: Harper and Row, 1971)pp. 111–112.

2. Joseph Stiglitz, *Economics of the Public Sector* (New York: W.W. Norton and Co., 1986).

3. S. N. S. Cheung, "The Fable of the Bees: An Economic Investigation," *Journal of Law and Economics* 16 (April 1973):pp. 11–33.

4. G. Hardin, "The Tragedy of the Commons," *Science* 162 (1968): 1243-48.

5. Roy A. Rappaport, "Ritual Regulation of Environmental Relations among a New Guinea People," in *Ecology, Meaning and Religion* (Berkeley, Calif.: North Atlantic Book, 1979).

6. Adam Smith, *An Inquiry into the Nature and Causes of the Wealth of Nations* (1776; reprint, Chicago: University of Chicago Press, 1976),p.213.

7. John Rawls, *A Theory of Justice* (Cambridge: Harvard University Press, 1971).

8. See Aldi J. M. Hagenaars, "The Definition and Measurement of Poverty," in *Economic Inequality and Poverty*, ed. Lars Osberg (New York: M. E. Sharpe, 1991).

9. The statistics used in this discussion are taken from the annual *Statistical Abstract of the United States* (Washington, D.C.: GPO, 2000). Means-tested programs are those that require some evidence that recipients are below the poverty line before benefits are distributed.

10. Amartya Sen, *Poverty and Famines: An Essay on Entitlement and Deprivation* (New York: Oxford University Press, 1984).

11. Simon Kuznets, "Economic Growth and Income Inequality," *American Economic Review* 45, no. 1 (March 1955):pp. 1–28.

12. Patricia Ruggles, "The Impact of Government Tax and Expenditure Programs on the Distribution of Income in the United States," in *Economic Inequality and Poverty*, ed. Lars Osberg (New York: M. E. Sharpe, 1991).

13. For an interesting discussion of this and other regulatory takings cases, see Leigh Raymond, "The Ethics of Compensation: Takings, Utility and Justice," *Ecology Law Quarterly* 23, no. 3 (August 1996): 577–622.

14. N. Kaldor, "Welfare Propositions of Economics and Interpersonal Comparisons of Utility," *Economic Journal* 49 (September 1939): pp. 549–552.

15. W. Baumol, *Economic Theory and Operations Research* (New York: Prentice-Hall, 1977), 526–31.

16. A more detailed description of utilitarianism and other philosophical approaches to ethics is found in Chapter 6.

5

Public Choice and Political Economy

INTRODUCTION

In 1993, Joseph Stiglitz, a highly regarded academic economist (Princeton and Stanford), joined the Council of Economic Advisers (CEA), a group of economists appointed by the president of the United States to analyze economic policy issues and provide guidance on alternative policy initiatives. Stiglitz was a member of the CEA from 1993 to 1995 and served as chair from 1995 to 1997. Based on his experiences at the CEA, he presented a lecture highlighting the problems of using economic analysis to improve public policies.[1] He noted that there were a great many policy changes that could have been made that would have satisfied the Pareto criterion, either with or without compensation, or that would have come very close to satisfying it in the sense that only a very small group of relatively wealthy individuals would have suffered small losses from the change. Although he was able to report a few successes, the thrust of his lecture was that most of the potential Pareto improvements were blocked by various informational and strategic problems raised by interest groups and other political constituents.

In the last chapter, a fairly clear and compelling role for the government was identified: implement policies that correct market failures and leave the parts of the economy in which there are no market failures alone. The existence of market failures opens the possibility for changes that will increase the well-being of some without reducing the well-being of anybody else, although achieving this Pareto outcome will often require compensation. At the end of the last chapter, we raised some concerns about the Pareto criterion as a guide for making and evaluating public policies. It may turn out that redistributive policies that violate Pareto optimality will be judged socially beneficial on the basis of other criteria. But even if the Pareto criterion is inadequate for evaluating all conceivable policy options, it is hard to imagine why it would ever be in the best interests of society not to make changes that benefit some while leaving no one worse off. And yet, according to Stiglitz, such outcomes are common. The purpose of this chapter is to explore this problem in greater detail.

One easy explanation for the difficulties associated with realizing Pareto optimal changes might be that politicians and bureaucrats are misguided, venal, stupid, or irrational. The problem with this explanation is that much of economic theory is based on the expectation that people are rational, reasonably honest, and well informed, and there is some evidence that this expectation is generally met. If people, including voters, bureaucrats, and politicians, are rational in their private economic decisions, why would they suddenly become irrational or misguided in their public decisions? Rational voters would be expected to vote ineffective or dishonest politicians out of office, and rational politicians, realizing this, would be expected to search for optimal policies to keep the voters happy.

On the other hand, the public sector might be plagued with situations similar to those described by the prisoners' dilemma, in which rational pursuit of self-interest leads to suboptimal outcomes. Because much of politics involves bargaining and strategic behavior under conditions of uncertainty and incomplete information, it may not be too surprising that the government frequently fails to take advantage of the opportunities for Pareto improvements. In fact, there is a branch of economics, known as "public choice economics," that is devoted to this precise question. Public choice economists emphasize the difficulties of ascertaining the true preferences of citizens who may have incentives not to reveal their desires in the hopes of free riding on the provision of public goods. Even if the analyst can accurately discern preferences, finding a way to aggregate conflicting preferences is generally problematic. These problems lead to the notion that government failure can be as significant in public policy analysis as market failure.

In the next section of the chapter, we explore a problem with the basic conclusion of the last chapter, that correcting market failures will increase social welfare. This is known as the problem of the "second best." Following discussion of the second best, we turn to the insights of public choice economics, which include the analysis of different types of voting rules, rational voter behavior in democracies, preference revelation, lobbying and special-interest groups, and preference aggregation. In the final part of the chapter, we return to the experiences Stiglitz describes in his lecture and his explanations for the difficulties of making apparently sensible policy changes.

THE THEORY OF THE SECOND BEST

The hypothetical economy described in Chapter 3 achieved allocative efficiency and Pareto optimality automatically as long as producers and consumers behaved rationally and the basic conditions of perfect information, perfect competition, and no market failures were met. As shown in Chapter 4, however, the conditions required for the invisible hand to perform its magic are extremely demanding and unlikely to be met in any real-world economy. Market failures are common, and it may not be possible for the government to correct them all or to insure that the other conditions are met. It would appear that the best that can be done is to correct those market failures that can be corrected while ignoring those that cannot. The *theory of the second best*, however, shows that this conclusion is not necessarily justified. In other words, in an imperfect economy with numerous market failures, welfare may not be increased by correcting some of the market failures while leaving others in place.

A mathematical demonstration of the theory of the second best is presented in Appendix 5.1 at the end of this chapter. The basic intuition behind these results is that a market failure that cannot be eliminated may lead to distortions in related markets so that the conditions for a second-best optimum may require systematic distortions in other markets to compensate for the market failures that cannot be corrected. Suppose, for example, that there is a monopoly that also produces a negative externality. Recall that in the case of negative externalities, firms are likely to produce more than is socially optimal. A monopoly restricts output to maximize profits, however, and this restrictive effect may offset the impact of the negative externality. If the monopoly were broken up but nothing was done about the negative externality, the competitive firms would produce more total output with more of the negative externality. In this case, it may be better to leave the monopoly in place rather than breaking it up using antitrust laws.

One of the best examples of the application of the theory of the second best comes from international trade theory. Basic trade theory shows that the first-best optimum is free trade among all countries. However, individual nations often have incentives to pursue protectionist policies (for reasons that will be addressed later in this chapter). Perhaps welfare could be increased by setting up zones of free trade (known as customs unions or free trade areas), even though relations between these zones would continue to be governed by protectionism. Customs unions are made up of several countries that agree to eliminate all trade barriers applied to goods traded among the members of the union while adopting a common external tariff with respect to countries not belonging to the union. The European Union (EU) is an example of such an arrangement. Note that the North American Free Trade Agreement (NAFTA) between the United States, Mexico, and Canada is not a customs union because there is no common external tariff.[2] Free trade agreements have many of the same effects as those generated by the formation of a customs union, and the two are often analyzed in a similar manner.

Customs unions are an example of a second-best solution, the first-best arrangement being universal free trade. The question for countries contemplating the formation of a customs union is whether such an arrangement will increase welfare, assuming that universal free trade cannot be attained. Given what has been said about the theory of the second best, it is not surprising that the answer to this question is ambiguous. Welfare may or may not increase. The formation of a customs union is likely to lead to increased trade among the members of the union and reduced trade with countries outside the union. Suppose that a country imports agricultural goods from a country that has a comparative advantage (see Chapter 2) in agriculture. The formation of the customs union between these two countries leads to expansion of this trade, and this *trade creation* will lead to increased welfare for the importing country as the deadweight losses from the trade barrier are recovered.

On the other hand, if the country had traditionally imported food from a country outside the union, the formation of the union can lead to *trade diversion* as trade shifts from the low-cost producer with a comparative advantage outside the union to a higher-cost producer within the union. Trade diversion may reduce welfare because the importer is now obtaining its food from a higher-cost producer, and the government earns no tariff revenue on the imported food. Thus, the welfare effects of creating a customs union depend on whether there is more trade creation than trade diversion and the precise nature of the trade diversion that is generated. It is conceivable that greater welfare would be achieved by maintaining a generalized protectionist structure as opposed to setting up free-trade zones.

There are many other examples where second-best theory gives rise to somewhat counterintuitive results. For example, the simple prescriptions about trade liberalization are somewhat problematic in the presence of international environmental externalities.[3] This theory raises some serious problems for policy analysis. It suggests that detailed analysis of the entire economy would be required before deciding whether interventions that would ordinarily be justified on the basis of arguments about market failures are warranted. Because the informational requirements for such analyses could be substantial, the theory of the second best would seem to be a prescription for inaction. While recognizing the significance of these theoretical results, most analysts do not attempt to locate second-best optima, relying instead on straightforward application of the standard results of welfare theory. In most real-world cases, it appears that correcting market failures will enhance welfare.

It is in this sense that Stiglitz expresses his frustration at the inability of government policy makers to realize simple Pareto improvements. For Stiglitz and most other analysts, the government has a responsibility for certain public goods, redefinitions of property rights to correct externalities, and a variety of other efforts that are expected to improve efficiency, and such actions can be recommended without worrying greatly about the full implications of the theory of the second best. Still, the theory sounds a cautionary note about these efforts[4] and provides ammunition for those who prefer less government intervention for whatever reason.

The Political Setting for Public Policy

Much of the discussion so far has focused on the conceptual foundations for policy analysis, with little reflection on the actual processes through which policies are designed, selected, implemented, and evaluated. The models developed in earlier chapters include consumers, firms, and an abstract government capable of implementing policies leading to a set of expected improvements in economic performance. Such models are useful for identifying some general principles, but they are far too simple to capture the full complexity of public policy making. To begin with, the government is not a monolithic entity that functions as if driven by a single consciousness. Governments usually include legislative, judicial, and executive branches. In democracies, the legislative branch is elected and charged with establishing the laws and regulations that are at the heart of public policy. The executive branch is then charged with implementing these legal institutions. In the United States, the executive branch is led by an elected president who is responsible for overseeing the implementation of policies and working with the legislature in the design and selection of these policies. In parliamentary systems, a prime minister is chosen from among the successful legislative candidates to form and head the government. In both cases, actual policy implementation is carried out by appointed officials as well as the permanent staffs of various government agencies. The judicial branch is less directly connected to policy making but has important roles in enforcing the rules created by the other branches and in interpreting the law and refining its ambiguities both at the constitutional level and at the level of the criminal and civil codes.

For our purposes, it is useful to think of the legislative and executive branches as being made up of politicians and bureaucrats. Politicians are elected by the voting public to design and select policies aimed at accomplishing broad public purposes. Because the public is normally divided over what constitutes appropriate public policy, politicians in the legislative and executive branches are likely to be similarly divided. Such divisions can be made more complicated by the existence of political parties that aim to promote broad agendas that will attract more voters. The implementation of the policies created by politicians is done by bureaucrats, who are permanent employees of the various government agencies and as such are not subject to elections. The producers and consumers in our earlier models are also voters whose policy preferences are influenced by the nature of their particular roles as consumers, producers, and citizens. Of course politicians and bureaucrats are also voters and consumers with their own sets of preferences. There is no reason to expect that any of these agents would behave rationally only when making economic choices, becoming irrational when faced with political decisions. Public choice economics begins with this insight, treating all these agents as rational pursuers of self-interest. In this sense, public choice is the application of economic thinking to politics.[5]

If politicians and bureaucrats are rational utility maximizers, what exactly is it that determines their utility? As consumers, of course, these individuals would have the same kind of utility functions as described in previous chapters. In their roles as politicians and bureaucrats, however, one might expect that other arguments would enter into the utility functions. For politicians, a primary source of increased utility is probably the ability to be reelected. If reelection depends on campaign contributions from *special-interest groups*, politicians are likely to pursue such contributions with great vigor. At the same time, it is possible that politicians genuinely wish to adopt policies that will serve the interests of their constituents and of the nation as a whole. In other words, a politician's utility may depend not only on the narrow self-interest of retaining her job but also on the realization of policies that serve the public or national interest. One problem is that local interests to which the politician must be the most responsive if she wishes to be reelected may not actually be in the broader national interest. It is easy to imagine circumstances where satisfying local desires will lead to reductions in general social welfare. Such situations are often referred to by the pejorative term *pork barrel* politics. In some cases, however, policies that please the voters or special-interest groups may also be in the national interest so that the advancement of a politician's interests in retaining her or his job may not conflict with the effort to promote the public interest, at least as she understands it. Likewise, both private and public interests may serve to motivate the individuals working in government bureaucracies. Although bureaucrats are often seen as inefficient, self-serving individuals protected by membership in a civil service that does not require accountability, it is not difficult to find people working in government agencies who are genuine public servants in the best sense of that expression. It is also not difficult, of course, to find bureaucrats who are lazy and unproductive, but it should be noted that the same could be said of some workers in the private sector. It is often thought that bureaucrats seek to increase their power through working to obtain better positions, larger budgets, or an expansion of the responsibilities and staff assigned to their agency. This type of motivation could lead to growing demands for public resources even if the expansion of the agency is not in the public interest. In addition, some bureaucrats may attempt to use their positions to advance personal preferences that are inconsistent with the decisions and directives generated by the political system.

It should be noted that similar problems can arise in large private-sector firms, many of which are almost as bureaucratic as the government. Workers in these firms may behave in much the same way as public bureaucrats, competing for additional influence within the firm, using the inability of the firm to detect shirking as cover for inefficiency or incompetence, or attempting to move the firm in directions that are not consistent with the interests of the shareholders or the policies of the firm's management. Problems such as these are often referred to as *principal-agent problems*. They arise in circumstances where one individual or agency (the principal) makes decisions on the policies the organization is to pursue, while others (the agents) are charged with implementation. Because the principals never have complete control over the actions of the agents, there will frequently be some divergence between the intentions of the principals and the actions actually carried out by the agents.

A classic example of the principal-agent problem is the conflict between the interests of a firm's shareholders (the principals) and its management (the agent). The shareholders may wish maximum profits consistent with growth in the value of the firm along with generous distribution of the profits in the form of dividends. The management, on the other hand, may wish to maximize growth by sacrificing profits or to maintain a peaceful climate in its relations with its employees by offering generous wages and bonuses out of the firm's revenues, thereby reducing the size of the profits and dividends for shareholders.

In the public sector, similar agency problems can arise in the implementation of policies designed by politicians but put into practice by bureaucrats. At one time, the U.S. Congress passed legislation requiring the United States Department of Agriculture (USDA) to monitor the state of "family farms." To carry out this directive, the USDA had to develop a definition of family farm so that such farms could be identified and monitored. Whether this definition corresponded to the image in the minds of the members of Congress, or whether they even had a coherent image of the family farm, is impossible to determine. In 1993, the Navajo nation sued the U.S. government and a private coal company over royalty payments that had been negotiated by the U.S. Department of the Interior on their behalf during the 1980s. The Navajos claimed that the secretary of the interior at the time had not represented their best interests. Their case against the U.S. government was thrown out on the grounds that U.S. law does not actually require the Interior Department to act in the best interests of the Navajos. However, the judge reprimanded the former interior secretary for making a biased decision following private meetings with the coal company.[6] In this case, the agent (the secretary of the interior) did not act in the best interests of the principals (the Navajos) who owned the coal deposits that were being mined by the coal company.

Policies to protect *food safety* in the United States are another example of potential divergence between the intentions of a policy and its actual effects as implemented. The passage of the Meat Inspection Act in 1906 was stimulated by the publication of Upton Sinclair's novel, *The Jungle*, in which a wide range of unsanitary and unhealthy practices of the meatpacking and processing industries were described. Since then, Congress has added new policies to modify the coverage and scope of food safety legislation related to meat products. To implement these policies, the USDA established the Food Safety Inspection Service (FSIS), which employs a large number of inspectors charged with visually examining slaughtered carcasses and meat products for signs of contamination and disease. In recent years, problems with bacterial infections of meat (salmonella and *E. coli* bacteria) have increased. These pathogens cannot be detected by visual inspection. Policies designed to address this issue have been largely stymied by an extremely effective lobbying campaign by the meat industry.[7]

In 1996, in an effort to more effectively control bacterial infections in meat, the FSIS introduced a system known as Hazard Analysis Critical Control Point (HACCP). Under HACCP, packing plants are required to identify critical points in the processing procedures where contamination may occur and set up monitoring systems at these points to insure that the procedures are functioning correctly. This system aims to prevent contamination before it happens and places greater responsibility on the industry itself for controlling unsanitary meat. The role of the inspectors has shifted from visual examination of carcasses to oversight of the monitoring processes at the critical points.

Many of the inspectors believe that this system will be ineffective in insuring food safety, and there have been numerous conflicts between the inspectors and the managers of plants.[8] The inspectors believe that HACCP will allow the industry to avoid the economic costs of producing safe products. There has long been an adversarial relationship between meat inspectors and the managers of meatpacking plants. The inspectors can reject meat and even close down the plant if they find unsanitary conditions. Such actions can be very costly to the packers. At the same time, the inspectors are in somewhat precarious positions in that they are subject to pressure from the packers not to interpret the standards so tightly that plants are shut down frequently and meat products rejected as unsafe. Without strong support from the USDA, the FSIS, and the legal system, the inspectors may fear that conscientious implementation of the inspection rules could be dangerous for their personal well-being. This adversarial relationship complicates the implementation of the HACCP system. Note that none of these issues has anything to do with the benefits and costs of the new rules, which, according to one study, would favor the new system under a wide range of assumptions about its actual impacts on public health.[9]

In this case, the beliefs and experiences of the agents charged with implementing the new procedures run counter to the beliefs of the principals (FSIS, USDA) about effective control of meat safety and about the reliability of the meat industry to police itself. Eventually, a compromise of some sort will be worked out and the new rules will be implemented in some way. It is likely, however, that the precise nature of the final practices will be quite different from the original intentions of those designing this new policy. Note that HACCP is a new procedure developed by an executive agency to address a problem that has not been dealt with through congressional legislation. Several attempts have been made to introduce legislation aimed at improving the safety of the U.S. meat supply, which has been brought into question by an increasing number of food poisoning cases. The bills introduced by these legislators have never been voted out of the agricultural committees charged with their initial review. Lewis believes that this inaction is due to the influence of the meat industry lobby.[10]

Groups such as the meat industry lobby are important actors in the policy process. Lobbies and special-interest groups attempt to influence both the politicians designing new policies and the bureaucrats charged with implementing them. In many democracies, there are large numbers of these groups which seek to advance policies favorable to their members and prevent policies perceived to be unfavorable to them. A special-interest group provides a public good to its members in the sense that obtaining favorable legislation benefits all those associated with the group, whether they support it or not, and the amount of the benefits available is the same for everyone. In other words, the benefits are characterized by difficult exclusion and nonrivalry. The public-good nature of the efforts of special-interest groups would normally be expected to give rise to free riding that would reduce their effectiveness. Olson has shown, however, that relatively small groups with very intense interests are often able to overcome the free-rider problem.[11]

On the other hand, large, diverse groups (e.g., consumers) have great difficulty in developing effective organizations to advance their interests, particularly as the individual benefits are likely to be small. Experience in the United States shows that special-interest groups, such as the meat industry associations, are generally quite effective at influencing public policy, often at the expense of larger, less-organized groups. It is not uncommon, in fact, for some of these groups to work directly with Congress and the bureaucracy on the design and implementation of public policies.

The picture of special-interest groups that seems to emerge from the preceding discussion is one of selfish individuals working to advance their interests at the expense of the rest of society. In many respects, this image is quite accurate. It should also be noted, however, that special-interest groups have much valuable information about the details of their particular activities, and this information can be very important in making policies. It would be impossible for members of Congress to have expert knowledge on everything from the biochemical processes in meat contamination to the engineering and physics of nuclear waste disposal. In a sense, lobbying activities are a lot like advertising. Advertising almost always contains a large dose of misinformation and persuasive propaganda designed to convince people to buy the advertised product, but it may also contain information on prices or particular product characteristics that may be of interest to consumers seeking to make the best choice. In a similar manner, lobbies may provide useful technical information to both politicians and bureaucrats, whose expertise necessarily has to be concentrated in broader areas.

The stylized model developed in this section includes politicians and bureaucrats who make decisions on the basis of their own preferences and judgments as well as under the influence of consumers and producers in their roles as voters or members of special-interest groups. Politicians usually belong to a political party which aims to organize diverse voters in support of the party's agenda. From a public choice perspective, all of the participants in the policy process attempt to advance their self-interest. Because the interests of these individuals vary greatly, the process is characterized by conflict and stalemate, bargaining and compromise.

VOTERS AND POLITICIANS IN THE CHOICE OF PUBLIC POLICIES

As shown in Chapter 4, consumers of public goods may have an incentive to hide their true preferences in the hope of avoiding payment for a good they expect to be provided in any case. Why should one contribute to National Public Radio during the all-too-frequent pledge drives, when the broadcasts can be heard whether one contributes or not? In a similar manner, citizens in their role as voters may not be able or willing to reveal their true preferences for public goods provided by the government. In the case of private goods, a price for each unit is established by market supply and demand, and the total quantity produced and consumed is determined by aggregating all the individual supplies and demands that are generated at this price. In the case of a public good, consumers cannot buy units of the good at some price per unit because the good is indivisible. The unit of public defense that is consumed by each individual is the same and equal to the total amount supplied. Once it is decided that a public good is to be provided, itself a political decision, the amount of that good to supply is usually determined through political processes.[12] Political arguments about the size of the defense budget or the amount of overall government spending are staples of election-year politics.

In nondemocratic regimes, dictators can avoid the problem of preference revelation by substituting their preferences for those of the citizens of the country. In representative democracies, however, voters elect representatives who are supposed to advance policies reflecting the voters' preferences. This can give rise to the kinds of principal-agent problems described before. In general, political candidates propose policies they hope will be consistent with the preferences of a large enough number of voters to insure their election. Their problem is to determine what the voters prefer in a setting where there is little other than public opinion polls to serve as a guide and where voter preferences are fluid and subject to change and manipulation. As accurate as survey polls may be, they may not capture the precise preferences of the public for the kinds and amounts of public goods that they would like to see provided.

In addition, once elected, politicians spend most of their time on relatively obscure technical issues (e.g., how to inspect meat in packing plants or whether a particular pesticide should be controlled) that are of interest to only a tiny segment of the voting public. The main source of information for determining a position on these issues is the research of the politician's staff and the information provided by concerned constituents and special-interest groups. For broader public issues, polls may provide some guidance, but there are almost always sharp divisions among voters about the policy questions that are brought up for public debate. Faced with polls indicating that 40 percent of her constituents favor increased defense spending while 40 percent oppose the increase and 20 percent are undecided, a particular politician may be hard-pressed to determine an appropriate course of action. Somehow, diverse public preferences have to be detected, aggregated, and translated into specific policy decisions. In democracies, such tasks are accomplished through voting.

Voting Rules

It is often thought that a *simple majority* (i.e., more than 50 percent) of the voters is enough to insure that the results of an election are democratic and that democratic results are sufficient for appropriate aggregation of voter preferences. In the modern world, there are few examples of *direct democracy* in which every citizen votes on whatever issue is before the public. In *representative democracies*, voters elect a relatively small number of individuals to represent their interests in the legislature. The election of these representatives is usually determined by a simple majority. However, there are other *voting rules* that could be used. For example, a *supermajority* of two-thirds or three-quarters of the voters could be used in place of a simple majority voting rule. In general, such voting rules are used in legislative or constitutional deliberations rather than for the popular election of representatives.

Thus, in the United States, constitutional amendments have to be approved by majorities of two-thirds of the members of both houses of Congress and ratified by three-quarters of the states. In the European Union, decisions often require agreement by supermajorities based on a system of weights for the different countries. A coalition of a large country such as Germany with two or three small countries such as Belgium, Austria, or Greece may be enough to block a law that is favored by the other eleven or twelve countries belonging to the EU. At the extreme, some decisions require *unanimity*, that is, agreement by all of the parties. This is the case for decisions taken by the World Trade Organization, for example, where rules on international trade are adopted only with full consensus.

A unanimity voting rule means that each party can veto any decision. If a decision requires everyone's agreement, a single individual can block it by withholding support. Unanimity is the only voting rule that is consistent with the strict application of the Pareto criterion. Recall that Pareto improvements make at least one individual better off without reducing the well-being of anyone else. If a policy is proposed that improves the lot of some at the expense of others, the Pareto criterion requires that those who lose out from the change be able to veto it. In the last chapter we noted that redistribution violates the Pareto criterion because the benefits to individuals with low incomes are transferred from the wealthy, whose utility is consequently reduced. If the Pareto criterion is taken as the exclusive guide to policy, such changes would be disallowed even if the utility loss of the wealthy is less than the utility gain of the poor. A unanimity voting rule allows this criterion to be enforced.

But it is not only on questions of redistribution that vetoes might be exercised under a unanimity rule. Virtually every decision taken by legislatures is opposed by someone, because any government intervention is likely to have an impact not only on efficiency but also on the distribution of income-earning opportunities. A requirement for unanimous consent would mean that very little would ever be decided. Unanimity is clearly a more stringent voting requirement than the majority rules. Likewise, supermajorities are more difficult to assemble than simple majorities. It is for this reason that these more demanding voting rules are generally reserved for such extremely important decisions as constitutional change or the adoption of laws that pertain to areas thought to be fundamental to the integrity of organizations or states.

The advantage of stricter voting rules can be seen by noting that almost half the voters may be dissatisfied with the outcome if the decision is made by simple majority. Suppose that democratically elected representatives must decide whether to provide government-subsidized disaster insurance to farmers. If we assume that the public's preferences on this issue are precisely reflected by the representatives (a highly dubious assumption), a simple majority of 52 percent leaves 48 percent of the public dissatisfied. While technically a minority, 48 percent of the voters may constitute a very large number of people. If instead of deciding on crop insurance, the legislature is considering a law that would affect a fundamental right such as religious freedom or the ownership of private property, it may be unwise or unjust to allow the majority to determine the outcome. Suppose that 45 percent of a country's population is Muslim and 55 percent is Christian. If the constitution provides that any issue proposed is to be settled by a simple majority rule, the Christians could propose the expropriation of all of the property of the Muslims and use their majority to enrich themselves at the expense of their fellow citizens. A voting rule requiring a super-majority might preclude this outcome (unless a substantial number of Muslims agreed that their property should be expropriated). For fundamental issues, minorities may need protection from the *tyranny of the majority* through either stringent voting rules or legal side constraints such as the Bill of Rights in the United States, which establishes certain rights as inviolable even if a majority wishes to override them.

The disadvantage of these more stringent voting rules is the inverse of their advantage. If a three-quarters majority is required for an action to be taken, 26 percent of the population can prevent its implementation. This *tyranny of the minority* is a recipe for inaction because it is generally not too difficult to muster enough opposition to block virtually anything that might be proposed. It is for this reason that most ordinary public business is decided on the basis of simple majorities rather than the more stringent rules, which are reserved for cases where fundamental principles and minority rights need protection.

The tyrannies of the majority and the minority are not the only problems with democratic decision making. Over two hundred years ago, a French nobleman, the Marquis de Condorcet, identified an interesting problem with majority voting. Suppose that three individuals (Long, Tall, and Sally) are faced with deciding how to divide up a sum of money (say, $300) on the basis of majority voting. It is likely that their initial inclination would be to vote for an equal division, $100 for each. However, Sally may recognize that she could do better for herself by striking a deal with one of the others, say Tall. Suppose Sally and Tall propose a division of $120 for each of them and $60 for Long. Sally and Tall would vote for it and Long would vote against it. Based on majority voting, Sally and Tall's proposal would win because two out of three is a majority. Realizing this, Long might try to tempt Tall by proposing a division of $130 for Long and Tall and $40 for Sally. This proposal would also garner a majority with Long and Tall in favor and Sally opposed. And so would an infinite number of other proposals that the three might devise. The problem identified by Condorcet (and later by C. L. Dodgson, a mathematician who, under the pen name Lewis Carroll, wrote the classic book *Alice in Wonderland*) is often referred to as *cycling*. One can imagine a series of votes on different pairs of alternatives where the new alternative always defeats the one approved in the previous round so that there is never a definitive outcome. Voting cycles seem to undermine the effectiveness of majority voting in aggregating voter preferences into a precise decision.

Another possible problem with majority voting can arise if it is possible for representatives to trade votes, a process known as *logrolling*. Suppose that there are three representatives (Urban, Rural, and Suburban) who have to decide on two proposals, one to provide food stamps to low-income urban consumers and the other to provide direct income supplements to farmers. Urban strongly favors food stamps and is mildly opposed to the income supplements for farmers. Rural strongly favors the income supplements and is mildly opposed to food stamps. Suburban strongly opposes both. If logrolling is impossible, neither proposal will be approved because neither can command a majority. On the other hand, suppose that Urban contacts Rural and offers to support the proposal for income supplements if Rural will agree to support the food stamp proposal. Given that neither is strongly opposed to these proposals, they may be able to make a bargain, with the result that both proposals are approved even though neither actually has the support of the majority.

Logrolling is a common feature of most modern democracies. The U.S. food stamp program has historically been administered by the USDA as part of overall farm policy. The strategy has been to tie this program, which is popular with urban legislators, to farm programs, which might not otherwise obtain the necessary support in a country that is predominantly urban. It is sometimes thought that logrolling leads to excessive growth of government programs as policies that would not normally be approved garner sufficient votes through bargaining and vote trading. While it is possible that logrolling could lead to such results, it may also serve to resolve stalemates in cases where there is a consensus that something should be done but there is insufficient support for any of the alternatives. Likewise, logrolling is often thought to protect intensely felt minority interests, as in the case of farmers who may feel that their interests will be neglected unless they can offer programs such as food stamps to gain support for farm programs from urban lawmakers. The key in all these cases is the intensity of the voter preferences involved. If a group of voters or their representatives feel very strongly that a particular policy needs to be adopted, they may be able to bargain with others, offering the only currency of value in political arenas, their votes, to those whose mild opposition can be overcome by the promise of support for their concerns.

Differences in *preference intensity* are common in both the economic and political arenas. In a market, consumers can express the intensity of their preferences by allocating more dollars to the purchase of preferred items. Voters, however, generally have only one vote to allocate, which makes it difficult to determine how strongly they favor a particular position. Logrolling is one way to allow preference intensities to count. There are other ways to more fully exploit the information contained in the differing intensities of voter preferences.[13] One of the most famous was suggested by Jean-Charles de Borda in the eighteenth century. Suppose that there are five candidates or policy proposals. The *Borda procedure* allows each voter to rank the five alternatives, assigning five points to the most desirable, four to the second most desirable, and so on until reaching the least desirable, which receives one point. The winner is the one with the largest number of total points. A similar procedure is to assign ten votes to each voter, who can then allocate them among the alternatives, putting all ten on one if that alternative is strongly preferred or distributing them among the alternatives if there are no strong favorites. These procedures are rarely used in the election of representatives or in policy decisions within legislatures even though they could potentially increase the accuracy of the expression and aggregation of voter preferences. There are circumstances, however, in which simple majority voting can lead to a fairly accurate picture of consumer preferences as well as determinate outcomes in the sense of avoiding the problem of cycling. One such case is examined in the next section.

The Median Voter

One of the earliest and most influential results in the literature on public choice is the *median-voter theorem*, usually attributed to Downs, although, according to Mueller, it was first described by Hotelling in 1929.[14] Figure 5.1 shows a continuum that includes all the possible political positions between those on the far left and those on the far right. Point *M* represents the mid-point or median of the continuum. This means that half of the possible political positions are located to the right of *M* and half are to its left. Suppose that each voter has a preference for one of the positions on this continuum and will vote for candidates who are closest to that position. Voters whose preferences are located at the point marked *A* and to its left will choose a candidate whose position is at *A* because that position is closer than any other to the right of *A*. Voters with preferences at *B* or to its right will favor candidates with positions at *B* for the same reason. Voters between *A* and *B* will choose the candidate who is closest, which means that those to the left of the midpoint of the segment between *A* and *B* (point *X*) will vote for the candidate at *A*, while those to the right of the midpoint will vote for the candidate at *B*. If voters are evenly distributed along the continuum, the candidate defending a position at *B* will win because there are more voters from *B* to the right than from *A* to the left and the voters between *A* and *B* are equally divided between the two.

Left	A X M B	Right

Figure 5.1: The median voter.

   Clearly, both candidates have an incentive to position themselves as close to the median as possible. Candidates who take extreme positions can never win as long as the assumptions underlying this model are met. It turns out that even if the distribution of voters along the continuum is skewed, candidates will still be able to claim the group that is more extreme than they are and divide the rest so that a movement toward the median will capture voters from the other candidate. In situations where candidates have incentives to position themselves at the median, the problem of deciding what position to take when confronted with a divided public is made much easier.

   Suppose that the 40 percent of the voters who favor more defense spending are on the right while the 40 percent who favor the status quo are on the left. If the remaining undecided voters have preferences that lie somewhere between these two positions, a candidate on the left could propose modest increases in defense spending that would be more acceptable to the 40 percent who favor the status quo than the opponent's proposal and would attract a greater share of the median voters. The candidate on the right might also find it advantageous to temper his or her proposals in an effort to gain undecided votes. The result may be that both candidates end up with similar positions so that the final outcome depends on other factors, such as who was better able to get out the vote, than any real differences in their positions on defense spending.

The median-voter theorem offers one explanation why many voters in the United States see little difference between Republicans and Democrats. Although it is frequently the case that the overall political philosophies of the two parties are different, the actual policies pursued by successful candidates usually are quite similar. Because the party activists are often more extreme than the bulk of the electorate, candidates may have to adopt fairly extreme positions in order to secure the party's nomination. Once nominated, however, these candidates must move as quickly as possible toward the center if they are to have any chance of winning the election. In the United States, the 1964 Republican presidential candidate, Barry Goldwater, and the 1972 Democratic presidential candidate, George McGovern, suffered crushing defeats because they were seen by many voters as too far from the center. In terms of Figure 5.1, winning Republican candidate Richard Nixon was at B in 1972 while McGovern was at A.

For some, the median-voter theorem proves the superiority of two-party democracies. Because the candidates are forced to the center, there is continuity in the policies a country follows, and the danger of extremism is reduced. In addition, if there are only two candidates, the winner must necessarily have received a majority of the votes. Two-party elections may help maintain a kind of consensus that appeals to the largest number of voters, even if the voters on the extremes begin to feel unrepresented or if many feel that there is no real difference between the candidates and therefore no particularly good reason to bother voting. A disadvantage to the two-party system is that voters on the extremes may choose not to vote if the candidates in the middle seem too far from their positions. Likewise, the entry of additional candidates could lead to the choice of a candidate at some distance from the median. In the United States, some elected presidents have been weakened by the entry of third-party candidates who siphon off enough votes from the extremes to ensure that the winner does not obtain a majority of the popular vote. George Wallace in 1968 and Ross Perot in 1992 received 14 percent and 19 percent of the popular vote, respectively, with the result that neither Richard Nixon nor Bill Clinton began his presidency with a majority of the popular vote.

In the U.S. presidential election of 2000, the third-party candidacy of Ralph Nader led to an even stranger outcome. U.S. presidential elections are decided by an electoral college made up of delegates chosen by each of the fifty states. The number of delegates from each state is equal to the number of congressional representatives and senators from that state. This system actually gives a greater weight to votes from small states than from large ones because every state has two senators. With two senators and 52 representatives, California has 54 electoral votes. Wyoming has three based on its single representative and two senators. California's population is over 33 million so each of its electoral votes represents about 613,000 people compared with 160,000 people in Wyoming, which has a population of only 480,000. Because the distribution of electoral votes is not proportional to each state's population, it is possible for a candidate to win the national vote but lose the election in the electoral college. This is precisely what happened in the 2000 election in which

George W. Bush obtained more electoral votes than Al Gore who won the popular vote. Gore might have won in the electoral college as well if Ralph Nader had not been a candidate. In the key state of Florida, the outcome was a virtual tie with Bush declared the winner by a few hundred votes. Many political analysts felt that if Nader had not been in the race, several thousand of his voters would have cast their ballots for Gore, changing the outcome of the election. In any case, the winner of the 2000 election not only received less than a majority of the total vote but fewer votes than his opponent; this may compromise his ability to govern.

In parliamentary systems, it is common for numerous parties to contend for votes in electing the representatives from whom the prime ministers and their cabinets are chosen. Multiparty systems such as those found in Israel or Italy allow for representation of the full range of positions, from the extreme left to the extreme right. Many of these parties are able to seat only a few of their candidates, but their influence can be magnified by the fact that they may be able to enter a coalition with the party receiving the largest share of the vote and play a role in governing the country. For example, if the party with the largest share of the votes ends up with only 35 percent of the members of parliament, it will need to strike a deal over policy with parties that share at least some of its positions in order to put together a majority coalition and form a government. Under these circumstances, all voters may feel that they have a reasonable chance of being represented in the government, but there is a greater risk that a government will be held captive by a small faction whose defection would cause it to fall.

Some Israeli governments appear to have followed policies that did not reflect the views of the majority of the population because the ruling party was beholden to small religious parties on the extreme right. In 1999, elections in Austria led to the inclusion in the government of a politician who has made comments that seem to show more sympathy for Adolf Hitler than is usually acceptable in Europe, much to the consternation of Austria's partners in the EU. Jörg Haider's Freedom Party won 27 percent of the vote, which was equal to the percentage won by the center-right People's Party but less than the Social Democrats, a center-left party that won 33 percent of the vote. The two centrist parties were unable to form a government, so the center-right People's Party and the far-right Freedom Party joined together to do so despite much widespread opposition.

Rent Seeking

As noted earlier, voters generally have only one vote to allocate, and even under such alternative voting procedures as the Borda count, their influence on the final outcome is likely to be small. Some have suggested that it is irrational to vote. After all, the act of voting carries some cost in terms of learning enough about the issues and candidates to vote one's interest and in the time required to complete the voting act itself. If the impact of the vote is virtually nil, the costs of voting may be thought to outweigh the benefits. Such an account makes sense only if one believes that the only motivation for voting is to determine the outcome. In fact, voters may have complicated motivations related to civic duty and participation in public decisions, so that most choose to vote even though they believe their vote would hardly be missed if they stayed home instead. Refusal to vote can be seen

as a form of free riding on the provision of an important public good, democratic governance. For many, free riding in these circumstances is seen as morally wrong, and such
beliefs generally guarantee a sufficient turnout for the elections to be meaningful. Still, in
the United States at least, voter turnout is often quite low. An alternative to free riding as
an explanation for low voter turnouts is that the nonvoters are generally satisfied with the
way things are going and are content to let others determine the election results.

At the same time, some individuals may feel very strongly about particular issues and
experience dissatisfaction at their inability to insure that the policies or candidates they favor
are chosen. With a single vote, such individuals have almost no influence on the final outcome. Moreover, many issues never come up for a direct vote. If one believes strongly in the
need to protect endangered species, it may be frustrating to be limited to casting one vote for
candidates who share that concern but only as one small part of a full agenda of public issues.
One recourse for such individuals is to magnify their influence through joining with others
who have the same concerns to form a special-interest group. In modern democracies, such
groups are often extremely influential in policy design and implementation.

As noted earlier, a special-interest group provides a public good to its members and is
unlikely to be able to overcome the free-rider problem unless it is relatively small and its
members have very intense interests. Such groups often focus on a single issue, which
becomes the basis for judging all political activity. Thus, committed environmentalists
belonging to such groups as Greenpeace or the Sierra Club argue for subordination of all
other human activities (trade, agriculture, business, jobs, etc.) to their cause. When the
cause that is defended by a special-interest group is self-serving in the sense that it concerns advancing the economic interests of its members, the actions of the group are often
referred to as *rent seeking*. Rent seeking is defined as the use of productive resources
(labor, capital) to convince the government to introduce distortions into the economy in
such a way as to advance the economic interests of the rent seeker. For example, an industry association might conclude that its members will be able to realize greater profits by
using some of their resources to convince the government to implement a trade barrier as
opposed to using the resources to invest in improved technology or a new plant. Although
protectionism reduces the well-being of the majority, it is difficult for policy makers to
resist the pressure for protectionist policies from groups that are capable of mobilizing
resources and votes for those who support their agenda.

The use of resources to lobby the government for legislation favorable to a group's economic activities is likely to reduce the general welfare, because the government distortions
generally result in inefficiencies that show up as higher consumer prices, increased taxation, or some other negative economic effect.[15] Because international trade often seems
somewhat removed from people's daily lives, it has long been a prime target for rent-seeking groups. Protection from foreign competition allows firms to increase their prices and
earn higher profits than would be the case if they had to confront foreign firms in the
domestic market. In the 1980s, the U.S. government instituted a trade barrier known as a
voluntary export restraint (VER) on imports of Japanese automobiles. It has been estimated that this policy cost U.S. consumers $3.2 billion in higher automobile costs.[16] Much of
this was transferred to Japanese automobile makers, but a substantial portion was retained

by U.S. producers, whose profits increased dramatically while this policy was in place.

Rent seeking is a wasteful activity that clearly violates Pareto optimality. The increased welfare of the rent-seeking group is obtained at the expense of the general public, and because this type of change redistributes income, it is not possible to use compensation to achieve a Pareto outcome. It is also frequently the case that successful rent seeking contributes to increased income inequality because the organizations and firms with the resources to conduct effective lobbying efforts are often wealthier than many of the consumers who end up paying higher prices for goods or higher taxes. Stiglitz cites U.S. dairy policies that keep milk prices above the levels that would prevail in a competitive market, with particularly deleterious effects on the welfare of low-income children.[17]

Not all special-interest group lobbying is aimed at rent seeking. In some cases, groups seek to influence policies that have little economic impact. For example, the National Rifle Association seeks to prevent any regulation on the sale or use of firearms in the United States, not as a way to increase the returns to firearm sellers (although its efforts may have that effect as well) but, at least ostensibly, to protect a cherished principle (the "right" to bear arms). In other cases, special-interest groups may carry out multiple activities that include, but are not limited to, rent seeking. For example, in addition to its rent-seeking activities (e.g., lobbying for farm subsidies), the American Farm Bureau Federation provides a variety of services to its members (e.g., insurance) as well as lobbying for national policies that may be in the public interest as well as the interests of its members (e.g., NAFTA). Some lobbying may actually serve a useful purpose by providing helpful information to lawmakers. The general problem of special-interest groups, however, is that they are motivated by the particular interests of their members, which may only occasionally coincide with the interests of the wider society. It is much more common for the efforts of special-interest groups to advance the welfare of their members at the expense of the welfare of nonmembers, and this is probably true whether they are engaged in explicit rent seeking or defending a principle supported by a minority of the population. The imperfections associated with the various voting rules and the negative influences of rent-seeking special-interest groups raise serious questions about the likelihood that democratic processes will converge on a socially optimal set of public policies. This is the topic of the next section.

SOCIAL WELFARE AND THE POSSIBILITY OF DEMOCRATIC DECISION MAKING

In Chapter 4, we introduced the concept of a social welfare function based on individual utilities. This social welfare function was used to pick out the initial distribution so that the second fundamental theorem of welfare economics could be applied in reaching a general social optimum. This abstract representation is useful in making a general point about the effectiveness of markets in reaching Pareto optimality and allocative efficiency. However, it is impossible to make it operational for use in practical policy analysis. In the real world, we are constrained to rely on the voting mechanisms described earlier rather than using some measure of utility to specify a social welfare function that will solve the problem of locating a socially optimal initial distribution and provide guidance on the supply of other public goods. There seem to be a great many problems with voting and democratic decision making, however. Depending on the circumstances, voting may not provide a clear picture of policy preferences, minority interests may fall prey to the tyranny of the majority, minorities may be able to sabotage broadly beneficial policies, and policy decisions may run counter to the best interests of society as a result of the actions of special-interest groups.

In 1951, Kenneth Arrow published a small book on democratic decision making for which he won the 1972 Nobel Prize in economics.[18] Recognizing the problems with the concept of a social welfare function as a guide for social policy, Arrow explores the possibility of arriving at a social optimum through democratic decision making. Arrow identified five conditions that one might expect to hold if democratic decision making is to lead to a social optimum. These conditions are specified as axioms, and Arrow proves mathematically that arriving at a social choice without violating any of the five axioms is impossible. Before describing the full system of axioms, consider the *voting paradox* identified earlier in the discussion of cycling. Suppose that there are three voters (or groups of voters) faced with three options. These options are to be voted upon in successive pairwise contests, that is, option 1 against 2 followed by 2 against 3 and, finally, 1 against 3. The three voters, Youth, Midlife, and Retired, have the following preferences:

> Youth prefers option 1 to option 2 and option 2 to option 3.
> Midlife prefers option 2 to option 3 and option 3 to option 1.
> Retired prefers option 3 to option 1 and option 1 to option 2.

If the first vote is option 1 versus option 2, Youth and Retired will vote in favor of 1 and it will win. The second vote pits option 2 against option 3 with option 2 winning as it receives the votes of Youth and Midlife. Finally, option 3 wins against option 1 with votes from Midlife and Retired.

The social preference ordering that results from this pairwise voting would indicate that option 1 is preferred to option 2, which is preferred to option 3, but that option 3 is preferred to option 1. In other words, the preference ordering is not transitive. Intransitivity, or cycling, is a problem because the outcome is indeterminate. Note that this problem is not resolved by replacing the pairwise voting procedure with a single vote on all three options. The single vote would mean that each option receives one-third of the votes as Youth, Midlife, and Retired each choose their preferred option. One implication of these results is that transitivity is required if democratic decision making is to lead to a determinate outcome.

Not surprisingly, therefore, transitivity is one of the axioms Arrow included in his analysis. The complete system of axioms Arrow believes should be satisfied in democratically reaching a social optimum is as follows:

1. *Transitivity.*
2. *Unrestricted domain.* This axiom means that all possible preference orderings are allowed. In the context of the preceding example, this axiom prevents Youth's preference ordering, for example, from being excluded a priori.

3. *The Pareto principle*. If one person's preference ordering is unopposed by all the others, that preference ordering is retained in the set of social preferences. In contrast, any preference opposed by one or more is excluded. The Pareto principle is equivalent to a unanimity voting rule. This is also the principle that guarantees that the final decision will be a social optimum.

4. *Nondictatorship*. This axiom prevents the imposition of one individual's preferences on all others. Dictatorship is inconsistent with democratic decision making.

5. *Independence of irrelevant alternatives*. The choice between any two options is not affected by preferences on other options not being considered. This axiom rules out strategic behavior such as vote trading.

Arrow shows that these five axioms cannot all hold at the same time. The mathematical proof is fairly difficult, but it may be possible to develop an intuitive explanation for this result. Because of the unrestricted-domain axiom, any possible set of preferences is admitted for consideration by the voters. If all the voters prefer a particular option over the others, democratic voting will lead to the selection of that option. If there is no unanimous choice among the alternatives, however, then some method will have to be used to select one of them. The final option selected also has to satisfy the Pareto principle in the sense that it is acceptable to all so that no one will wish to veto it.

Independence of irrelevant alternatives means that there can be no bargaining or strategic behavior. The voters can consider only the choice that is before them. Suppose that Youth strongly desires a resolution of this issue. She might be willing to vote for option 2, even though her first choice is option 1, in an effort to forestall the cycle. This means that option 2 would win on the first and second votes. The preference ordering would show 2 preferred to both options 1 and 3. The victory of option 3 over option 1 on the third round of voting would be irrelevant. The axiom on irrelevant alternatives rules out this sort of strategic behavior. It also means that the options have to be compared sequentially, and that leads inevitably to the transitivity problem identified in the voting paradox. However, a choice among the three alternatives has to be made. The only way to do this is to violate the nondictatorship axiom, allowing one individual to impose her preferences. The five axioms cannot all be satisfied at the same time, and this shows that, under what appear to be fairly reasonable restrictions, it may be impossible to realize a democratic decision.

This result is referred to as Arrow's general possibility theorem or, more accurately, as his *impossibility theorem*. It has generated a great deal of discussion, much of which focuses on how one might avoid the impossibility result. It turns out that the five axioms are really quite restrictive, and it may be possible to realize democratic decisions that are socially optimal (that is, that satisfy the Pareto principle) if one or more of the axioms is eliminated. The most common candidate for elimination is the axiom on independence of irrelevant alternatives. Without this axiom, bargaining and vote trading become possible, and the voters may be able to make a democratic choice that is acceptable to all. An alternative voting rule that allows voter intensities to come into play, such as the Borda count, might lead to similar results. The problem with relaxing this axiom is that it is not clear that bargaining and strategic behavior will give rise to the best choice for society as a whole. Strategic behavior suggests that individuals may misrepresent their true preferences in an effort to influence the final result. As long as the Pareto principle is retained, the final choice has to be one over which no one wishes to exercise a veto, even if that choice is the product of logrolling. But this choice may still be a poor reflection of the social preferences of the society as a whole.

In the absence of bargaining or rules allowing preference intensities to count, it appears that it would be difficult to obtain consistent, democratic social choices that would be acceptable to all. One possibility is to abandon the Pareto principle and seek a social choice that would be generally satisfactory even if not fully optimal in the traditional sense. Amartya Sen, 1998 Nobel laureate in economics, proved another impossibility theorem to illustrate a case in which it might make sense to abandon the Pareto principle.[19] For this case, Sen introduces a new axiom that is used in conjunction with the Pareto principle and the axiom of unrestricted domain. The new axiom is that each individual is decisive over his or his own choices and cannot influence the decisions of the other. With these three axioms in mind, Sen tells a story of two individuals, Prude and Lewd, who are faced with the choice of either reading or not reading *Lady Chatterley's Lover*, a novel that was quite controversial at the time it was published because of its focus on female sexuality. There are four possible outcomes:

1. Both read the book.
2. Lewd reads it.
3. Prude reads it.
4. Neither reads it.

Let P stand for "is preferred to." Lewd's preference ordering is represented as: 1 P 3 P 2 P 4. Lewd believes that the best outcome would be for both of them to read the book, but if only one can read it, she prefers that Prude be the one to do so because she believes it would be good for him. The worst would be to waste a perfectly good book by having no one read it. Prude's preference ordering is represented as 4 P 3 P 2 P 1. Prude thinks the book is obscene and should be read by no one. However, if someone has to read the book, he would prefer it to be himself because he thinks it would be a bad thing for Lewd to pollute her mind further with this kind of literature. The worst for Prude would be for both to read it.

In this story, there is a Pareto optimal solution, which is for Prude to read the book while Lewd does something else. The problem is that each individual is decisive over only his or her own decisions. That means that Lewd can only decide to read or not to read the book herself. She cannot decide what Prude is to do. The result is that Lewd will decide to read the book and Prude will decide not to read it, and this outcome is Pareto inferior: both could move from their third preferences to their second if the result were for Prude alone to read it. Sen solves this paradox by calling for elimination of the Pareto principle. Without the Pareto principle, Lewd faces a choice between reading or not reading the book and it simply does not matter what Prude does. In a sense, this example is a case that really should not be a public choice anyway. Most of us would agree that an individual's choice of reading material is not something that should be decided by the general public.

On the other hand, there has been a long history of censorship, so perhaps the story is not as irrelevant as it might seem.

Sen's story has the structure of a prisoners' dilemma. If the two could enter into a binding contract specifying that Prude would read the book while Lewd would not, they could reach the Pareto optimal choice. But note what this implies. The contract would give substance to what Sen calls *nosy preferences*, that is, one individual's preferences with respect to the preferences held by another. Lewd and Prude both have preferences about the tastes and preferences of the other. Should such preferences be allowed to enter into social decision making? Many people would argue that to allow nosy preferences into the social arena would be to risk oppression. Suppose that the majority is highly offended by the fact that some people love classical music. Why would we want that preference to carry any weight at all in social decision making? What if the classical-music haters used their electoral power to force the classical-music lovers to listen to the latest rap group? Sen's example suggests that there may be many situations in which people should mind their own business, and in such situations, the Pareto principle should not be allowed to count.

So what does this mean for democratic decision making? My own conclusion is in line with Winston Churchill's observation that democracy is the worst form of government—until one considers the alternatives. For the choice of representatives and for the public policy decisions made by these representatives, there is really no alternative to some form of voting procedure as practiced in modern democracies. The results of democratic voting are not likely to satisfy everyone (so they will violate the Pareto principle), and it may be necessary to allow procedures such as vote trading, strategic behavior, and lobbying to arrive at determinate decisions. As noted in the last chapter, Pareto optimality may not be an appropriate guide for policy decisions that have distributional consequences. Because most policy problems include both efficiency and distributional dimensions, it is almost always the case that any action to resolve the problem will violate the Pareto principle. In other words, strict attention to the Pareto principle is a recipe for inaction. For Sen and many others, abandoning such a principle is a small price to pay for effective public policy. Of course, without the Pareto principle, some other guide for locating the best course of action will be needed. Such guidance is probably best found in the study of ethics, the topic of the next chapter.

CONCLUSION

Much of the literature on public choice seems to indicate that locating a social optimum through democratic voting procedures is just about impossible. If that is the case, the idea that the way to overcome market failures is to rely on collective actions guided by rational voters, politicians, and bureaucrats may be misguided. Such a conclusion seems consistent with Stiglitz's observations on the difficulty of making sensible economic policy in the political setting of Washington, D.C. Stiglitz offers four explanations for the inability to realize Pareto optimal changes that can provide some additional insights into the problems of political decision making. He notes, for example, that it is often difficult for the government to make credible commitments on policy issues. In the case of dairy policy, the CEA wished to replace the arrangements that increase dairy prices with a system based on direct payments to dairy producers that would be equal in value to the revenue earned through the higher dairy prices. Dairy farmers would be no worse off and milk consumers would benefit from lower milk prices. For technical reasons, the social cost of a program of direct payments would be less than the total cost to society of the price-support system. Direct payments, however, are highly visible, and this visibility makes them vulnerable to subsequent political attack. Even though this policy change would be socially beneficial, it was strongly opposed by the dairy lobby because the government was unable to make a credible commitment that the direct payments would be maintained in the future.

Stiglitz also notes that bargaining in the political arena is often carried out with imperfect information, which can lead to unstable coalitions. He describes a case where the CEA developed a new approach to cleaning up toxic wastes in the United States, in close consultation with environmental, labor, and business groups. Legislation was drafted and approved by the relevant congressional committees. The participants had agreed to this proposal because it seemed a reasonable compromise given political realities. Before the proposal could be brought before the full Congress, however, the 1994 elections took place, changing the majority party in Congress. Business leaders, expecting that the new Republican Congress would look more favorably on their interests than its predecessor, defected from the coalition supporting the new approach. The result, contrary to the expectations of the business groups, was not a different approach that favored their interests but rather a stalemate that simply left the status quo ante in place. Stiglitz also describes the effects of political rivalry. Republican congressional leaders often undermined policy initiatives, even those that had broad bipartisan support, simply to prevent the president from being able to claim a political victory. It is likely that Democratic congressional leaders have behaved in a similar manner when the political roles were reversed.

A final explanation for the inability of the government to realize Pareto optimal changes is the uncertainty about the true consequences of an action. Stiglitz points out that complicated explanations that may lead to better predictions of consequences are ineffective because the public and the politicians do not have the time or patience to work through the complexities. This is a particularly acute problem for economic issues, because good economic analysis is often difficult, complicated, and somewhat counterintuitive. In the case of debates about trade agreements, for example, the main political discussion focuses on the impact of trade liberalization on jobs. The economic demonstration that the gains from more liberal trade show up in increased efficiency and greater welfare for both producers and consumers is a complex argument. Likewise, showing that the number of jobs in an economy is determined by macroeconomic policies rather than trade policies is not only complicated but seems to fly in the face of common sense as well. In these circumstances, reliable economic analysis does not impress the decision makers who are more interested in trying to compare the likely number of jobs gained or lost as a result of a potential trade agreement.

Along with the theory of the second best and the problems of public choice highlighted in earlier discussions, Stiglitz's experiences seem to suggest that one should be pessimistic about the ability of the government to implement policies that are beneficial to society. Such a conclusion is often used to support arguments for reducing the role of the government. From this perspective, even if there is an obvious social problem, the best thing that can be done is to ignore it, because trying to find a solution through public policy can only make things worse. In this context, it is interesting to consider Stiglitz's conclusion on the nature and role of government in society:

> Those who said that I would leave the White House with a more jaundiced view of the role of government were only partly correct. While special interests do often dominate over the general interests and while seeming near-Pareto improvements are often resisted, these failures do not undo the great achievements of the public sector, from mass education to a cleaner environment. These failures should focus our attention on re-examining both how and what the government should do.[20]

The fact that government failure is a frequent occurrence does not mean that there is no role for the government in the solution of the many collective action problems that society has to face. The theory of the second best and the potential for government failure should be taken seriously in both the design and evaluation of public policies. But they should not be taken to mean that the government should be severely limited in the scope of its activities. The fact that public decision making and government action are imperfect does not mean that we should give up on them altogether. Instead, we should seek to find ways to improve these activities so that important social goals are more precisely realized. The kinds of judgments that are required for improvements in both the process of political decision making and the ultimate goals being targeted are ethical, as we shall see in the next chapter.

SUMMARY

1. The theory of the second best shows that the simple rule that policies should be implemented to correct market failures may not lead to increased efficiency and welfare.

2. The economic assumption that individuals are rational can be applied to their political behavior as well. Public choice is the application of economic principles to political behavior.

3. In democracies, voting is used to determine appropriate action on many collective action problems. Unfortunately, voting may not result in an accurate aggregation of social preferences because of cycling and other problems.

4. Voting rules range from simple majorities to unanimity, which is equivalent to a Pareto voting procedure. Supermajorities and unanimity are usually reserved for fundamental issues because they are more difficult to reach than simple majorities. Simple majority rules can lead to the tyranny of the majority, while supermajorities can lead to tyrannies of the minority.

5. One problem with democratic voting is that preference intensities are not included. Logrolling and alternative voting procedures such as the Borda count allow preference intensities to influence the outcome.

6. Rent seeking is the effort by special-interest groups to influence government policy in directions favorable to the group but at the expense of general social welfare.

7. Arrow and Sen have developed impossibility theorems that call into question the possibility of achieving socially optimal outcomes through democratic voting. One implication of their results is that procedures such as logrolling may allow voting to reach a determinate outcome and that the Pareto principle (which implies unanimity) may not be appropriate for collective choice.

8. Public choice shows that politics complicates the simple rules derived from welfare economics and the search for efficiency.

KEY CONCEPTS

theory of the second best	unanimity
trade creation	tyranny of the majority
trade diversion	tyranny of the minority
special-interest groups	cycling, voter paradox
pork barrel	logrolling
principal-agent problems	preference intensity
food safety	Borda procedure
simple majority	median-voter theorem
direct democracy	rent seeking
representative democracy	impossibility theorems
voting rules	nosy preferences
supermajority	

DISCUSSION QUESTIONS

1. Explain why the creation of a free trade agreement (FTA), such as NAFTA, may not increase welfare in the countries forming the FTA.

2. Do you believe that people behave rationally when faced with political questions? Do political results usually seem to reflect broad social preferences fairly accurately or not? Why or why not?

3. What are the advantages and disadvantages of parliamentary systems as compared to the U.S. political system?

4. Explain why a unanimity voting rule is consistent with the Pareto principle. Do you think more political decisions should be based on unanimity or supermajorities than is presently the case?

5. For what kinds of public decisions would it be advisable to require more stringent voting rules than simple majorities?

6. How do the Borda count and other alternative voting procedures as well as logrolling allow preference intensities to count?

7. Do you think that there are political reforms that could improve the performance of the U.S. political system? Explain.

SUGGESTIONS FOR FURTHER READING

Buchanan, James M., and Gordon Tullock. *The Calculus of Consent*. Ann Arbor: University of Michigan Press, 1974.

Mueller, Dennis. *Public Choice II*. New York: Cambridge University Press, 1990.

Olson, Mancur. *The Logic of Collective Action*. Cambridge: Harvard University Press, 1971.

———. *Power and Prosperity*. New York: Basic Books, 2000.

———. *The Rise and Decline of Nations*. New Haven, Conn.: Yale University Press, 1982.

Olson, Mancur, and Sati Kähkönen, eds. *A Not-so-Dismal Science*. New York: Oxford University Press, 2000.

Sen, Amartya K. "The Possibility of Social Choice." *American Economic Review* 89, no. 3 (June 1999): 349–78.

APPENDIX 5.1. THE THEORY OF THE SECOND BEST

The basic structure of the theory of the second best can be illustrated with a simple general equilibrium model. Assume that there are three goods denoted — x_1, x_2, and x_3 — and two individuals U_1 and U_2 — and that the output side of the economy is represented by an implicit production function — $F(x_i) = 0$. Consumption of x_1 by U_1 is denoted x_{11}, consumption of x_2 by U_1 as x_{12}, of x_2 by U_2 as x_{22}, and so on. Using this notation, we have:

Utility functions
$$U_1 = U_1(x_{11}, x_{12}, x_{13})$$
$$U_2 = U_2(x_{21}, x_{22}, x_{23})$$

Production
$$F(x_{11} + x_{21}, x_{12} + x_{22}, x_{13} + x_{23}) = F(x_1, x_2, x_3) = 0$$

The problem is to maximize the utility of the first individual while holding that of the second constant. Note that this is a mathematical representation of Pareto optimality in that we seek to maximize the well-being of one individual without reducing the well-being of the other. In addition to the constraint that individual 2's utility not be decreased below a prespecified level (U^*), the production possibilities of the society are constrained as given by $F(\bullet) = 0$. The problem is to

(1) Maximize
$$U_1 = U_1(x_{11}, x_{12}, x_{13})$$
subject to
$$U_2 = U^* \text{ and}$$
$$F(x_1, x_2, x_3) = 0$$

which is represented by the following Lagrangean equation:

$$\mathscr{L} = U_1 - \lambda_1(U_2 - U^*) - \lambda_2 (F(x_1, x_2, x_3)).$$

The solution to this problem is to differentiate the equation with respect to the three goods and the two lambdas, setting the partial derivatives equal to zero (assume the second-order conditions for a maximum are satisfied by the nature of the utility and production functions and ignore the derivatives for the two lambdas). Let double subscripted U

indicate marginal utilities and single subscripted F marginal products:

(2) $\partial U_1/\partial x_{11} = U_{11}$ $\partial U_1/\partial x_{12} = U_{12}$ $\partial U_1/\partial x_{13} = U_{13}$
     $\partial U_2/\partial x_{21} = U_{21}$ $\partial U_2/\partial x_{22} = U_{22}$ $\partial U_2/\partial x_{23} = U_{23}$
     $\partial F/\partial x_1 = F_1$ $\partial F/\partial x_2 = F_2$ $\partial F/\partial x_3 = F_3$

The first-order conditions from this system are:

(3) $U_{11} - \lambda_2(F_1) = 0$ $U_{12} - \lambda_2(F_2) = 0$ $U_{13} - \lambda_2(F_3) = 0$
     $U_{21} - \lambda_2(F_1) = 0$ $U_{22} - \lambda_2(F_2) = 0$ $U_{23} - \lambda_2(F_3) = 0$
     $U_2 - U^* = 0$ $F(x_1, x_2, x_3) = 0$

These results can be expressed in terms of marginal rates of substitution (MRS) and marginal rates of transformation (MRT). Recall that the MRS is the ratio of the marginal utilities of two goods for a given individual (e.g., U_{11}/U_{12}), and the MRT is the ratio of the marginal products (e.g., F_1/F_2). In this case, there are three MRS for each individual and three MRT on the production side. Consider the first-order conditions given by $U_{11} - \lambda_2(F_1) = 0$, and $U_{12} - \lambda_2(F_2) = 0$. These imply that $U_{11} = \lambda_2(F_1)$ and $U_{12} = \lambda_2(F_2)$. When these two equations are expressed as a ratio, the λ_2 cancel and we have the expected result that the MRS is equal to the MRT (see Chapter 4). Let MRS_{112} represent the ratio of the marginal utilities (U_{11}/U_{12}), that is, the marginal rate of substitution between goods 1 and 2 for individual 1, and use similar notation for the other five MRS in the system. Also let MRT_{12} represent F_1/F_2 and use similar notation for the other two MRT. Then the general equilibrium is given by:

(4) $U_{11}/U_{12} = U_{21}/U_{22} = MRS_{112} = MRS_{212} = MRT_{12}$
     $U_{11}/U_{13} = U_{21}/U_{23} = MRS_{113} = MRS_{213} = MRT_{13}$
     $U_{12}/U_{13} = U_{22}/U_{23} = MRS_{123} = MRS_{223} = MRT_{23}$

These conditions are the standard general equilibrium results that the competitive optimum is reached where the MRS are equal to the MRT (see Chapter 3, Figure 3.6). These are the conditions that will be realized in an economy with no market failures, perfect information, and rational consumers and producers. The problem posed by the theory of the second best is to determine the social optimum if some of the basic conditions shown before are violated. Suppose, for example, that there are two monopolies in the economy, one of which is a natural monopoly that cannot be eliminated, while the other can be broken up under antitrust laws. The theory of the second best shows that it may not be the case that welfare will be increased by breaking up the second monopoly while leaving the first in place. The first-best outcome would be to eliminate both monopolies. If the first-best optimum cannot be attained, would the second-best optimum be achieved by fixing the market failure that can be corrected? The answer depends on the particular circumstances of the economy.

To see the reason for this uncertainty, suppose that one of the marginal conditions shown in equation (3) is violated. Specifically, let the condition ($U_{11} - \lambda_2(F_1) = 0$) be replaced by:

(5) $U_{11} - k(F_1) = 0$ where $k \neq \lambda_2$

If it is assumed that this violation cannot be changed or corrected, it can be added into the original Lagrangean as a new constraint. The second-best problem is to maximize U_1 subject to the constraints that the utility of the second individual not be altered, that the output provided by F not be surpassed, and that the violation shown in equation (5) be met. The new Lagrangean is:

(6) $\mathcal{L}' = U_1 - \lambda_1(U_2 - U^*) - \lambda_2(F(x_1, x_2, x_3)) - \lambda_3(U_{11} - k(F_1))$

Following the same procedures, derive new first-order conditions. Let $\tilde{U}_{11} = \partial^2 U_1/\partial x_{11}^2$, and similarly for the other derivatives of the marginal utilities, and $\epsilon_{11} = \partial^2 F/\partial x_{11}^2$, and similarly for the other derivatives of the marginal products. Then,

(7) $U_{11} - \lambda_2 F_1 - \lambda_3(\tilde{U}_{11} - k\epsilon_{11}) = 0$
$U_{12} - \lambda_2 F_2 - \lambda_3(\tilde{U}_{12} - k\epsilon_{12}) = 0$
$U_{13} - \lambda_2 F_3 - \lambda_3(\tilde{U}_{13} - k\epsilon_{13}) = 0$
$\lambda_1 U_{21} - \lambda_2 F_1 - \lambda_3 k\epsilon_{11} = 0$
$\lambda_1 U_{22} - \lambda_2 F_2 - \lambda_3 k\epsilon_{12} = 0$
$\lambda_1 U_{23} - \lambda_2 F_3 - \lambda_3 k\epsilon_{13} = 0$

These first-order conditions for the second-best optimum lead to different results from those presented in equation (4). For example,

(8) $U_{11}/U_{12} = F_1/F_2 - \tilde{U}_{11}/\tilde{U}_{12} - \epsilon_{11}/\epsilon_{12}$
and $U_{21}/U_{22} = F_1/F_2 - \epsilon_{11}/\epsilon_{12}$

The optimum is reached not where $MRS_{112} = MRS_{212} = MRT_{12}$ but rather where $MRS_{112} = MRT_{12} - \tilde{U}_{11}/\tilde{U}_{12} - \epsilon_{11}/\epsilon_{12}$ and $MRS_{212} = MRT_{12} - \epsilon_{11}/\epsilon_{12}$. There are similar distortions in all the other optimum conditions. The basic result is that in the presence of a violation of the conditions defining the competitive optimum, the second-best optimum requires that the marginal rates of substitution for the two individuals not be equal to each other or to the marginal rate of transformation for the two goods.

To return to the example described earlier, if one of the two monopolies in the economy can be eliminated but the other cannot, the second-best optimum would not necessarily be reached by eliminating the second monopoly. Economic welfare may well be greater if both monopolies are left in place, although this state would be inferior to the first-best optimum in which all monopolies would be eliminated. Note that the results in equation (7) imply that all of the original marginal conditions would need to be modified by the addition of other terms. Thus, achieving a second-best optimum may require the systematic distortion of the entire economy. Because the informational needs required to derive the optimal conditions in a second-best world would be immense, it is unlikely that it would be possible to figure out just how the economy should be distorted to achieve this second-best optimum. This is what makes the theory of the second best such a problem for policy analysis. Given that we will never be able to eliminate all market failures, the simple notion that the government should eliminate those failures that it can, leaving everything else alone, may not be the best guide to policy.

NOTES

1. Joseph Stiglitz, "The Private Uses of Public Interests: Incentives and Institutions," *Journal of Economic Perspectives* 12, no. 2 (spring 1998): 3–22.

2. Free trade agreements (FTAs) are subject to a problem known as trade deflection. Suppose the United States applies a 15 percent tariff to imported textiles while Mexico has a tariff of 5 percent. Textile exporters in Southeast Asia can send goods destined for the U.S. market first to Mexico, where the tariff is lower, and then to the United States without further duties because of the FTA. To avoid this problem, the NAFTA agreement includes elaborate rules of origin to distinguish products that are truly Mexican or Canadian from those that are simply deflected through these countries to avoid higher tariffs. A common external tariff prevents trade deflection (all countries have the same tariff structure) but requires that trade policies be set at the regional rather than the national level. In fact, the fifteen members of the European Union (EU) do not negotiate separately at international trade talks organized by the World Trade Organization (WTO). Instead, the EU negotiates on behalf of all its members, which have the same trade policies with respect to non-EU countries in any case.

3. See William Baumol, "Environmental Protection, International Spillovers and Trade," Wicksell Lectures (Stockholm: Almqvist and Wicksell, 1971); and Michael Rauscher, *International Trade, Factor Movements and the Environment* (New York: Oxford University Press, Clarendon Press, 1997).

4. It is not clear, for example, that welfare in the United States increased following the breakup of the AT&T telephone monopoly in the 1970s, and many question whether the antitrust suit against Microsoft will result in greater efficiency and enhanced welfare.

5. According to Mueller, "Public choice can be defined as the economic study of nonmarket decision making, or simply the application of economics to political science. The subject matter is the same as that of political science: the theory of the state, voting rules, voter behavior, party politics, the bureaucracy, and so on. The methodology of public choice is that of economics, however. The basic behavioral postulate of public choice, as for economics, is that man is an egoistic, rational utility maximizer." Dennis C. Mueller, *Public Choice II* (New York: Cambridge University Press, 1990), 1.

6. See "Navajos' Suit over Royalties Is Thrown Out," *New York Times*, Section A, 8 February 2000.

7. See Charles Lewis, "Congress Is a Captive of the Meat Industry," *USA Today*, March 1999, 10–11.

8. See Pan Demetrakakes, "The HACCP Hassle," *Food Processing* 59, no. 9 (September 1998): 24–29.

9. S. R. Crutchfield, J. C. Buzby, T. Roberts, and M. Ollinger, "Assessing the Costs and Benefits of Pathogen Reduction," *Food Review* 22, no. 2 (May–August 1999): 6–10.

10. Lewis, "Congress Is Captive," 10–11.

11. Mancur Olson, *The Logic of Collective Action* (Cambridge: Harvard University Press, 1974).

12. Theoretically, it would be possible to compute the socially optimal amount of a public good to supply using information on consumers' marginal rate of substitution (MRS). Recall from Chapter 4 that the optimum for a public good is found where the sum of the MRS of all consumers is equal to the marginal rate of transformation. In reality, carrying out such calculations in a world of many goods (instead of the two-good example illustrated in Box 4.2) in which information on consumer utilities is difficult to discern would be virtually impossible.

13. For a complete discussion of alternative voting procedures, see Dennis C. Mueller, *Public Choice II* (New York: Cambridge University Press, 1990), chaps. 7, 8.

14. Ibid., p. 180.

15. A. O. Krueger, "The Political Economy of the Rent-Seeking Society," *American Economic Review* 64 (June 1974): 291–303.

16. Paul Krugman and Maurice Obstfeld, *International Economics: Theory and Policy*, 5th ed. (New York: Addison-Wesley-Longman, 2000), 204. Voluntary export restraints are agreements by an exporting country to limit exports to a specific quantity. They are usually entered into when the exporter fears that other kinds of trade barriers are about to be implemented. Because VERs are administered by the exporting country, the rents generated by the higher prices for its products are retained in the exporting country rather than in the importing country. A tariff, on the other hand, has the effect of retaining these rents in the importing country so it is less desirable from the point of view of the exporter than a VER.

17. Stiglitz, "Private Uses," p. 10.

18. Kenneth J. Arrow, *Social Choice and Individual Values* (New York: John Wiley, 1951). The book was revised in 1963, but the axioms appear in both editions.

19. Amartya K. Sen, "The Impossibility of a Paretian Liberal," *Journal of Political Economy* 78 (January–February 1970) 152–57.

20. Stiglitz, "Private Uses," p.21.

6

Ethical Considerations in Public Policy Analysis

INTRODUCTION

In previous chapters, a complex conceptual framework for policy analysis based on economic principles has been developed. This framework provides a foundation for practical decisions concerning the best course of action to follow when faced with collective action problems. Many economists would suggest that an economic optimum defined as Pareto optimality and allocative efficiency is a good state of affairs. From this perspective, the best thing to do when faced with collective action problems is to adopt policies that move society closer to the optimum so defined or at least do not move society away from it. This account has the appearance of ethical neutrality in that it focuses on the technical concept of economic efficiency. We saw in Chapter 4, however, that the emphasis on Pareto optimality and efficiency left open several ethical puzzles related to economic justice, equality, and human flourishing. The purpose of this chapter is to bring these ethical considerations into the analytical framework.

Questions about the best thing for individuals or groups of individuals to do are fundamentally ethical. Many philosophers have suggested that ethics is the study of how individuals ought to live and what they ought to do.[1] This understanding can be divided into two parts, the first concerning questions of individual morality and the second having to do with ethical courses of action for collective entities such as nations or states. It is, of course, the second that is of greatest relevance for a book on public policy analysis. Economic analysis can provide insights into the pertinent facts related to a collective action problem, as can other disciplines such as biology, engineering, political science, law, and so on. To move from an understanding of the facts to a decision about what to do, however, requires some judgment on the questions of what constitutes a good state of affairs and what the right way to attain that state of affairs entails. As noted earlier, economists often employ Pareto optimality and allocative efficiency as criteria for policy recommendations with little or no reflection on their ethical content. These economic criteria, however, as well as any other rules for making decisions on the best course of action to follow, contain implicit value judgments, and it is clear that practical policy analysis is as much an exercise in ethics as it is in economics, law, engineering, or politics.

It could still be argued that technical economic analysis is value free, generating factual information that can be turned over to policy makers who will combine this information with ethical judgment in choosing particular policies. This position, however, fails to recognize that even technical economic analysis is based on ethical presuppositions. It has been suggested that economists share a worldview that includes the propositions that people are self-interested, that the good life is one in which an individual is free to pursue his self-interest as he defines it, and that the best society is one that permits individuals to realize good lives so defined.[2] While the first of these assumptions could be interpreted as a factual observation and subjected to empirical testing, the other two clearly are value judgments about how we ought to live. Establishing the veracity of such judgments requires a great deal of further argument and debate. Even with well-reasoned argument in support of these propositions, universal agreement that they are true is likely to be much more difficult to attain than would be the case, for example, with the proposition that people will buy less of a good as its price increases. To this point, we have largely ignored such considerations, developing a conceptual approach to the economics of policy analysis as if these value propositions were self-evident. This chapter is designed to rectify that omission.

ECONOMICS AND ETHICS

The Enlightenment thinkers most closely associated with the invention of the discipline of economics were moral philosophers. For much of his career, Adam Smith held the chair in moral philosophy at the University of Glasgow. His first book, entitled *The Theory of Moral Sentiments*, preceded the publication of his most famous work, *An Inquiry into the Nature and Causes of the Wealth of Nations*, by some seventeen years. Best known as a philosopher, David Hume also laid the foundations for modern theories of international finance and monetary economics. John Stuart Mill, Henry Sidgwick, and Karl Marx are all counted among the great economists as well as among the great philosophers of the nineteenth century. Further back in time, Aristotle made important observations about exchange and the role of currency in determining equivalent values in Book V of his *Nichomachean Ethics*, and much subsequent philosophy in the Middle Ages dealt with such questions as "just" prices. Clearly, there is a long-standing link between economics and ethics, a link that was strengthened in the twentieth century by such recent Nobel laureates in economics as John Harsanyi (1994) and Amartya Sen (1998).

Current interest in the relation between ethics and economics, however, reverses a trend that had been prominent for much of the twentieth century. Beginning with the marginalist revolution in economics at the end of the nineteenth century, it became fashionable to try to make economics more scientific, along the lines of physics and other natural sciences dependent on deductive reasoning and hypothesis testing. In 1932, Lionel Robbins published an influential book entitled *An Essay on the Nature and Significance of Economic Science* in which he argued that *interpersonal welfare comparisons* are subjective value judgments that cannot be derived from the propositions of scientific economics.[3] For Robbins, economics is useful in analyzing alternative means for attaining ends that are chosen by others. Thus, it was argued, economists have nothing to contribute to discussions of distributional questions, which are hopelessly bound up in weighing increases in the well-being of one individual against decreases in the well-being of another. Since interpersonal welfare comparisons of this nature were deemed impossibly subjective, Robbins argued that economics could only provide information on the most efficient way to accomplish ends that were determined through political or other processes.

It was in this context that the Pareto criterion was rediscovered. Pareto optimality and the associated concept of compensation provide a basis for policy decisions without relying on what some see as subjective interpersonal welfare comparisons. Moreover, the Pareto criterion is closely related to measures of efficiency and thus is fully consistent with the notion that economics can address efficiency questions but has nothing to contribute to discussions of distributional justice.[4] This dichotomy between efficiency and distribution is seen in long-running debates concerning the possibility of achieving economic growth along with an equitable distribution of economic goods and whether redistribution reduces economic efficiency.

Interestingly, some of the most heated controversies in philosophical ethics mirror these economic debates. In judging whether a particular action is right or wrong, some philosophers look to the consequences of the action, judging that it is right if it brings about good consequences and wrong if it does not. This is similar to the economic proposition that the goals of the political-economic system are to maximize utility or otherwise bring about the best consequences and that such objectives are best met through the efficient use of resources. On the other side are philosophers who feel that there are rules that should never be violated and duties that should never be shirked regardless of the consequences that may flow from such behavior. From this perspective, it is wrong to break a rule against killing other people or to fail to carry out a duty to tell the truth even if doing so would lead to a better overall outcome. This position could support efforts to achieve a more equal income distribution, for example, even if such redistributive policies violate Pareto optimality and lower efficiency.

As has already been noted, the treatment of *efficiency* and *equity* as distinct issues that may be in conflict ignores the fact that both are determined simultaneously within the economic system. The Coase theorem shows that an efficient outcome will be realized regardless of how property rights are assigned if the individuals can bargain freely. But even if the rather demanding bargaining requirement is met, the way in which property rights are assigned determines the income streams of the two individuals and the distribution of income between them. Limiting economics to questions of efficiency is misleading, because any policy recommendation derived from efficiency criteria will carry with it implications for equity and the fair distribution of resources. (See Box 6.1). Likewise, philosophical debates about paying attention to consequences as opposed to rights and duties, or vice versa, suffer from the same defect. For practical policy analysis, it is unreasonable to ignore either consequences or nonconsequentialist considerations such as rights and duties. This proposition is further developed after the review of some of the main concepts associated with alternative ethical visions.

Box 6.1. Global warming: An example of the inherent link between economics and ethics

While there is still much controversy about global warming, there appears to be a growing consensus that the earth's climate is changing and that human activity, notably the burning of fossil fuels, is contributing to this process. Global warming is thought to result from higher concentrations in the atmosphere of "greenhouse" gases, notably carbon dioxide, methane, and nitrous oxide, which trap solar heat that would normally be radiated into outer space. The effects of global warming include rising sea levels that will disrupt the lives and well-being of both human and nonhuman animals residing in coastal areas; unstable weather patterns leading to droughts, floods, and other weather-related problems; and changes in natural ecosystems that could affect biodiversity. If it is correct that human behavior is contributing to global warming, the world is faced with a collective action problem, as described in previous chapters, and the practical policy issue is to determine what to do about this problem. The economic approach to the analysis of such a problem is to determine the benefits and costs of alternative courses of action for dealing with it. As will be explained in the next chapter, the first step in cost-benefit analysis is to identify and measure the impacts of alternative policies, including a policy of doing nothing, and to determine who will bear the costs and who will enjoy the benefits associated with each policy option.

In the case of global warming, measuring the effects of alternative policies is difficult and controversial. Although determining the precise magnitudes of these effects raises a host of technical issues, it is clear nevertheless that the distribution of the costs and benefits of any action will be highly uneven. There are two aspects to this distributional question. First, the costs of global warming and the benefits of slowing it accrue to people of different generations. If nothing is done now, those of us currently alive will not bear the cost of policies to reduce global warming, nor will we suffer the consequences, which are likely to be felt only in the distant future. If we do take action now, future generations will benefit from our efforts, but current generations will have to bear the costs. The second distributional dimension of global warming concerns the spatial impacts of climate change. It is generally thought that the economic costs of global warming will fall most heavily on low-income households in developing countries. Moreover, countries such as the United States are the main sources of greenhouse gases as a result of extensive energy consumption. Per capita energy consumption in the United States is ten times as high as it is in developing countries and each U.S. citizen is responsible for emissions of carbon dioxide that are twenty times those of a person living in Africa. If nothing is done to slow global warming, wealthy individuals in the industrialized countries will not have to bear the cost of curtailing their energy use, while people living in the low-income countries of Asia, Africa, and Latin America will bear the brunt of the growing costs of climate change.

These distributional issues cannot be ignored in analyzing policies to mitigate the effects of climate change. As will be explained in the next chapter, economists use a procedure, known as discounting, to compare costs and benefits that are spread over time. Discounting is based on the recognition that a unit of money received in the future is worth less than the same amount of money received today. There are several ways to explain this difference but the easiest way to understand it is to note that a dollar received today can be placed in an interest-paying bank account which means that it will be worth more than a dollar at some point in the future. The value today of costs and benefits arising ten years from now is less than their face value measured at that future time. Discounting is the procedure used to correct future benefits and costs for this time value of money. The rate at which these future impacts are discounted is essentially an exercise in determining how much weight should be attached to economic effects that occur at different times. High discount rates will reduce the value of distant events a great deal while low discount rates attach greater present value to these future costs and benefits. A discount rate of zero is equivalent to not discounting at all and means that a future impact is valued in the same way as one that happens today.

The choice of the discount rate to use in any cost-benefit analysis determines the weight to be attached to the interests of people who will be living in the future. This is particularly important for the analysis of global warming because the processes take place over such long periods of time and discounting can reduce the size of distant economic impacts quite severely. In any case, the

> Box 6.1. (Continued)
>
> discount rate is not just a technical parameter to be looked up in a table and applied to the costs and benefits generated by projects or policy changes. It carries with it implicit value judgments concerning the weight we wish to attach to the interests of future generations. Likewise, a cost-benefit analysis of climate-change policies that ignored the differential impacts of global warming on wealthy and low-income countries would reflect a value judgement that the interests of those in the high-income, polluting countries are more important than those of the people in developing countries. It is simply not possible to do an economic analysis of the benefits and costs of alternative public policies without confronting these kinds of distributional issues. Such issues are fundamentally ethical in that they concern what we should do and how we ought to live.
>
> *Source:* The factual information in this account can be found in E. Wesley F. Peterson, "The Ethics of Burden-Sharing in the Global Greenhouse," *Journal of Agricultural and Environmental Ethics* 11 (1999): 167–96.

CONSEQUENCES, RIGHTS, AND VIRTUES

As noted earlier, ethical questions about the right thing to do and the right way to live can be addressed at both the individual and the societal level. For many people, individual morality is defined by their religion or some other *comprehensive doctrine* that provides guidance on ethical behavior. At the social level, such comprehensive doctrines are often in conflict, as shown by the long history of wars involving competing religions or ideologies. One response to this plurality of comprehensive views is to search for a broader ethical theory that would be acceptable to all people despite their divergent personal moralities. As it turns out, most religions and political ideologies share a great many beliefs about what is good and what is the right thing to do. If there is enough agreement on basic ethical principles among these competing belief systems, an *overlapping consensus* about public morality may be possible.[5] The realization of this overlapping consensus is, of course, problematic. Much has been written on ethical systems for public entities over the centuries, but many fundamental differences remain. The following comments focus primarily on public ethics rather than individual morality and are not intended to cover all aspects of the debate. The goal is to summarize some of the philosophical discussions most relevant to public policy analysis without pretending to cover all the nuances of the discussion or all the schools of thought.

Consequentialist Ethical Theories

One can think of ethical theories as being made up of two parts, theories of the good and theories of the right. Theories of the good attempt to define and explain just what is meant by good. For example, the ethical theory known as *utilitarianism* would define the good as human happiness (utility).[6] For most ethical theories, the theory of the good also specifies the nature of a good state of affairs. In the case of utilitarianism, a good state of affairs would be one in which there is the greatest amount of human happiness possible. Alternative theories of the good could be built on the notion that the good is beauty or truth, for example, with a good state of affairs being one in which beauty flourishes or truth is promoted.

Theories of the right concern the kinds of action that should be taken with respect to whatever notion of the good is being defended. Thus, for example, utilitarianism would suggest that the right course of action is to maximize human happiness. Complete ethical theories will generally have both theories of the good and of the right, and the various combinations of these two elements give rise to a very large number of possibilities. One way to approach the question of the good is to see it as an end with *intrinsic value*, meaning that human happiness is not a means to an end but an end in itself. In contrast, economic efficiency is not valued as an end in itself but rather as a means to an end (human happiness or well-being). One problem encountered in attempting to define the good concerns the possibility that there might be multiple goods. For example, a possible theory of the good might identify both human happiness and the natural beauty of the earth as intrinsically valuable. Such a representation could lead to internal conflicts if it is believed, for example, that the right thing to do with respect to these two ends is to maximize them both.[7] In such a case, some method for comparing and aggregating the two goods would be required. In addition, it might turn out that increasing happiness would diminish nature so that some method for dealing with trade-offs between the two intrinsically valuable ends would also be needed.

Theories of the right might be seen as theories about which means are the best for accomplishing ends that have intrinsic value. In the Western philosophic tradition, the theory of the right has generated a great deal of controversy. The two main modern contenders are *consequentialism* and *nonconsequentialism* (also referred to as *rights-based* or *deontological* theories). Utilitarianism includes a consequentialist theory of the right leading to the suggestion that the right thing to do is to maximize the good defined as human happiness. From this perspective, ethical actions are those that contribute to good consequences defined as increases in human happiness. Economics is fundamentally consequentialist, although it is not always recognized that the constrained maximization problem described in Chapter 2 is an exercise in utilitarian ethical theory. Such problems are set up to determine principles for choosing courses of action that will maximize good consequences (profits, utility) given that there are resource constraints. Many economists believe that such principles describe not only what people actually do but what they should do as well. Consequentialism is even more evident in applied economics and policy analysis, where decisions on the best policy to adopt are driven by such concerns as whether particular courses of action will lead to good consequences defined with reference to the relative size of benefits and costs. Evaluating policies on the basis of the consequences they are likely to generate seems intuitively to be the most sensible way to choose among a set of policy options.

Despite the appeal of consequentialism, however, there are serious problems with the idea that the right course of action is to be determined in terms of how well it promotes good outcomes. The appointment of the Australian philosopher Peter Singer as professor of bioethics at Princeton University generated an outpouring of invective and condemnation from groups opposed to some of Singer's writings on euthanasia for severely disabled infants.[8] Singer has argued that parents of severely impaired newborns should be allowed to euthanize these infants in order to have a healthy child in their place. The reasoning behind this position is utilitarian in that replacement of the disabled infant, presumed to have less potential for happiness, with a healthy one would result in greater total utility. It could also be argued that the resources required to keep such severely impaired infants alive would be better employed for other purposes, such as finding cures for diseases that cause a great deal of misery. Many people find such ideas extremely objectionable, and numerous letters were sent to the newspapers to express outrage that anyone would even be allowed to make such suggestions. Steve Forbes, a multimillionaire with political ambitions, announced that he would discontinue his financial contributions to Princeton as long as Singer remained a member of the faculty. The problem with consequentialism illustrated by this example is that it may recommend actions felt to be immoral by most people in order to achieve the end of greater net human happiness. Singer's utilitarian conclusion probably would not be included in the overlapping consensus required for a public ethics.

Another problem with consequentialism is that the procedure may require weighing the gains of the many against the losses of a few in a manner that seems unfair. Bovine somatotropin (Bst) is a naturally occurring hormone in cattle that plays a role in milk production. Through genetic engineering, large quantities of this hormone (sometimes referred to a rBGH for recombinant bovine growth hormone) can be synthesized and injected into cows to increase their milk production and feed-conversion efficiency. The benefits of this technology include increased milk supplies and lower prices. The costs include the difficulties of some dairy farmers who, for whatever reason, are unable, unwilling, or slow to adopt the new technology. These producers will face lower prices and may go out of business. The aggregate loss in utility made up of what may be catastrophic losses for these farmers is almost certainly less than the total gain in utility made up of small increases in well-being of the very large numbers of people who will benefit from the lower milk prices.

Strict application of the utilitarian calculus would favor the introduction of Bst because the increase in utility of milk consumers is larger than the loss of utility of the producers who do not adopt the technology. This seems bizarre given that the individual utility increases of the consumers are likely to be so small that they would hardly be noticed while the misery of the small number of producers who lose their farms could be extreme. The insensitivity of consequentialism to such distributional questions should not come as a surprise given what we know about utilitarian economics and the common perception that there is a trade-off between efficiency and distributional equity. Some economists appear to favor maximizing the good consequence of efficient resource use even if such a maximization exercise leaves great inequalities and poverty intact.

A final criticism of consequentialism is that such a doctrine seems much too demanding in that it might require individuals to make large sacrifices in the name of achieving good outcomes for others. For example, utilitarian ethics might require that an individual sacrifice his life in order to save other people. The loss in utility of the person who loses his life may be less than the amount of utility that would be lost if several other people were to lose theirs. In this case, the appropriate action is to minimize the utility loss, so the individual would be required to make the supreme sacrifice to realize the best consequence. While we would normally consider someone who made such a sacrifice a hero for demonstrating the virtues of bravery and beneficence, we would probably not want to make such behavior a moral requirement. In other words, volunteering to make sacrifices for the greater good may be seen as virtuous, but failing to sacrifice one's life to save others is not generally thought to be immoral.

Nonconsequentialist Ethical Theories

A common element in these examples is that consequentialism appears to require some individuals to give up their personal well-being to achieve a greater good for the broader society. Such a result seems to run counter to commonly held beliefs about the kinds of burdens society can appropriately place on people. Many people feel that there are certain rights that all people have as a result of being part of the human family and that these rights should never be violated, even if their violation would result in good overall consequences. Moral theories that reduce the importance of consequences in determining the right courses of action are often referred to as *rights-based* or *deontological* theories or, more generally, simply as *nonconsequentialist* theories. One way to represent nonconsequentialist approaches to ethics is based on the idea that one should never do anything known to be wrong. Thus, for example, if it is wrong to kill another human being, one should never take another's life, even if the world would be a better place if one did. Such a situation might have arisen if someone had had the opportunity to kill Adolf Hitler or any one of several more recent despots (e.g., Pol Pot in Cambodia or Slobodan Milosevic in the former Yugoslavia) before these men were able to cause the immense harm that they did.

An obvious question for proponents of such theories concerns just what should be included in the list of rights, the violation of which is judged to be wrong. One of the main sources for deontological reasoning is Immanuel Kant, the eighteenth-century German philosopher who defined and argued for the *categorical imperative*.[9] The categorical imperative can be formulated in two ways. The first, presented as a theoretical imperative, calls on people to act only in ways that they would be willing to see generalized as universal laws. Thus, if we would be willing to see a rule that one should never commit murder become a universal law, we should take such a rule as a guide to our own behavior. This rule would then be part of the categorical imperative. If, on the other hand, we would not want to live under a universal law requiring that we always sacrifice our personal well-being for the greater good, such a rule would not be included.

The second way to represent the categorical imperative, which Kant saw as the practical equivalent to the first, is to call on people to act in ways that never treat themselves or others merely as a means to an end but rather as an end in themselves. The assassination of Hitler could be judged to be wrong on the grounds that to do so would be to treat Hitler merely as a means to an end (reduced human misery, death, and destruction). From the categorical imperative, Kant is able to derive certain duties one should always observe. Such duties include injunctions against murder, breaking promises, and telling lies, as well as requirements to help others, to express gratitude, and to develop one's abilities.[10]

The most prominent twentieth-century philosopher working in the tradition of Kantian deontology is John Rawls. Rawls views the good as the "satisfaction of rational desire," a formulation very similar to that of many utilitarian economists.[11] He follows Kant, however, in believing that the principles governing right behavior should have priority over the

good. This is one way in which Kant and Rawls are nonconsequentialists. The good is
largely a consequence or an end, and Rawls argues that while the good is of great importance in moral theory, the obligation to respect moral rights and duties is prior to pursuit
of the good. Rawls also draws on another great tradition in the history of philosophy, the
concept of a *social contract*. In Chapter 3, we reviewed some of the thoughts of Thomas
Hobbes on human nature and its implications for social organization. We noted that his
vision of people in constant conflict and strife led to the suggestion that it would be in the
interest of all concerned if people would enter into a social contract with an all-powerful
monarch who would act to control the passions that would otherwise tear society apart.
Hobbes's understanding of the content of the social contract is just one of many. The idea
that there is a social contract between the public and the sovereign (or in the modern world,
the state) implies that there is an agreement between the two parties that sets out their
rights, responsibilities, and duties. It could be argued that modern constitutions are social
contracts in this sense.

Rawls employs an ingenious thought experiment to derive a set of principles that he
believes would be included in a social contract. He asks us to imagine that members of a
"well-ordered society," not unlike modern North America or Western Europe, are assem-
bled behind "a veil of ignorance" to deliberate on principles of justice to be included in a
social contract. The *veil of ignorance* filters out virtually all knowledge about each individual's personal characteristics in the world in which she or he will come to live once the
elements of the social contract have been determined. In the "original position" behind the
veil of ignorance, the parties do not know whether they are male or female, short or tall,
straight or gay, black or white, rich or poor, and so on. Stripped of all information that
might cause them to seek rules that would favor their particular interests, they are expected to come up with unbiased principles that would serve the general good.[12]

The veil of ignorance is a way to operationalize Kant's categorical imperative. Behind
it, there is no temptation for rational individuals to misrepresent their beliefs about what
should be taken to be universal laws. As a result, a social contract developed under these
conditions would not be expected to be biased in ways that would favor some groups over
others. In the *original position*, Rawls suggests that rational individuals will adopt two
principles of justice. The *first principle of justice* states that "each person is to have an
equal right to the most extensive total system of basic liberties compatible with a similar
system of liberty for all."[13] The second, often referred to as the *difference principle*, calls
for an equal distribution of "social primary goods" unless an unequal distribution is to the
"advantage of the least favored."[14] According to Rawls, the parties would give priority to
the first principle in the sense that they would allow restrictions on liberty only if such
restrictions increase overall liberty in some way. This means that basic liberties are not to
be infringed in the name of the difference principle or of increases in efficiency, for example. Rawls believes that the parties to the contract would want as much freedom to realize
their goals as possible. However, once basic liberties are insured, he believes that they
would favor a fairly equal distribution of social *primary goods*, which he defines as liberty, opportunity, income, wealth, and the bases of self-respect. The parties in the original
position have certain facts about the society in which they will come to live. They know

that there will be people of different ethnic origins and religions, different physical and intellectual capabilities, and so on. What they do not know is who they will be and what characteristics they will have when they come out from behind the veil of ignorance. It seems reasonable to expect, therefore, that an individual in the original position would not agree to a system allowing a radically unequal distribution of the goods he or she will need to lead a meaningful life because the risk that he or she will wind up on the short end of the distribution is too great.

Rational individuals, however, would understand that a completely equal distribution of the social primary goods could mean an equal share of a pie that is smaller than it could be. In other words, the parties are presumed to understand that complete equality might reduce incentives to make the kinds of efforts that would increase the overall prosperity of the society. The difference principle allows for inequalities but only insofar as such inequalities make the least advantaged better off.

Rawls's theory of justice is hypothetical in the sense that it pertains to what one might expect in an ideal world as opposed to the messy real world in which we actually live. Nevertheless, it is possible to use it to help in thinking about the ethics of public policies. From this perspective, the principles of justice take priority over such consequentialist factors as efficiency or economic growth. Consistent with other deontological or rights-based theories, basic liberties are to be protected regardless of the consequences. Once institutions have been established to protect basic liberties, the next concern is how goods and other necessities of life are distributed. When one considers the real world, it seems evident that the inequalities observed in all countries are not structured to benefit the least advantaged. Since existing inequalities are not to the advantage of these people, justice will be enhanced by policies to reduce them. Of course, such policies should not infringe upon the basic liberties achieved under the first principle of justice, but redistribution consistent with the two principles of justice is allowed.

Other nonconsequentialist philosophers develop different lists of rights and duties that should take precedence over consequences. In all cases, these rights and duties can be seen as constraints on individual and collective behavior. Such deontological constraints could include the kinds of duties identified by Kant, Rawlsian principles of justice, or individual rights.[15] An important issue in specifying these deontological constraints is whether they can be violated by accident. For some, their violation can only result from an intentional action. Thus, a duty not to cause harm would not be violated if the harm is brought about unintentionally. A distinction is made in most legal systems between accidental and intentional violations of the law. Restitution may be required in cases where an individual unintentionally damages someone's property, but a jail sentence is usually reserved only for those who damage property on purpose.

On the other hand, if a person has a right not to be harmed, that right is violated regardless of the intentions of the one who violates the right. This suggests that rights and duties carry different implications for ethical behavior. Many have pointed out that possession of a right implies a duty for others not to violate that right. Likewise, a duty not to harm someone confers a right not to be harmed on that person. In this sense, the two words are like two sides of a single coin. However, as the preceding discussion suggests, the two concepts could be distinguished by paying attention to motivations or intentions. Rights are often thought to imply protection for an individual from both intentional and nonintentional acts. An individual's property rights in his or her automobile are violated whether the accident is due to negligence, intentional misbehavior, or a pure accident for which the other party cannot be held responsible. On the other hand, duties can be violated only through some sort of purposeful action. It may be possible to unintentionally mislead another person, but an unintentionally untrue statement does not qualify as a lie because the person making it believes it to be true. Lies are defined as intentionally misleading or untrue statements. Thus, while rights could conceivably be violated by accident, duties can be violated only by intentional action.

The distinction between rights and duties is sometimes represented by the terms *victim-based* (for rights) and *agent-based* (for duties).[16] Agent-based moral theories assign duties to individual agents and recommend that they carry out those duties. Victim-based approaches, in contrast, identify individual rights, the violation of which is immoral because of the harm it brings to the victim. One of the most prominent victim-based theories is the libertarian approach of Robert Nozick.[17] Nozick suggests that individuals can come to possess property in two ways, (1) initial acquisition and (2) trade or exchange. "Just acquisition" of property can occur, for example, when an individual uses her own labor to take possession of something that is not already owned by someone else. Once people own a piece of property, they can use it, sell it, bequeath it, offer it in barter, and so on. As long as these exchanges are voluntary and uncoerced, they can be considered just. If the distribution of goods in a particular society has been generated by just acquisition and just exchange, that distribution is just and should not be altered by the state even if it is highly unequal. This understanding would preclude redistributive policies, because they violate the rights of those whose income or wealth is transferred to others, and this rights violation represents an unjust exchange.

In the process of deriving these results, Nozick explores the question of whether human beings could live in *anarchy*. Recall that Hobbes used the concept of the *state of nature* to describe a world in anarchy with no authority to act as a check on the passions and desires of self-interested people. Nozick draws on the work of another great English philosopher, John Locke (1632-1704), to imagine a state of nature in which people are able to cooperate in general but in which, nevertheless, feuds, reprisals, and other forms of conflict are likely to arise. Nozick begins with the Lockean state of nature and argues that rational individuals would naturally form associations for mutual protection from the actions of others. Such associations would arise from voluntary agreements among the members with no violation of the principles of just acquisition and exchange. The protective associations would find that there are economies of scale in the provision of the public good of security for their members. In other words, a natural monopoly arises as protective associations attempt to take advantage of the scale economies. The end result of this process, Nozick suggests, would be a dominant protective association that would look a great deal like a state. Thus, a minimal state would arise through a sequence of voluntary agreements in which no individual rights are violated. Similar arguments have been made by James Buchanan, who won the Nobel Prize in economics in 1986. Instead of a Lockean state of nature, Buchanan follows Hobbes in describing the initial state of anarchy as a war of all against all. He represents this Hobbesian state of nature as a giant prisoners' dilemma. In Chapter 4, we found that the dominant strategy in a prisoners' dilemma is defection and that one way out of the dilemma is the creation of institutions that provide enforceable contracts. For Buchanan, the purpose of the state is to serve as the institution that enforces contracts, thereby allowing people to realize the benefits of cooperation. Buchanan's *protective state* is very similar to Nozick's *minimal state*. In both cases, its role is to provide a very particular public good (security, a system of law that enforces contractual agreements) without which social and economic activity would be impossible. In both systems, this role is seen as crucial, but its scope is also seen as very limited.[18]

Nozick and Buchanan are proponents of the political approach known as *libertarianism*. In general, libertarians prefer that governments be limited to enforcing *negative liberties*. The distinction between negative and *positive liberties*, originally made by Isaiah Berlin, is based on two senses in which terms like freedom or liberty can be understood.[19] In the first sense, an individual may be free from interference by others. In the second, one may be free to do or accomplish something. The first type of freedom is referred to as negative liberty because its realization requires that people refrain from doing things that reduce another's freedom of action. Classic negative liberties include the freedoms of speech, assembly, and religion as in the U.S. Bill of Rights. In contrast, to achieve positive liberty, one must be free to take control of one's life.[20] Some libertarian writers have interpreted positive liberty as being contingent on control over resources. Thus, to be free in this sense would require that one have enough resources to be able to accomplish goals and to control one's future.

Negative and positive liberties could be reformulated in terms of rights. A negative right would be a right to be free of interference while a positive right would be a right to pursue one's goals. This reformulation makes clearer the reason why some people may connect positive liberties with command over resources. If a person has the positive right to pursue his goals, he also must have the means to carry out this pursuit. Negative rights, on the other hand, can be realized by simply leaving people alone. For libertarians, the problem with positive rights is that their existence would extend the role of the state quite substantially. If individuals have positive rights to pursue their goals, it could be inferred that society has an obligation to insure that all citizens have at least the minimum set of resources required to accomplish such goals. Guaranteeing that everyone is capable of realizing positive freedoms would require redistributive policies that would violate individual negative liberties in the same way that they would violate Pareto optimality. For Nozick and Buchanan, the only rights that should be protected by the state are individual rights to be left alone. If poverty and inequality result from this system, there is nothing that can be done about it. Those who are able may elect to make voluntary, charitable contributions, but any state-sponsored scheme to reduce inequalities would involve unjust and coercive transfers and should be ruled out no matter how bad the consequences.

It is interesting to note that the Universal Declaration of Human Rights adopted by the United Nations in 1948 includes both positive and negative rights. The *human rights* enumerated in the declaration have no force of law until they are translated into agreements or covenants ratified by members of the UN. The two main covenants that have been adopted concern (1) civil and political rights and (2) economic, social, and cultural rights. The covenant on civil and political rights includes rights to free speech, freedom of religious belief, and other negative rights similar to those included in the U.S. Bill of Rights. This covenant has been ratified by the United States. The covenant on economic, social, and cultural rights, on the other hand, has not been ratified by the United States. This agreement includes positive rights such as rights to food, clothing, and shelter, the realization of which may require transfers from the wealthy industrialized countries to low-income countries. While the U.S. government is willing to make small voluntary contributions to development in these countries, it is not willing to accept the notion that they have rights that might generate duties in the wealthy countries.[21]

An important problem in nonconsequentialist moral theories is how to handle cases in which rights or duties conflict. It is not hard to imagine situations where telling the truth would cause harm to someone. In such cases, the duty to tell the truth is in conflict with the duty not to cause harm. Conflicts also arise within a particular right or duty. Thus, for example, observing an obligation not to cause harm to one individual could result in harm being caused to another, as in a case where the first individual intends to harm another and the only way to prevent that is to drop something heavy on his or her head. The earlier distinction between agent-based and victim-based theories is relevant to this question. The victim-based theories of Nozick and Buchanan focus on rights that should never be violated and are subject to the problem just outlined. If someone has a right not to be harmed regardless of the consequences, it would be wrong to harm him or her even if doing so would prevent injury to others. But the decision not to harm the first individual is simultaneously a decision to harm the others, so that it is impossible to determine a moral course of action in this case. Agent-based nonconsequentialist theories, on the other hand, remove consideration of the victims to focus on an individual's duties. Thus, some might argue that it is more important for an individual to do his duty not to cause harm than it is to prevent the harm to others. The culpable individual is the one who causes harm, and the duty of the bystander is not to prevent that harm but rather to avoid actively causing harm.

Virtue Theory

This kind of thinking is pretty hard to swallow for those who favor at least some attention to consequences. The death of numerous individuals seems a greater evil than the death of a single individual, even if that individual is perfectly innocent. It turns out that most nonconsequentialists do not rule out some attention to consequences. Rawls suggests that any ethical doctrine that took no account of consequences "would be simply irrational, crazy."[22] One way to include attention to both rights and consequences would be to make rights a part of the good consequences that are to be promoted. Thus, for example, one might argue that the rights to life, liberty, and the pursuit of happiness should be included as arguments in individual utility functions that also include goods and other utility-generating elements and that are to be maximized.[23] From this perspective, the deontological constraints can be made part of the state of affairs that one wishes to realize.[24]

Some philosophers, however, question whether states of affairs should have such a prominent role in guiding individual behavior. For example, Foot argues that a good state of affairs might be thought to include the well-being of others.[25] Virtuous individuals would wish to promote such a state of affairs because they are beneficent, that is, they wish to do good for others. But beneficence is only one of the virtues, and it may turn out that other virtues such as friendship or justice may override beneficence in a particular case. In these circumstances, states of affairs may be of secondary importance.

Foot is working from a tradition in ethics that is rooted in ancient Greek philosophy. This rich tradition is as varied and complex as anything that has been produced since, and it would be impossible to provide a full account of it in the context of this chapter. For our purposes, it is useful to focus on Aristotle, who arguably had the most influence on subsequent thinking about a great many topics, including ethics. For Aristotle, the good is something that is desired intrinsically, and happiness is the only thing that satisfies this condition.[26] Happiness stems from fulfilling one's functions, and the function that is central for human beings is reason, something Aristotle saw as uniquely human. But happiness is not all there is to it. For the function of an entity to be properly fulfilled, it must be fulfilled in the best way. Aristotle uses the example of a harp player, suggesting that the function of a harp player is to play the harp with the virtue of excellence. He concludes that human happiness is the exercise of reason in accord with virtue.

But what is meant by virtue? Much of the *Nichomachean Ethics* is an inventory of the virtues that would be associated with particular human experiences.[27] These experiences are generally portrayed as polar opposites, with virtue represented as the mean between them. Thus, for example, bravery is the mean between fear and its opposite, reckless overconfidence. Other virtues include temperance, generosity, truth telling, magnanimity, friendliness, and justice. The trick for practical morality is to locate the precise nature of the mean that is the virtuous position to take when confronted with these human experiences. Aristotle suggests that the way to discover this mean, and therefore what virtue requires, is through observation of the behavior of a person of good character.

It is this emphasis on character that distinguishes *virtue ethics* from both consequentialist and nonconsequentialist ethical theories. For many people, debates about consequences and rights are cold and abstract in contrast to the original question posed by the Greeks concerning "how I ought to live" and "what it means for me to have a good character." There is some affinity between the agent-based deontological theories described earlier and virtue theory in that both place the actions of individual agents at the heart of the ethical question. In both cases, lists of duties or virtues are offered as guides for individual moral behavior. The difference is that virtue theory motivates these lists with reference to the development of character.

From this perspective, Rawls's procedure of deriving moral principles from a hypothetical original position is not only absurdly abstract but also misguided in that it eliminates such determinants of individual character as gender, nationality, sexual orientation, religion, and so on. Because these attributes are central to the formation of character, they cannot be simply assumed out of existence in the study of ethics. For some virtue theorists, the reintroduction of these personal characteristics leads to a rejection of the idea that there can be a universal morality. From this perspective, good character is determined in local settings with reference to particular cultural traditions and values. A person of good character in Singapore might not be considered highly virtuous in California's Silicon Valley. It is often noted that Aristotle's ethics are derived for Greek culture in the third century before the Christian era, a culture that condoned slavery and the subjugation of women.

The emphasis on the local and the particular can lead to *cultural relativism*. This is the belief that behavioral norms are peculiar to a given culture and that there is no way to transcend one's own culture to evaluate the practices of others. In some parts of Africa and the Middle East, female genital mutilation (the excision of the clitoris, often accompanied by sewing up the vaginal opening) is common. This practice serves no medical purpose and in fact can lead to severe health problems in the young girls who are subjected to it. A strict cultural relativist would argue that female genital mutilation is a local tradition that cannot be criticized by those who do not share the cultural heritage of those who practice it. Both consequentialist and nonconsequentialist ethicists, on the other hand, would condemn this practice on the grounds that it violates human rights or that it causes a great deal more pain and suffering than can be justified by respect for tradition or anything else. The principles at the base of these criticisms are considered universal rather than dependent on a particular cultural context. Note that on some readings of Aristotle (e.g., that of Martha Nussbaum), virtue ethics would not condone such practices as female genital mutilation either.

On the other hand, it is clearly true that we are all products of our particular societies and cultures, and it seems strange to neglect this important context when thinking about ethical behavior. An emphasis on social context is a prominent feature of *communitarianism*, an approach to politics that is in many respects the polar opposite of libertarianism. Whereas libertarians focus on individual freedom, communitarians feel that individual rights should be subordinated to the broader interests of the community. Most communitarians of the nineteenth century were Marxists, socialists, or communists who believed that a good state of affairs would include requirements that individuals sacrifice their personal interests for the good of the proletariat, the commune, or some other social group. This tradition lives on in China, for example, where individual rights to have children are strictly controlled in order to reduce the problems of high population growth for the country as a whole. Many modern communitarians are more closely tied to the conservative right than were their nineteenth-century predecessors. These thinkers often believe that there was a greater sense of solidarity and brotherhood in the past and argue for a return to the less individualistic world they imagine existed in earlier times. For both left and right communitarians, there is a sense that the community itself has value independent of the value or well-being of those who make it up. To borrow from utilitarian constructs, one might say that communitarians believe that the total utility of a community is greater than the sum of the individual utilities of its members.

Communitarians often seem willing to accept violations of individual rights in the name of protecting the community. Thus, for example, a communitarian might favor banning pornography in order to protect the solidarity and cohesiveness of the community, even if such a ban violates the right to free speech. An example of communitarian thinking can be found in an exchange between some prominent philosophers concerning the right to assisted suicide. The U.S. Supreme Court had accepted a case concerning the right of states to make it illegal for physicians to assist terminally ill patients in taking their lives. A group of prominent philosophers, including both Rawls and Nozick, wrote a friend-of-the-court brief arguing that human dignity required that individuals be given this option when faced with that most personal of all human conditions, their own death.

Michael Sandel argued against this position. He noted that despite claims to represent a neutral, universalist position, the philosophers' brief implicitly contains the value judgment that the best way to live is as free and autonomous agents who are able to construct a life of dignity that has meaning and value for them. For Sandel, this is a mistake because

> [t]he philosophers' emphasis on autonomy and choice implies that life is the possession of the person who lives it. This ethic is at odds with a wide range of moral outlooks that view life as a gift, of which we are custodians with certain duties. Such outlooks reject the idea that a person's life is open to unlimited use, even by the person whose life it is.[28]

Sandel bases most of his argument on the work of Locke and Kant, the founders of the tradition that grounds much of the liberal perspective represented in the brief.[29] There appears to be a communitarian element in the arguments as well, however.

Sandel objects to seeing life as something that is possessed by the one who lives it. But if my life is not my possession, whose life is it? If it is a gift, from whom does this gift come? One possibility would be that human life is a gift of the gods, who retain ultimate possession of it. It could be that the duties entailed by this gift are duties to the community. Sanctioning suicide would undermine strongly held communal beliefs about life (i.e., that a person's life is a gift in which the community has an interest), and a decision to allow assisted suicide would favor one view of what life is (that of the writers of the brief) over this alternative. In the end, the Supreme Court ruled that states do have the right to ban physician-assisted suicide. This ruling does not prevent the legalization of assisted suicide, something that was done by Oregon voters who approved a referendum allowing assisted suicide in fairly well-defined circumstances. The U.S. House of Representatives passed legislation in 1999 banning assisted suicide. This bill, which would overturn Oregon's law, has not been taken up in the Senate as of this writing.

It is interesting to note that these three great ethical traditions (consequentialism, non-consequentialism, and virtue theory) overlap in many ways. Aristotle's belief that the good is human happiness clearly influenced later thinkers and appears to have been largely incorporated in such diverse theories as Mill's utilitarianism and Rawls's deontology. In addition, the kinds of behavior that are viewed as virtuous or obligatory by partisans of these different approaches to ethics are often quite similar. Traits such as generosity, beneficence, fairness, or justice in dealing with others are seen as desirable in virtually all these traditions. As with rights-based approaches, virtue theory can give rise to conflicting behavioral directives when two or more of the virtues are incompatible. Thus, an important virtue such as fidelity to one's friends and family could conflict with the virtue of not causing harm. The brother of the person known as the Unabomber faced this dilemma when he realized that he was in a position to prevent further harm to people to whom his brother might send letter bombs but that doing so would violate his brother's trust.

The problem of resolving conflicts between different duties, rights, or virtues is common to ethical theories that do not include one-dimensional goals as is the case, for example, with utilitarianism. As long as some way can be found to measure everything in terms of happiness, all behavior can be reduced to its contribution to or detraction from total happiness, and conflicts among the components of this happiness can be ignored. The problem with this, of course, is finding the appropriate common denominator so that increases in utility can be corrected for any associated decreases to obtain the net impact. If this is not possible, some way of treating trade-offs between moral goals will plague utilitarianism as well. One way to deal with these conflicts is to admit that they are unavoidable and to seek some ranking so that in cases of conflict, one would choose the higher-ranking duty or virtue. The difficulty of this approach can be imagined by asking yourself, for example, whether honesty is more important than generosity. Another strategy might be to draw on more than one ethical approach to sort out these conflicts. For example, in the case of the Unabomber's brother, it might be possible to draw on consequentialist ethics to help determine which of the two duties, fidelity to one's brother and prevention of harm, should be used as a guide. The consequences of keeping faith with his brother would have been further deaths or injuries, which might be deemed worse than the consequence of betraying him. Alternatively, the decision might be determined by figuring out which course of action would come closer to satisfying the categorical imperative or some other deontological rule.

Adam Smith provides another way to think about this issue. Smith, often seen as an advocate of a libertarian, laissez-faire economics, actually develops an ethical theory in the tradition of the ancient Greek Stoics. Much of *The Theory of Moral Sentiments* is devoted to discussions of virtuous behavior, which is represented as the mean between two extremes. Instead of using a person of good character to locate this mean, however, Smith introduces the idea of an *impartial spectator* who acts as a guide to virtuous behavior and who might also provide guidance in cases of conflicting virtues or duties:

> [T]he first of those three virtues [prudence, justice, and beneficence] is originally recommended to us by our selfish, the other two by our benevolent affections. Regard to the sentiments of other people, however, comes afterwards both to enforce and direct the practice of all those virtues; and no man, during the whole of his life, or that of any considerable part of it, ever trod steadily in the paths of prudence, justice, or of proper beneficence, whose conduct was not primarily directed by a regard to the sentiments of the supposed impartial spectator, of the great inmate of the breast, the great judge and arbiter of conduct.[30]

The impartial spectator plays a role similar to Rawls's veil of ignorance by removing self-interest from the determination of proper behavior. Smith's impartial spectator represents the point of view of others into which we can project ourselves through what Smith and other philosophers of his time refer to as *sympathy*.[31] Sympathy is a state that arises if we imagine what it would be like to be in others' shoes. Perhaps the Unabomber's brother could have asked which course of action would better satisfy an impartial spectator. In any case, he did report his brother to the police, putting an end to a long-running mystery. Although Smith's moral theory has many similarities to virtue ethics, his interest in the design of a good society places him in the tradition of the Enlightenment, as exemplified by Kant and Mill, for example. Smith's ethics are situated in a broader framework that includes concern not only for individual character but also for good states of affairs for society in general.

ETHICS FOR PUBLIC POLICY ANALYSIS

The preceding sketches of some of the main themes found in philosophical ethics may seem quite abstract and disconnected from the kinds of practical problems confronted in the analysis of public policy. The variety of conflicting ideas advanced by proponents of the different ethical approaches would seem to make the discovery of an overlapping consensus an impossibility. Such a conclusion would be strengthened if I had attempted to incorporate other philosophical traditions into the discussion. There are well-developed moral systems associated with all the great religions, for example. In addition, an important alternative to Enlightenment liberalism has developed from some nineteenth-century ideas, most notably as initiated by Nietzsche and others who object to the notions of rationality and individualism that are at its center. Existentialism and postmodernism have both grown out of this tradition which takes exception to some of the most central beliefs reported in the preceding discussion. Finally, alternative ethics have been proposed by feminists, ecologists, and others with visions that are often fundamentally opposed to the main precepts of the liberal ethical traditions discussed in this chapter. Full exploration of these alternatives would require a great deal more time and expertise than I can bring to bear on them. I mention them only in the interests of inclusiveness. While I claim no expertise in postmodern or feminist ethics, my sense is that they have less to contribute to policy analysis than do the great ethical traditions discussed here. But that conclusion may be due to my limited understanding of them.

Even within the traditions that have been discussed, wide-ranging disagreement seems highly likely. Consider the deficiency payments that were a major component of U.S. agricultural policy between 1972 and 1995 (see Chapter 10). These payments were direct transfers to farmers based on the difference between a politically determined target price and the prevailing market prices for commodities such as grains and cotton. These direct payments were financed through general taxation and were thus part of a policy to redistribute income. Such a policy would not find favor with libertarians, who, as noted earlier, see government redistribution of income as an unjust violation of individual rights. Communitarians, on the other hand, might favor such a policy on the grounds that it protects traditional farming systems from market forces that tend to undermine them. These traditional farming systems might be seen as central to the prosperity of important rural communities that have intrinsic value and are therefore worth protecting even if some individual rights are violated in the process.

It seems almost impossible to reconcile these conflicting moral positions. The problem is compounded by the fact that most countries include a plurality of diverse communities, some of which are made up of individuals who reject many of the beliefs of the broader culture. In the United States, for example, some religious communities see the secular instruction found in public schools as inimical to their religion. Since they cannot change the nature of public instruction, they often elect to withdraw their children from that system in favor of home schooling, for example. The ethnic tensions fanned into life by opportunistic politicians during the war in Bosnia and Herzegovina continued to simmer after the war ended. One result has been that all versions of the region's history have become highly contested, with each group advancing stories that highlight the innocence of its members and the perfidy of others. In some regions, the postwar organizations overseeing the transition to full statehood have directed that history not be taught at all in public schools until tempers have cooled enough to allow reasonably accurate versions of both recent and distant events to be presented.

Such profoundly contested worldviews would seem to preclude the possibility of crafting a public ethics that would be acceptable to all. On the other hand, it may be the case that in most countries, the majority of citizens would share enough values and beliefs to be able to find a workable compromise on public policy. Even in Bosnia and Herzegovina, people of different ethnic groups generally lived together in harmony for centuries, although the recent conflict has generated complicated patterns of anger and resentment that make the future of that country uncertain. Harbour suggests that there is a shared moral core that is likely to be accepted across widely divergent human populations. She cites cross-cultural studies that find broad, general acceptance of values such as justice, beneficence, special obligations to family and members of one's community, some limited sacrifice of one's interests to those of the group, honesty, courage, self-control, and condemnation of murder, incest, and rape.[32] The broad acceptance of these values does not solve the kinds of conflicts over specific rights and wrongs that have so divided many societies. The fact that most people seem to share a core of moral values does, however, provide a basis for discussion and debate. People may differ about the implications of these values for particular problems, but it is possible that agreement on many issues can be reached through democratic deliberation and a reliance on pragmatic common sense.

Harbour's list of basic moral values is quite similar to other lists that have been proposed throughout this chapter. This lends some support to the notion that these values are universal, making it more likely that an overlapping consensus can be discovered. Why does it seem that the same values seem to be prominent in almost all discussions of ethical behavior? In the 1930s, Ross offered a list of moral duties that he took to be self-evident although contingent on the particular circumstances surrounding their application.[33] Ross's duties include fidelity, correcting wrongs that one has committed, gratitude, justice, beneficence, self-improvement, and avoidance of harm to others. For Ross, the source of these values is intuition. Through examination of our conscience, we discover that there are certain values that are self-evident. If intuition is a faculty that all people have, independently of their cultural background, it is not surprising that similar sets of values seem to show up on most lists of duties or virtues.

The field of evolutionary biology may offer an explanation of how such an intuitive faculty might have come to be present in most human beings. Economists and biologists have long been puzzled about the existence of altruism. Economic reflections on altruism were briefly touched on in the discussion of economic rationality in Chapter 2. For biologists, altruism is hard to explain because it does not seem to contribute to the survival of the particular individual who risks her life altruistically. However, suppose that the real driving force in the behavior of human and nonhuman animals is the perpetuation of genetic sequences through time. It could be that the evolution of a gene for altruism would increase the chances of the overall survival of the species, even though a particular individual and her genes are sacrificed. Species survival is what is needed to insure the long-term survival of these selfish genes, which are widely shared among different individuals in a population.

Morality and altruism are not the same thing, and it is not immediately obvious why selfish genes would program us to believe that honesty, for example, is a duty we should observe. There is interesting evidence, however, of a link between our physical selves and moral behavior. Researchers have found a region of the brain that seems to control social behavior. Individuals who have suffered damage to this part of the brain often demonstrate antisocial behavior, including lying, harming others, and so on.[34] It may be wrong to conclude that all our values and behavior are programmed into us by our genetic code. There has been a long-running debate about whether behavior is primarily determined by genes (nature) or by society and culture (nurture). For many people, it seems obvious that both nature and nurture influence the way we behave, and while different societies can be expected to nurture their children in different ways, our common genetic humanity may make us all fundamentally much more similar than is often thought. And these genetic similarities could provide a basis for common moral beliefs that are activated in some way by the particular cultural experiences of each individual.

Of course, even if all this is true, there is still a problem of translating these broad values into practical decisions on the best course of action to follow when confronted with a particular collective action problem. The great ethical traditions all contain useful ideas about the selection of the best course of action, but they do not include specific instructions on deficiency payments or assisted suicide. It does seem, however, that most people would like to behave virtuously and that they prefer a society organized to accomplish the best outcomes without violating rights or shirking important obligations.

Many scholars have attempted to work out ethical systems that take account of all of these considerations. We noted earlier Amartya Sen's suggestion that both consequences and rights be included in good states of affairs.[35] One way to conceive of this marriage of rights and consequences would be to imagine that the policy goal is to maximize good consequences, subject to the constraint that the rights and duties included in Ross's list, for example, not be violated. Another approach might be to use rights-based ethical approaches to evaluate basic social institutions with consequentialist approaches used to evaluate particular policy initiatives.[36] Approaches such as these could be complemented by reference to virtue theory. I have argued elsewhere that public policies that require sacrifices of national, group, or generational interests might be justified by noting that virtuous behavior requires that people accept their responsibilities and obligations toward others even if this means making modest personal sacrifices.[37] Approaches such as these imply that the analyst would want to examine the broad consequences of alternative policies, including their impact on widely accepted rights, duties, and virtues. Thus, for example, in the case of deficiency payments described earlier, one would want to know how this policy affects income and other indicators of well-being for farmers, consumers, taxpayers, people in other countries, and future generations, as well as how it affects the environment, individual rights, rural and urban communities, and so on. It may well turn out that the policy has little or no impact on some of these variables, and that would allow dropping them from further consideration. It is also possible that the policy will have a positive impact in one area and a negative impact in another, raising questions that may be resolved only through democratic debate and argument. At the end of such a process, it may be the case that the participants will remain divided as to the best course of action to take. In that case, the only way out of the dilemma is to allow the political system to resolve the issue, recognizing that this will surely leave some dissatisfied.

Policy analysis is not a process of using recipes to select the best course of action. It is necessarily riddled with conflict and disagreement. Yet decisions are made and courses of action are charted. The goal of the policy analyst should be to shed light on the advantages and disadvantages of alternative public interventions. Accomplishing that goal will require attention to economic consequences, political realities, and ethics. In a sense, the analyst should be drawing upon all the concepts discussed in this and preceding chapters to develop a case for the course of action that seems to be the most sensible, with emphasis on making clear the considerations that have led to and support these conclusions. The next part of the book uses this conceptual framework to examine the basic tools of policy analysis and their application in specific cases.

SUMMARY

1. Decisions on public policies always reflect some sort of value judgment and thus are inherently ethical. Economic analysis of public policies also requires ethical reflection, and there has been a long-standing link between economics and ethics.

2. Three great ethical traditions that are relevant for public policy analysis are consequentialism, nonconsequentialism, and virtue theory. Consequentialism, of which utilitarianism is a prime example, is the doctrine that the ethical course of action is that which generates the best consequences.

3. Consequentialism seems to require unreasonable sacrifices by individuals for the greater good. Nonconsequentialism (also called rights-based or deontological theory) holds that the right thing to do is to avoid violating rights or duties regardless of the consequences. Rawlsian liberalism, the libertarianism of Nozick and Buchanan, and Kantian deontology are examples of this ethical approach.

4. Virtue ethics is an older tradition that holds that ethics is about individual character. Individuals of good character behave in accordance with such virtues as beneficence or justice.

5. An emphasis on individual character can lead to rejection of the idea that there are universal values. This rejection can take the form of cultural relativism, which holds that values are specific to a given culture and cannot be criticized by those who are not of that culture.

6. A related doctrine is communitarianism, which rejects the individualism of both consequentialism and nonconsequentialism. Communitarians believe that the interests of the

community should take precedence over individual interests.

7. While it is surely very difficult to locate an overlapping consensus on the values that should inform policy analysis, the common humanity of different people may offer some hope that workable, democratic compromises can be found.

KEY CONCEPTS

interpersonal welfare comparisons
efficiency and equity
comprehensive doctrines
overlapping consensus
utilitarianism
intrinsic value
consequentialism
nonconsequentialism
rights-based or deontological theory
categorical imperative
social contract
veil of ignorance
original position
first principle of justice (Rawls)

difference principle
primary goods
victim-based
agent-based
anarchy
state of nature
protective/minimal state
libertarianism
negative/positive liberties
human rights
virtue ethics
cultural relativism
communitarianism
impartial spectator
sympathy

DISCUSSION QUESTIONS

1. Nonconsequentialism calls for respecting rights and duties no matter what the consequences are. Is it possible that the reason nonconsequentialists emphasize rights is that they believe that respecting rights will lead to good consequences? Is there really a distinction between nonconsequentialism and consequentialism?

2. Economics is inherently utilitarian because it focuses on benefits and costs. Explain.

3. Describe the two versions of the categorical imperative and explain how this principle is nonconsequentialist.

4. Discuss the similarities between Rawls's veil of ignorance and Adam Smith's impartial spectator.

5. How might conflicting values be reconciled in an effort to select policies that solve collective action problems? Is there any way to resolve conflicts over such profoundly contested issues as abortion, genetic research, homosexuality, and so on?

SUGGESTIONS FOR FURTHER READING

Dasgupta, Partha. *An Inquiry into Well-Being and Destitution*. New York: Oxford University Press, Clarendon Press, 1993.

Hausman, D. M., and M. S. MacPherson. *Economic Analysis and Moral Philosophy*. New York: Cambridge University Press, 1994.

Nozick, Robert. *Anarchy, State and Utopia*. New York: Basic Books, 1974.

Rawls, John. *A Theory of Justice*. Cambridge: Harvard University Press, 1971.

Sen, Amartya K. *On Ethics and Economics*. New York: Basil Blackwell, 1987.

Singer, Peter. *Practical Ethics*. New York: Cambridge University Press, 1999.

Singer, Peter, ed. *A Companion to Ethics*. Cambridge, Mass.: Basil Blackwell, 1994.

NOTES

Much of this chapter was originally developed in collaboration with George C. Davis in our article "Consequences, Rights, and Virtues: Ethical Foundations for Applied Economics," *American Journal of Agricultural Economics* 81, no. 5 (1999): 1173–80.

1. See Peter Singer, ed., *A Companion to Ethics* (Cambridge, Mass.: Basil Blackwell, 1994), xi.

2. Charles K. Wilbur, ed., *Economics, Ethics and Public Policy* (New York: Rowman and Littlefield, 1998), 4.

3. Lionel Robbins, *An Essay on the Nature and Significance of Economic Science* (London: Macmillan, 1932).

4. Brennan suggests that Robbins wrote his book as a response to economists who were arguing for more equal income distributions on the grounds that the marginal utility of an additional dollar to a poor person is greater than the loss of utility of the wealthy person whose dollar is transferred to the poor person. On this basis, total utility would be maximized by equalizing the distribution of income. Robbins wanted to show that such an analysis rests on subjective value judgments. To do so, he denied that the changes in utility between the rich and poor individuals after redistribution could be compared. Geoffrey Brennan, "Economics," in *A Companion to Contemporary Political Philosophy*, ed. R. E. Goodin and P. Pettit (Cambridge, Mass.: Basil Blackwell, Cambridge, 1996), 133.

5. The concept of an overlapping consensus is due to John Rawls, who provides extensive development of the idea in *Political Liberalism* (New York: Columbia University Press, 1996), 133–72.

6. The classic statements of this theory are *An Introduction to the Principles of Morals and Legislation* (1823) by Jeremy Bentham and *Utilitarianism* (1863) by John Stuart Mill.

7. Note that the natural beauty of the earth could be seen as valuable because it contributes to human happiness. Such an interpretation would change the beauty of the earth from something with intrinsic value to something that has instrumental value, that is, something that can serve as a means to an end. Many environmentalists would argue that the beauty of the earth has intrinsic value and should be preserved for its own sake regardless of any effects it may have on human happiness.

8. See John Leo, "Singer's Final Solution," *US News and World Report*, 4 October 1999, 17; Katha Pollitt, "Peter Singer Comes to Princeton," *Nation*, 3 May 1999, 10; editorial, *Philadelphia Inquirer*, 26 September 1999; and letters to the editor, *Philadelphia Inquirer*, 3 October 1999. Singer, well known for his utilitarian writings on ethics and animal rights, is a highly regarded philosopher, although many take exception to some of his conclusions.

9. Immanuel Kant, "Grounding for the Metaphysics of Morals," in *Classics of Moral and Political Theory*, ed. M. L. Morgan (Indianapolis: Hackett Publishing, 1992), 1013–18.

10. Much of this description of Kant's ethics is based on notes from a workshop presentation by Professor Robert Audi, University of Nebraska, Lincoln, 8–13 June 1998.

11. John Rawls, *A Theory of Justice* (Cambridge: Harvard University Press, 1972), 30–31.

12. Ibid., 11.

13. Ibid., 302.

14. Ibid., 303.

15. The word *deontological* has its root in the Greek word for duty and refers to ethical theories

that focus on the study of moral obligation in contrast to teleological theories that emphasize pursuit of good consequences. The word *teleology* is from the Greek word telos, for "ends" or "purposes." See also Nancy David, "Contemporary Deontology," in *A Companion to Ethics*, ed. Peter Singer (Cambridge, Mass.: Basil Blackwell, 1993).

16. Samuel Scheffler, ed. *Consequentialism and Its Critics* (New York: Oxford University Press, 1991), 10.

17. Robert Nozick, *Anarchy, State and Utopia* (New York: Basic Books, 1974).

18. It is interesting to compare the concepts of the protective or minimal state with the arguments made by Thomas Paine in *Common Sense*. Paine suggested that government is a "necessary evil" that is required to control the inherent wickedness of human beings. Government promotes happiness "negatively by restraining our vices." Thomas Paine, *Common Sense* (1776; reprint, Mineola, New York: Dover Thrift Edition, 1997), 3.

19. Isaiah Berlin, "Two Concepts of Liberty," in *Four Essays on Liberty* (New York: Oxford University Press, 1979).

20. This distinction is somewhat vague, and many philosophers have criticized its usefulness. See Chandran Kukathas, "Liberty," in *A Companion to Contemporary Political Philosophy*, ed. R. E. Goodin and P. Pettit (Cambridge, Mass.: Basil Blackwell, 1995), 534–37.

21. In 1997, U.S. foreign aid amounted to only 0.09 percent of the U.S. GNP. Denmark contributed 0.97 percent of its GNP and Norway 0.86 percent. The average for all industrialized countries was 0.40 percent. Official development aid from the United States was less than $7 billion in 1997. Japan contributed $9.4 billion in that year, even though its economy is only about half the size of the U.S. economy. Foreign aid statistics can be found through the Development Aid Committee (DAC) of the Organization for Economic Cooperation and Development (OECD) at the Internet Web site www.oecd.org/dac/.

22. Rawls, *Theory of Justice*, 30.

23. Partha Dasgupta makes this argument eloquently in *An Inquiry into Well-Being and Destitution* (New York: Oxford University Press, Clarendon Press, 1993), 30.

24. Amartya Sen, "Rights and Agency," in *Consequentialism and Its Critics*, ed. Samuel Scheffler (New York: Oxford University Press, 1991).

25. Philippa Foot, "Utilitarianism and the Virtues," in *Consequentialism and Its Critics*, ed. Samuel Scheffler (New York: Oxford University Press, 1991).

26. Much of the discussion of Aristotle's ethics is based on notes from a workshop led by Professor Robert Audi, University of Nebraska, Lincoln, 8 June 1998. The primary source is Aristotle's *Nichomachean Ethics*.

27. Martha Nussbaum describes Aristotle's approach to his list of virtues in "Non-Relative Virtues: An Aristotelian Approach," in *The Quality of Life*, ed. Martha Nussbaum and Amartya Sen (New York: Oxford University Press, Clarendon Press, 1993).

28. Michael J. Sandel, "Last Rights," in *New Republic*, 14 April 1997, 27.

29. The word *liberal* has taken on a peculiar sense in recent political discourse in the United States. Originally, liberalism meant that individual rights and liberties were to be respected, including such civil and political rights as free speech and freedom of religion as well as freedom to pursue one's ends in the market. This is still the primary sense of the word in Europe and in academic discussions, although the popular meaning in the United States has come to be more or less synonymous with "left-wing."

30. Adam Smith, *The Theory of Moral Sentiments* (1759; reprint, Indianapolis: Oxford University Press, 1982), 262.

31. See Stephen Darwall, "Sympathetic Liberalism: Recent Work on Adam Smith," *Philosophy and Public Affairs* 28, no. 2 (spring 1999): 139–64.

32. Frances V. Harbour, "Basic Moral Values: A Shared Core," in *Ethics and International Affairs*, ed. J. H. Rosenthal, (Washington, D.C.: Georgetown University Press, 1999), 112–13.

33. William David Ross, "Many Self-Evident Obligations," in *The Right and the Good* (New York: Oxford University Press, Clarendon Press, 1930).

34. See "Morality and the Brain: Enfants Terribles," *Economist*, 23 October 1999, 94–95.

35. Sen, "Rights and Agency."

36. An approach based on constrained maximization is developed by Gauthier in *Morals by Agreement* (New York: Oxford University Press, 1986). Sen ("Rights and Agency"), Dasgupta (*Inquiry into Well-Being*), and Ralph D. Ellis (*Just Results: Ethical Foundations for Policy Analysis* [Washington, D.C.: Georgetown University Press, 1998]) implicitly use such a procedure to incorporate rights or justice into the states of affairs to be promoted. Paul B. Thompson, Robert D. Matthews, and Eileen O.van Ravenswaay (*Ethics, Public Policy and Agriculture* [New York: Macmillan, 1994]) outline an approach based on applying deontological constraints at the level of institutional structure and using consequentialist measures to evaluate economic performance.

37. E. Wesley F. Peterson, "The Ethics of Burden-Sharing in the Global Greenhouse," *Journal of Agricultural and Environmental Ethics* 11, no. 3 (1999): 167–96; also Peterson, "Time Preference, the Environment and the Interests of Future Generations," *Journal of Agricultural and Environmental Ethics* 6, no. 2 (1993): 107–26.

Part II

Analytical Methods and Applications

Part II

Analytical Methods and Applications

The theoretical framework described in the first part of this book provides a foundation for analyzing and evaluating public policy interventions. In this part, these insights are translated into practical methods for the economic analysis of public policies. The basic analytical approach commonly employed in such analyses is to compare benefits and costs of alternative policy initiatives. If the benefits are greater than the costs, a case can be made that the policy should be implemented. Benefit-cost analysis, also called cost-benefit analysis, is clearly in the consequentialist tradition in that public projects and policies are evaluated in terms of the good and bad consequences they are likely to generate. Moreover, conventional benefit-cost analysis focuses on the search for the most efficient uses of scarce resources, with less concern for the fair distribution of economic goods. In line with the conclusion from Part I, efficiency and other good consequences should not be ignored simply because they are only a part of what is significant to human beings. Efficiency is important and should be part of any policy assessment even though political, ethical, and other concerns are also important. There are both formal and informal ways to bring such concerns into the analysis.

This part of the book includes three core chapters, each of which is followed by one or two case studies. The case studies are designed to illustrate the practical application of the analytical methods discussed in the core chapters. In many instances, it is in the case studies that some of the most interesting political and ethical aspects of public policy analysis are introduced. The three core chapters set out various methods for estimating and projecting the consequences of alternative policies. The first of these, Chapter 7, concerns benefit-cost analysis as applied to the evaluation of public projects. The other two core chapters set out analytical methods often employed to model market relationships and the impact of policies on important economic variables. Partial equilibrium models are described in Chapter 10, while Chapter 13 sets out some basic principles for economy-wide modeling.

7

Benefit-Cost Analysis of Public Projects

INTRODUCTION

The purpose of this chapter is to present the basic methods of *benefit-cost analysis* for public projects. Recall that a case can often be made for government action when there are market failures. For some people, public policy is the conscious effort to create institutions to regulate markets and behavior when confronted with public goods, common property externalities, other external effects generated by the way in which property rights are defined, income distribution, and so on. In many of these cases, appropriate policies involve laws and statutes that can be partially evaluated in terms of their economic impacts on market performance and the correction of market failures.

In addition to interventions that change regulations, property rights, laws, and statutes, governments also undertake *public projects* that produce physical outputs. A policy to support milk prices will affect the operation of dairy markets and is likely to have an impact on milk producers, consumers, and taxpayers across the country. A government project to build infrastructure to deliver irrigation water to an arid region will affect water markets in the project area, and it will also generate a wide range of indirect effects on other markets, such as those for labor or agricultural products. It differs, however, from such policies as the milk price supports, which establish legal institutions designed to influence behavior. The irrigation project will actually create physical structures.

This contrast between projects and policies may be somewhat artificial in that publicly funded infrastructure projects usually have fairly broad impacts on human behavior both within the region most directly affected by the project and outside it. As noted earlier, the construction of the interstate highway system in the United States generated a wide range of windfall gains and losses, sometimes in parts of the country at great distance from the actual highway construction. Nevertheless, for this initial presentation of benefit-cost analysis, it is useful to maintain the distinction between projects that are more directly concerned with the production of infrastructure or other public goods and policies that are intended to influence the functioning of markets.

Most public projects are undertaken only after studying the anticipated benefits and costs of the project. This is the case for the construction of dams and highways in the United States, for example, as well as for development projects in low-income countries, such as irrigation schemes, land development, or the establishment of new industries. The obvious question is why such projects should be carried out by governments using tax revenues, bond issues, or development loans to finance them rather than left to the private sector. One answer is that these projects are public goods that would be undersupplied by the private sector. For example, the construction of a dam may provide such benefits as flood control, low-cost electric energy, and recreational opportunities. A private firm might be able to capture the benefits associated with electricity generation and would thus have an incentive to make the investment in constructing the dam. The firm might not be able to charge for the flood-control and recreational benefits of the dam, however, so there could be a tendency for less than an optimal supply of these public goods. It is also possible that the private energy company would end up with some degree of monopoly or market power so that less than optimal amounts of electricity may be generated.[1] There may also be complications related to land acquisition and compensation for flooded land that can be handled only by governments. In the United States, for example, the private sector generally does not have the right to condemn property.

There are other reasons why governments are motivated to carry out public projects such as these. For example, a great deal of public investment has been made by the U.S. government to develop water delivery systems in the western United States. This region is characterized by relatively low rainfall but extensive natural resources such as minerals and scenic beauty. Along with needs for water to develop the mining and agricultural potential of the region, urbanization in California—and later in inland states such as Arizona, Colorado, Nevada, and Utah—has added to the region's water demand. Government investments in dams and water delivery systems have been crucial for the growth of the region. The motivation for these investments would appear to be to promote regional development in the western United States.

But is such regional development a public good? Do the government interventions correct externalities, monopolies, or other market failures, or are they driven more by a desire to redistribute income and opportunities from the East to people and industries in the West? Some might argue that western growth has been a public bad rather than a public good and that we would be better off if less public infrastructure had been built. Such development projects are undertaken for a variety of reasons in contexts where political interests and power relationships, as well as inefficiencies and market failures, shape the options that are placed before the public. Still, the general rationale for such projects is the perception that there is some sort of collective action problem that will not be corrected by individuals in the private sector. The expectation is that government action will solve the perceived problem.

EVALUATING PUBLIC PROJECTS

Collective action problems are often identified through political processes involving special-interest groups, political constituencies, or government agencies that have an interest in the particular problem. Some governments and government agencies have attempted to formalize part of this process through *planning* procedures aimed at anticipating potential problems.[2] The goal of such planning exercises is to identify solutions to anticipated problems and to locate needed resources for dealing with them before the problems actually arise. Planning efforts may indicate that there is an important issue to which government resources should be committed even if questions have not been raised by constituents or other special interests. In many developing countries, foreign aid donors play a role in identifying and selecting public policy problems. During the 1980s, the World Bank and International Monetary Fund (IMF) made financial support for many low-income countries conditional on policy reforms that might not have been implemented without this outside pressure. Sometimes policy issues are generated through processes that are repeated at regular intervals, such as annual government budget deliberations or periodic international trade negotiations.

Whether the issue has been identified through pressure from special interests, planning, or some other mechanism, the first step in determining what to do is to define the nature and scope of the problem. It is possible that a cursory review of the situation will reveal that the problem is simply not sufficiently important to warrant any further attention. Alternatively, it may appear that the problem is related to some other issue in a way that would require both to be addressed if public intervention is to have any hope of success. For example, the government of a developing country might identify low agricultural productivity as a problem requiring public intervention, only to discover, after further reflection, that the low productivity is part of a complex development problem involving poor roads, lack of effective storage facilities, lack of standardized weights and measures, illiteracy, and poverty. It may turn out that all these problems need to be attacked at once or that, given limited government resources, the best strategy would be to work on the road and storage problems first before initiating policies more directly connected to agricultural productivity. Sometimes an initial review will reveal that a problem is so large that the only sensible approach is to work on smaller parts of the problem, with a view toward gradually developing an overall solution.

Once the nature and scope of the problem have been defined, an initial set of alternative solutions can be developed. There are two important points in this context. First, the alternatives should always be seen as subject to revision. Suppose that the problem is insuring adequate water for agricultural development in the western United States. An initial set of alternatives might include building dams and canals to distribute existing surface water, tapping underground aquifers, or investing in research on drought-resistant crops. Each of these strategies could be implemented in dozens of different ways, or they might be used in combination, and there may be other alternatives that come up as further study is done. The alternatives being considered should be seen as provisional, subject to modification in light of further information on needs, resources, and constraints.

The second point is that the initial and final list of alternatives should always include the option of doing nothing. To evaluate alternative projects aimed at solving some public policy problem, it is necessary not only to compare the projects among themselves but also to compare them with the no-policy option to see if any action at all is warranted. In 1998, the city of Lincoln, Nebraska, commissioned a study to determine which of four alternative routes for a beltway around the city would be the best. The results of the study indicated that the costs of all three alternatives were greater than the benefits, suggesting that all four were inefficient as specified.[3] The option of not building a beltway did not seem to be up for consideration. As it turned out, the city council chose the least efficient of the four alternatives, a choice that may have been linked to local political realities. A better course of action in this case might have been to do nothing at all. Alternatively, it is possible that there were unrevealed political, ethical, or other concerns that overrode the monetary benefits and costs presented in the analysis. If that were the case, however, it is curious that the city council chose to spend $2 million on a study it then ignored because of these other considerations.

This second point will be raised frequently in this and subsequent chapters. The basic idea is that good public decisions can be made only on the basis of comparisons with the relevant alternative. It is not uncommon to hear commentators assert that a policy has failed because some economic variable such as the trade deficit or the unemployment rate has changed in a way deemed unfavorable. Such assertions often ignore the changes that would have occurred if the policy had not been implemented. It is possible that the unemployment rate or trade deficit would have been even worse without the policy or that the policy itself had no real impact on these variables, which were driven by some other set of events. This question is sometimes referred to as the problem of finding the appropriate *counterfactual*, defined as the situation that would have prevailed if no action had been taken. Comparisons of circumstances today with conditions in the past ignore the fact that many things will have changed between these points in time, making the past an inappropriate counterfactual for evaluating a given project or policy initiative.

The need for an appropriate base of comparison in policy analysis is similar to the need for controls in scientific experiments. Suppose that a new irrigation technique has been developed and agronomists are interested in determining whether this method is better than existing alternatives, as indicated, for example, by yield per acre. The standard procedure is to apply the new method to plants in one or more plots while using existing methods in other plots receiving identical treatment in all respects except for the type of irrigation. Determination of the best irrigation system is then based on comparing these results. This procedure is necessary because many factors will affect yields. An accurate evaluation of the new technique requires that these other influences be controlled so that the only difference between the plots is the irrigation method. Under these circumstances, any yield differences that are found can be attributed to the particular irrigation system that was applied. Although this kind of controlled experiment could not be done for most public projects or policies, one can obtain similar kinds of information by paying close attention to the specification of an appropriate comparison. In the case of the alternative approaches to water supplies in an arid region, an evaluation of what is likely to happen if nothing is done serves as the control, just as existing irrigation methods can be used as controls in the experiment evaluating the new irrigation technology. In both cases, if the alternative does not give better results than the existing systems, it may not be a good solution to the problem.

The third step in this analytical process is to predict the consequences of each of the alternatives and to attach monetary values to these consequences so that they can be aggregated into a measure of the net benefits. In the case of the beltway study for Lincoln, Nebraska, for example, the analysts measured the direct construction costs of the three alternatives as well as their impacts on traffic flows, traffic congestion, traffic accidents, residential property values, agriculture, school districts, parks and recreation, water quality, noise, wildlife habitat, wetlands, archaeological sites, and several other variables. The study did not include costs associated with disturbing a historic farm listed on the National Register of Historic Places in 1979 and located near the path of the option favored by the city council. Because this project is to receive federal funding, an environmental impact study will be required and it is likely that the effects of the different alternatives on the historic farm will be included in this analysis.[4] To determine the net effects of each alternative, the costs and benefits have to be expressed in common units that can be added up and compared across the alternatives. The most common way to do this is to attempt to express all of the impacts in monetary terms. The results can be presented in the form of a benefit-cost ratio (monetary benefits divided by monetary costs) or as net benefits (monetary benefits minus monetary costs). This information can then be used to help in deciding which alternative (including the alternative of doing nothing) is the best choice or as guidance in redesigning one or more of the alternatives in order to eliminate inefficiencies and improve the likely impact. The final step is to decide which course of action makes the most sense, all things considered.

There are, of course, many complications in carrying out these procedures. For example, determining monetary values for the impacts associated with traffic noise, wetland or wildlife habitat protection, the potential destruction of archaeological sites, and many of the other variables of interest is difficult because there are no markets for these goods and therefore no easily observed market prices. Even when market prices are available, externalities or other types of market distortion may mean that they are not appropriate measures of the true social value of the project impacts. In addition, the benefits and costs of a project do not occur all at once, so it will be necessary to take account of the time value of money, as noted in Box 6.1.

Because evaluation of public projects requires estimates of future impacts, there is always the possibility that unforeseen events or uncertainties about the way in which the project will actually play out will lead to mistakes in the projected impacts. Nothing can be done about unforeseen events, but it is possible to use estimates of the probabilities associated with risky future incidents to refine the analysis. Finally, it is important to include consideration of the incidence of benefits and costs as well as other ethical and political aspects that should be brought into the analysis. A project that generates large net benefits may not be acceptable if the benefits flow to wealthy individuals while all the costs are imposed on those who already have limited opportunities. The rest of this chapter is devoted to describing methods for handling problems associated with the time value of money, market distortions and missing markets, risk, and distribution.

DISCOUNTING AND THE TIME VALUE OF MONEY

We often think of investing as the action of purchasing stocks, bonds, or other financial instruments. In the present context, however, a broader definition is of greater use. In this sense, investments include the creation of physical capital—defined as plants, equipment, and other types of capital goods—as well as financial investments. Consider, for example, the case of a firm that is contemplating an expansion of its production facilities through the construction of a new building and the purchase of machines and equipment. Before undertaking such expenditures, the firm will want to know what it can expect to earn from the investment. If the returns in the form of increased sales of the firm's product are not large enough to cover the costs of the investment plus a return that is at least equal to the interest the firm could earn if it put its resources in the bank, the investment is not likely to be made.

In computing the returns to a potential investment, firms recognize that the costs and benefits will occur at different times. If the value of money is not constant over time, some adjustment to the future benefits and costs will have to be made in order to insure that they are measured in the same units as those arising in the present. As noted in Box 6.1, the procedure for making this adjustment is known as *discounting* and is based on the observation that a dollar received today is worth more than a dollar received a year from now because it can be put into a bank account where it will earn interest. Suppose, for example, that a firm has a choice between investing $50 million in a project that will return $200 million in twenty years or putting the money into a project that will return $100 million in eight years. The first thing to note is that the firm can always put the money in a bank account where it will earn interest. If the interest rate is 8 percent, the $50 million will be worth about $95 million in eight years and $247 million in twenty years. These figures are based on continuous compounding, using the formula $V = Ae^{rt}$, where V is the final value, A is the initial value, e is the exponential, r is the interest rate (0.08), and t is time (eight or twenty years). The return of $100 million in eight years for the second project is higher than what could be earned from the bank, while the other project returns only $200 million compared to the $247 million the bank will pay out. Thus, although the first project returns a higher total, it is the second that is a better use of scarce resources when the *time value of money* is taken into account.

In the preceding comparison, the value of $50 million today was projected into the future through compounding. It was also assumed that interest was compounded continuously. Suppose instead that the bank simply computed the interest once at the end of either eight or twenty years. The formula for this computation is $V = A(1 + r)^t$. Thus, assuming that the interest rate is 8 percent, $50 million dollars at the end of eight years would be worth $50 \times (1.08)^8 = \$92.5$ million, slightly less than the result obtained through continuous compounding. Discounting is simply a process of running these calculations in the opposite direction. Instead of finding out the future value of something one has today, it is used to determine the present value of something one will have in the future. The present value of $100 million in eight years is $54 million, computed as $100 million/$(1.08)^8$, which is consistent with the previous conclusion that the second project returns more than the initial investment. In the other case, however, the present value of the $200 million in twenty years is only $42.9 million, and this is less than the initial investment.

These examples illustrate the importance for private firms of taking time into account when computing the costs and benefits associated with investments. The same considerations apply to public investments. Suppose the government is considering a project that would benefit a particular group of people, say, shrimp farmers along the Gulf Coast. An alternative to financing this project would be to place the funds needed to implement it in a bank account in the name of the shrimp farmers, who could then share the interest paid by the bank. The government would probably not want to implement the project if the discounted present value of the benefits is less than the present value of the costs. This is another way of saying that the project may be ill advised because it generates returns that are less than could be earned by placing the funds in a bank.

Of course, the government may have different objectives from those of private firms, so that although the basic procedures are the same, the nature of the benefits and costs that are included may not be. For example, the private firm only needs to worry about actual market prices in computing the benefits and costs. If it can get away with discharging pollution into a river, it need not consider the cost of this pollution in determining the return on its investment. The government, however, would want to include this cost in its analysis, and this is one way in which benefit-cost analysis of public projects differs from private investment decisions.

Suppose that the government is considering a project designed to increase the amount of irrigation water available in an arid region. The western United States is such a region, as is the West African Sahel, which lies along the southern border of the Sahara Desert.[5] A major river flows through the region, but water from the river is not widely distributed, so the agricultural potential of the region is not fully realized. Assume that there is only one project under consideration so that the problem is to compare this project with the alternative of doing nothing. The proposed project involves the construction of canals and pumping stations to move the water through a large, relatively flat area with good agricultural soils. Construction is expected to take two years and will require expenditures for labor (construction workers and design and management technicians), equipment (pumps, bulldozers), raw materials, and land acquisition. Some of the equipment purchased for the project will be sold when the project is completed. There will also be annual maintenance and operations costs for the life of the project. The benefits will consist of increased agricultural output, primarily of cotton, the main crop in the region. The life of the project is taken to be twenty years, after which it will be turned over to a private corporation that will charge fees to cover the costs of management and water delivery.

The basic costs and benefits of this project are shown in Table 7.1. All figures are expressed in real dollars, that is, dollars that have been corrected for inflation.[6] Initial costs for labor, raw materials, equipment (net of $15 million in the salvage value for the equipment sold at the end of the construction period as recorded in the second row), and land acquisition for the project amount to $64 million in the first two years. This figure is obtained by adding the figures in the Net Benefits column from the first two rows of the table (by convention, the years begin at zero, so the first two years appear in the rows labeled Year 0 and Year 1). In Year 10 of the project, it is assumed that major repairs and some rebuilding will be required, costing a net total of $15 million based on expenditures of $22 million for labor, raw materials, and equipment less $7 million in salvage value for equipment that is resold. Annual operations and maintenance costs are $3 million per year except in Year 10, when they are doubled as part of the repairs and rebuilding.

Table 7.1. Benefit-cost analysis for hypothetical irrigation project (in millions of dollars)

Year	Managers and Technical Workers	Construction Workers	Raw Materials	Equipment	Land Aquisition	Operations and Maintenance	Δag Income	Net Benefits	Present Value @10%	Present Value @3%
0	-1	-3	-20	-20	-5	-3	0	-52	-52	-52
1	-1	-3	-15	+15	-5	-3	0	-12	-10.9	-11.7
2						-3	10	7	5.8	6.6
3						-3	25	22	16.5	20.1
4						-3	30	27	18.4	24
5						-3	35	32	19.9	27.6
6						-3	35	32	18.1	26.8
7						-3	35	32	16.4	26.0
8						-3	35	32	14.9	25.3
9						-3	35	32	13.6	24.5
10	-1	-1	-10	-10		-6	25	-3	-1.2	-2.2
11				+7		-3	35	39	13.7	28.2
12						-3	30	27	8.6	18.9
13						-3	25	22	6.4	15.0
14						-3	20	17	4.5	11.2
15						-3	20	17	4.1	10.9
16						-3	20	17	3.7	10.6
17						-3	20	17	3.4	10.3
18						-3	20	17	3.1	10.0
19						-3	20	17	2.8	9.7
20						-3	20	17	2.5	9.4
Total	-3	-7	-45	-8	-10	-66	486	356	112.3	249.2

In the first two years of the project, there are no benefits—that is, there is no income—because the water delivery system is not yet in operation, although there are maintenance costs in those years. These maintenance costs might be thought of as training and testing costs. In Year 2 of the project, the system comes on-line, and some increased agricultural output, valued at market prices, is recorded. In the next two years, the benefits increase as more of the system is brought on-line, until they reach $35 million in the Year 5 row. These benefits continue until Year 10, when the repairs and rebuilding disrupt production, reducing the benefits slightly. Beginning in the Year 12 row, the increased agricultural output begins to fall. It is assumed that the planners for this project recognize that it will eventually cause increased salinity, which will reduce yields to some extent, and this is reflected in the benefits for the final years of the project.

The total cost of the project, including the initial investment, the repairs and rebuilding in Year 10, and the annual operations and maintenance costs, is $152 million, less the salvage value of the used equipment ($22 million). The income over the life of the project is $486 million. Without taking the time value of money into account, the returns are almost four times as large as the net investment of $130 million. There are several ways to adjust these results to include the time value of money. The method illustrated in the table is based on the figures in the column labeled Net Benefits. These figures are simply the horizontal summation of each row. The next step is to discount these net benefits using the following formula:

$$PV = \frac{NB_0}{(1+r)^0} + \frac{NB_1}{(1+r)^1} + \frac{NB_2}{(1+r)^2} + \ldots + \frac{NB_{20}}{(1+r)^{20}}$$
$$= \sum_{T=0}^{20} \frac{NB_T}{(1+r)^T}$$

where PV is present value, NB is net benefit, the subscripts and superscripts indicate the year ($T = 0, 1, \ldots, 20$), and r is the discount rate. The results of this exercise are shown in the last two columns, for discount rates of 10 percent and 3 percent. The figures at the bottom of these columns are the summations of the discounted values and are referred to as the *net present values* of the project. At both discount rates, these values are positive, indicating that this project is an efficient use of resources.

The net present value of a project aggregates discounted benefits and costs. Another way to present this information is in the form of a *benefit-cost ratio*. If we treat the operations and maintenance costs as part of the net benefits, a benefit-cost ratio for this project can be calculated fairly easily. Using a discount rate of 10 percent, the initial investment costs are $62.9 million (52 + 10.9). In Year 10, repair and rebuilding costs of $22 million have a present value of $8.5 million. Total costs thus are $71.4 million (62.9 + 8.5). Benefits can be found by adding up the rest of the figures in the column for the 10 percent present value, with an adjustment for the benefits in Year 10. The $22 million repair and rebuilding cost have already been included, so net benefits in Year 10 should be $19 million, with a present value of $7.3 million. Total benefits thus are $181.3 million, giving a benefit-cost ratio of 181.3/71.4 = 2.54.

A third way to present the information on this project would be to compute an *internal rate of return* (IRR). The IRR is the discount rate that would make the net present value of the project equal to zero. For the hypothetical project illustrated in Table 7.1, the IRR is around 30 percent.

Net present value, the benefit-cost ratio, and the IRR are different ways of presenting the same results. If the present value of the difference between benefits and costs (the net present value) is positive, the estimated benefits of the project are greater than its costs. A benefit-cost ratio greater than 1 means that the discounted present value of the benefits is greater than the discounted present value of the costs. The IRR can be compared with the rate of return on alternative investments to determine whether the project has a higher return than the alternatives do. Note that the favorable results for this project do not mean that it should be implemented. They do indicate that implementing this project is an efficient use of resources.

So far the analysis of the hypothetical investment project has been done as if it were a private investment where market prices and interest rates are the appropriate measures to use. For public-sector investments, it is generally necessary to adjust the data used in the analysis if, for example, there is reason to believe that market prices do not accurately reflect the social opportunity cost of the resources used in the project. In subsequent parts of this chapter, we will investigate methods for adjusting the benefit-cost analysis so that it reflects social values rather than purely private values. The important question here concerns one of the values used in the analysis, the discount rate. Private firms clearly will use market interest rates to discount benefits and costs of potential investments, because these rates show the true opportunity cost of the firm's investment funds. But are market rates the right choice for public projects? Is there a need for a *social discount rate* that more accurately reflects society's time preferences?

Appendix 7.1 at the end of this chapter includes a discussion of the reasons why money has different values at different times. In large part, these differences are due to preferences between current and future consumption, driven by perceptions of the risk and uncertainty of future events. If financial markets function reasonably well, there is every reason to expect that market interest rates reflect *social time preferences* in the same way that prices in perfectly competitive markets reflect aggregate social preferences. Of course, if financial markets are not competitive, there could be a divergence between social time preferences and market interest rates. A particularly interesting way in which such a divergence might arise stems from the fact that private market rates are determined by people presently participating in markets. They do not reflect the interests of people who will be born in the future. For short-term projects, this problem may not be severe because the present generation will have enough interest in its children and grandchildren to include them in its investment decisions. In addition, short-term projects are less subject to the effects of discounting, which has a greater impact the further one is from the present.

For projects likely to generate impacts that will occur in the distant future, however, the results of a benefit-cost analysis will be highly sensitive to the discount rate chosen. Consider, for example, a project that will generate modest net benefits for the next five hundred years, at which point a major disaster will occur. The present value of this future disaster, using a 10 percent discount rate, is calculated by dividing its monetary value by 1.1^{500}, or 496,980,000,000,000,000,000, a number that is over 62 million times the size of the 1998 U.S. GNP of $8 trillion. Even if the disaster is enormous, it will hardly matter if our decision is based on its present value at a 10 percent discount rate. Things are quite different, however, if a discount rate of 3 percent is used instead. At this rate, the discount factor is 1.03^{500}, or 2,621,880, and the future disaster will carry much greater weight in the analysis. Because discounting is a process of dividing finite numbers by a number greater than 1 raised to ever higher powers, lower discount rates reduce the value of future events less than higher ones. This can be seen in Table 7.1, where all the numbers in the column under the 3 percent discount rate are larger than the corresponding values discounted at 10 percent. Not surprisingly, the net present value of the project is larger ($249.2 million as opposed to $112.3 million) at the lower discount rate.

To see the significance of these differences, imagine that the hypothetical irrigation project leads to such high salinity in the river that it becomes necessary in Year 21 to build a series of desalination plants downstream from the project to clean the water for other users. If the project planners were aware that this would happen, it would be necessary to include this future cost as part of the project. Suppose that the cost of the treatment plants is $700 million and that everything else in Table 7.1 stays the same. Then we can subtract the discounted present value of this future cost from the net present values in Table 7.1 to determine the overall value of the project. At 10 percent, the present value of $700 million is $94.6 million, which, when subtracted from the earlier result of $112.3 million, leaves a net present value for the project of $7.7 million. In other words, even with the large future cost, the project still passes the benefit-cost test. At 3 percent, however, the present value of the plant construction is $376.3 million, which means that the net present value for the project as a whole is -$127.1 million ($249.2 million minus $376.3 million).

Many scholars have argued that the appropriate social discount rate to use for public projects is substantially lower than market interest rates.[7] This would seem to be of particular importance for projects that have very long-term effects that could reduce the living standards of future generations. Using a lower discount rate for public projects, however, means that more of these projects will pass the benefit-cost test. Projects rejected in the private sector could be judged viable at a lower social discount rate. Whether this would be good for future generations depends on whether one believes the government will make better investment choices than the private sector will or that there are significant market failures and other considerations that make government investment preferable to private. Note that if one is particularly concerned about the future disaster, it would be possible to set up a bank account and let the magic of compound interest generate a fund that would cover the future costs. Suppose, for example, that a project will lead to a social cost of $100 trillion (over three times the size of today's world economy, as measured by the combined GNPs of all the countries of the earth) in one hundred years. At 10 percent, the present value of this cost is about $7.3 billion ($100 trillion/1.1^{100}), a fraction of a percent of the current U.S. GNP. If it seems unfair to future generations to burden them with this large cost, one could simply invest $4.5 billion in a bank account paying 10 percent for the next one hundred years ($4.5 billion $\times e^{100 \times 0.10}$ = $100 trillion).

Of course, one can never be certain that the funds in the bank account will still be there in one hundred years or that the bank will still be in business at that time. Most economists believe that the ordinary workings of a growing economy will cover future problems so that it would be redundant to set up specific funds to cover the costs of possible future disasters. The expectation is that if current generations save and invest at an appropriate level, the future can be assured without worrying much about social discount rates. On the other hand, if a particular course of action seems likely to lead to extremely serious consequences in the distant future, the analyst might elect to simply offer this fact as a reason to reject the course of action without trying to figure out a social discount rate or worrying about compensatory funds.

Global warming is a case in point. Current actions may lead to climate changes that would seriously impair the quality of human life in the distant future. In analyzing the benefits and costs of actions to reduce global warming, William Cline used a social discount rate of 2 percent to allow the distant events to carry more weight in the analysis.[8] An alternative is to argue that it is unacceptable to impose these costs and so actions should be undertaken to prevent them from arising, even if the costs to current generations are high. A third alternative is to note that there is so much uncertainty attached to distant events that it would be foolish to worry about them at all. Technological innovations may lead to the ability to correct future warming. Besides, future generations will almost certainly be richer than we are today (it has always been the case that standards of living increase through time), so sacrifices by the current generation for the benefit of the wealthier future people amounts to a redistribution from the poor to the rich, something that violates common sense as well as Rawls's difference principle.

The choice of discount rates for public projects is often controversial. In addition to the intergenerational issue described earlier, there are other reasons why market interest rates may not accurately reflect social time preferences. First, market rates are influenced by factors such as taxation and imperfections in capital markets that distort the relation between marginal rates of substitution between current and future consumption in the same way that market failures in goods or factor markets introduce distortions in the marginal rates of substitution between different goods or factors. Second, there are many market interest rates, and it is often unclear which, if any, of these rates is to be preferred. Interest rates charged to borrowers are higher than the rates paid to savers. Moreover, interest rates usually vary substantially according to the riskiness of the investment or the credit rating of the borrower.

Some economists have argued that the social discount rate should be chosen on the basis of the type of financing for the project being undertaken. A project financed by government borrowing should be evaluated using the interest rate charged to borrowers with similar credit ratings to that of the government agency. If a lower rate is used in these circumstances, there is a risk that government borrowing will crowd out private-sector borrowing because more public projects would be profitable. This effect would lower social welfare if useful private investments were crowded out by marginal government projects. On the other hand, if the project financing is more likely to affect consumption, it may be preferable to use a discount rate that is closer to the marginal rate of substitution between current and future consumption—in other words, a rate that more closely reflects social time preferences. Perhaps the best way to approach the problem of choosing a social discount rate is to make explicit arguments to justify the rate chosen and to use several rates to illustrate the range of outcomes that might be expected. As noted in Box 6.1, the choice of a social discount rate is largely an ethical question. The answers to such questions are best made on the basis of explicit arguments and justification.

SHADOW PRICES: MARKET DISTORTIONS AND MISSING MARKETS

Discount rates are obviously only one of many prices in the economy that might be distorted by market imperfections. In many developing countries, lack of information and other structural rigidities in labor markets have resulted in high unemployment and extensive underemployment. In such circumstances, the true opportunity cost of the labor hired to work on a project may be less than would be indicated by market wages. If some of the materials or equipment for the hypothetical irrigation project presented in the preceding section were supplied by monopolies, the market prices would be higher than the marginal value society attaches to them. In this case, a more appropriate price for the social evaluation of the project would be lower than the observed market price. In describing the hypothetical irrigation project, we noted that it was expected to cause increased salinity in the water. This cost was initially accounted for by reducing the agricultural benefits in later years to reflect the potential impact of salinity on production. But salinity may also affect other uses of the water. For example, the higher salinity could make the river unsuitable for recreational or commercial fishing. This externality is a cost for society that should also be included in evaluating the project.

In some cases, no prices for particular goods and bads can be observed because there is no market in which these items are traded. Suppose, for example, that the irrigation project will disturb the habitat of an animal that is in danger of becoming extinct. Society may attach a value to the continued existence of this species, and its increased endangerment (or its extinction) should be included as part of the cost of the project even if there is no market price to measure the value of this animal's continued existence. There are many examples of situations where missing markets raise the problem of determining the appropriate values to use in evaluating projects. There is no market for human life, but any public highway project, for example, must account for the costs of highway accidents and deaths related to the project. As a society, we may value a wide range of nonmarketed goods, including, for example, beautiful sunsets or historical sites such as graveyards, buildings, and archaeological sites, and so on. Both public and private projects can have an impact on these nonmarketed goods. In the case of public projects, some way of accounting for these impacts is generally incorporated into the evaluation of the project itself. The evaluation of a public project to build an airport would have to include some measure of the social cost of increased noise even though market prices for noise do not exist. In the case of private projects, it is common for governments to implement regulations or policies to force private firms to take account of such impacts. Thus, for example, a project to build a private airfield would have to comply with laws on noise pollution.

Government policies can also give rise to *price distortions*. Many developing countries have implemented polices that cause the exchange rate of their currencies to be overvalued. The exchange rate is the price of foreign currencies in terms of the domestic currency. The Mexican new peso had an exchange rate with the U.S. dollar of 9.36 on November 17, 1999, meaning that one could trade one dollar for 9.36 new pesos on that date. An overvalued exchange rate has the effect of lowering the cost of a country's imports while raising the cost of its exports.[9] If the government of a country with an overvalued exchange rate is evaluating a project that requires imported inputs, it will be necessary to adjust the price of these inputs to reflect their true value to the country. In the United States and other industrialized countries, government intervention in agriculture is common. Many of these policies are designed to increase the prices received by producers above the level that would prevail in competitive markets. If the hypothetical irrigation project described earlier is located in the western United States, it is likely that users would not be charged the full price of the water. For a social benefit-cost analysis, it would be important to correct the value of the benefits for these government subsidies.

The goal of *social benefit-cost analysis* is to evaluate public projects from the point of view of society as a whole. Suppose that the government uses a tariff to protect domestic producers from foreign competition. The effect of such a policy, as illustrated in Chapter 2, is to increase the price received by producers. But this increased revenue (producer surplus) is paid for by consumers, who are forced to pay a higher price for the products of the protected industry. Suppose that a firm is able to discharge pollution into a river, thereby saving the cost of disposing of the pollution itself. Such a procedure obviously does not eliminate the cost; it is simply shifted to someone else. In cases such as these, there is a social cost that is greater than the private costs faced by the producers. Where social and private costs diverge, the appropriate values to use in analyzing a public project are not the prices observed in the market but rather adjusted prices that reflect the marginal values society attaches to the good or service. These adjusted prices are often referred to as *shadow prices*.

To illustrate the use of shadow prices in benefit-cost analysis, consider the hypothetical irrigation project shown in Table 7.1. For that analysis, no adjustments were made for differences between social and private values other than to raise the issue of social discount rates. In Table 7.2, the analysis is repeated using hypothetical shadow prices to account for various subsidies and distortions. The adjustments shown in Table 7.2 are based on the following assumptions:

1. The prices for cotton, the primary crop in the irrigated region, are supported with a tariff of 10 percent paid for by consumers. In addition, it is assumed that the original figures for the agricultural benefits include a subsidy for the price of the water. The social value of the cotton is less than the private value due to these subsidies. In Table 7.2, the agricultural returns are adjusted using imaginary world cotton prices and market rates for irrigation water.

2. The column for land acquisition in Table 7.1 has been modified to include land and environmental, ecological, and cultural costs. The costs of land acquisition in the first two years are the same. It is assumed, however, that the construction of the irrigation system leads to a onetime cost of $1 million in Year 2 as compensation for disturbing and relocating a cemetery and some historic buildings. In addition, the project leads to some loss of commercial and recreational fishing due to the reduced flow in the main river channel. Over time, the environmental losses increase as the salinity in the river further disrupts

fishing and other uses and the reduced water flow leads to sedimentation. The project also causes some habitat destruction, reducing biodiversity. All these costs have been aggregated in the column labeled Land and Environmental Costs. They are assumed to increase dramatically over time.

3.   Finally, there are some adjustments in the initial costs. It is assumed that some of the raw materials are supplied by a monopolistic firm charging prices above marginal costs. It is also assumed that the region has extensive unemployment so that the opportunity wage of the construction workers is lower than market wages. Construction labor costs and raw material costs have been lowered to correct for these market failures.

These adjustments to the original values change the results of the analysis. The net present value of the project using a discount rate of 10 percent is negative (-9.2 in the Total row for the column Present Value at 10 percent), suggesting that this project is not socially efficient. While the project is viable given existing market distortions and government subsidies, it is not beneficial to society when costs associated with market failures, government interventions, and the impact on nonmarket goods are included in the analysis. At a lower discount rate, the project still passes the benefit-cost test, although the benefit-cost ratio is only 1.06. This example illustrates one of the dangers of using lower discount rates in social benefit-cost analysis. Even though this project generates substantial costs throughout its life, particularly in the later years, the positive returns early on are enough to generate a positive net present value at the 3 percent discount rate. This result is consistent with the observation made earlier that lower social discount rates will filter out fewer projects, leading to more public investments. This may not be a problem if one believes that the government generally makes better investment decisions than the private sector. In this example, however, many would question the wisdom of undertaking a project with such severe environmental costs in order to support a few farmers who want to grow cotton in the desert.

These results are based on assumed values, and no explicit documentation describing actual market or shadow prices has been included. Further details on the adjustment of market prices to more accurately reflect social values will be discussed in the case studies in the next two chapters. A few comments are in order here, however, concerning the realities of valuation in social benefit-cost analysis. There are two categories of problems that must be confronted in determining the appropriate values to use in such analyses. The first category includes all the situations in which market prices do not accurately reflect social opportunity costs. Determining the extent to which market prices diverge from social values may require a great deal of information about the particular circumstances that have given rise to the distortion. In many cases, it is quite difficult to form a reasonable estimate of the appropriate shadow price. Price distortions caused by government intervention, however, can often be removed by using an appropriate reference price. For example, world prices are good estimates of the social value of domestic goods protected by trade barriers. Thus, in the illustration presented earlier, the cotton is appropriately valued at the world price rather than the internal price, which was assumed to have been raised by a tariff. Suppose that the world price for a pound of cotton is $0.60 and that the tariff raises this price to $0.66 per pound. The difference between these two prices is a cost to consumers, which should be subtracted from the market value of the cotton to obtain its true social value.

Table 7.2. Benefit-cost analysis for hypothetical irrigation project, with corrections for distortions and subsidaries (in millions of dollars)

Year	Managers and Technical Workers	Construction Workers	Raw Materials	Equipment	Land and Environmental Costs	Operations and Maintenance	Δag Income	Net Benefits	Present Value @10%	Present Value @3%
0	-1	-1	-17	-20	-5	-3	0	-47	-47	-47
1	-1	-1	-13	+15	-5	-3	0	-8	-7.3	-7.8
2					-4	-3	7	0	0	0
3					-3	-3	23	17	12.8	15.6
4					-3	-3	26	20	13.7	17.8
5					-3	-3	26	20	12.4	17.3
6					-3	-3	25	19	10.7	15.9
7					-6	-3	25	17	8.7	13.8
8					-7	-3	25	15	7.0	11.8
9					-7	-3	25	15	6.4	11.5
10	-1		-10	-10	-10	-6	15	-23	-8.9	-17.1
11				+7	-14	-3	25	15	5.3	10.8
12					-14	-3	25	8	2.5	5.6
13					-14	-3	14	-3	-0.9	-2.0
14					-14	-3	14	-3	-0.8	-2.0
15					-17	-3	14	-6	-1.4	-3.9
16					-17	-3	14	-6	-1.3	-3.7
17					-18	-3	14	-7	-1.4	-4.2
18					-21	-3	14	-10	-1.8	-5.9
19					-26	-3	14	-15	-2.5	-8.6
20					-31	-3	14	-20	-3.0	-11.1
Total	-3	-3	-40	-8	-216	-66	345	-2	-9.2	6.8

Likewise, the effects of domestic taxes and subsidies on market prices can sometimes be eliminated by simply subtracting the tax or subsidy rate from the observed prices. In other cases, information on market supply and demand may be available for estimating equilibrium prices in the absence of the intervention. Alternatively, it may be possible to find prices in similar markets where there is no government intervention. The fees charged by the U.S. government on public grazing lands are lower than the fees charged on similar private land. In these cases, the appropriate price to use in social benefit-cost analysis is the private grazing fee rather than the subsidized public fees. Some government policies raise more complex measurement problems. Minimum wages raise wages above the market equilibrium. Because minimum-wage laws are in force throughout the entire country, there is no domestic reference market to serve as a guide for valuing the social opportunity cost of unskilled labor being paid the minimum wage. Moreover, there is no international benchmark that could be used, as was the case for traded goods such as cotton. Adjustments to the costs of unskilled labor would have to be based on studies of labor supply and demand and an estimate of the equilibrium wage in the absence of minimum wage laws.

Similar problems often arise in dealing with distortions brought about by market failures. For example, if markets are imperfectly competitive and the analyst suspects that firms are using their market power to charge prices above marginal costs, information on the firms' cost schedules could be used to estimate the equilibrium price that would prevail if the markets were competitive. The firms in this case would probably treat such cost information as a trade secret, however, adding to the difficulties of estimating a competitive price. On the other hand, information on the social cost of an externality may be fairly easy to find. The costs of cleaning up air or water pollution may be well documented. In other cases, information about the costs of an externality may be available. Consider the case of an industrial plant that generates air pollution that damages nearby fields used to produce grain or some other kind of agricultural good. It may be possible to directly evaluate this damage through study of the affected crops. Or, information on yields and crop quality in similar fields not affected by the air pollution could be used to estimate the value of the damage. In all these cases, the goal is to remove biases introduced by using values that do not reflect true social benefits and costs.

The second category of valuation problems arises when there is no market and therefore no price of any sort. Determining precise social benefits and costs in these cases is difficult, and the best one can do may be to provide a range of estimates. We have noted several examples of goods and services that have value to human beings but are not traded in markets. Beautiful scenery, biodiversity, peace and quiet, sites that have historical value, and life itself are examples of such goods. Two broad methods for handling these valuation problems have been developed: hedonic pricing and contingent valuation (see Chapter 9).

Hedonic pricing relies on the fact that there may be a market for other goods that can either serve as a proxy for the nonmarketed good or shed light on its value. Suppose that a project involves construction that will screen out the view some homeowners have of beautiful sunsets that occur regularly in that area. The value of the beautiful sunsets may be incorporated into the price of houses that have a good view. The prices of similar houses where there is no view of the sunset would be lower, and the value of the sunsets could be taken to be the difference in the prices of the two sets of houses.

Although the value we each attach to our own lives and the lives of those who are close to us is, for all practical purposes, infinite, we do not behave as if it were. This is shown by the fact that we choose to do things that will shorten our own lives—smoking, for example—and we undertake collective projects that are bound to result in additional human deaths—such as highway construction. If all human life has an infinite value, we would be unable to do almost anything. Any nonzero risk of being killed while crossing the street would be enough to keep us home if we attach an infinite value to our lives. While some may find the procedure objectionable, it is often the case that some value for human life is needed in benefit-cost analysis, and hedonic pricing may provide a way to discover this value. For example, if one expected that a project would lead to deaths that would not have occurred without the project, expected lifetime earnings lost due to these deaths could be used to develop an estimate of the value of these lives. Alternatively, some analysts exploit the fact that there are people who are willing to take jobs that carry above-average risks of death in return for a higher wage. The wage differential could be used with other statistical information to estimate a value for human life.

Hedonic pricing is limited by the fact that it is not always possible to find market prices that can be used as proxies or as the basis for statistical estimation of the value of non-marketed goods. *Contingent valuation* is an alternative approach based on survey techniques. An estimate of the social value of preventing the extinction of an endangered species might be obtained by conducting a survey in which a representative sample of the relevant population is asked how much they would be willing to pay to prevent this extinction. Many respondents might answer that they are indifferent to the continued existence of whooping cranes or spotted owls and would be unwilling to pay anything to prevent their extinction. Others, however, might feel that it would be a bad thing for these animals to become extinct and would be willing to contribute to their protection. Note that traditional demand curves for marketed goods are basically measures of consumers' willingness to pay for particular goods. This idea was used in Chapter 2 to develop the concept of consumer surplus. In the case of nonmarketed goods, measures of willingness to pay serve as estimates of demand for the good, allowing the analyst to compute the consumer surplus for use as a measure of the benefits of providing the good.

Contingent valuation and hedonic pricing rely on creativity in the formulation of the models used and the interpretation of the results. Despite the best efforts of the analyst, however, the estimated social values of nonmarketed goods are generally less reliable than valuations based on actual markets. As in the case of social discount rates, it may be best when faced with such problems to use a range of values for the nonmarketed goods and to present results reflecting different assumptions about these values. In some cases, it may be impossible to derive any reasonable estimate of the social value of the nonmarketed good, and the best procedure may be simply to note the impact of the project without trying to attach a monetary value to it. Some public projects have been canceled or substantially redesigned because of their likely impacts on nonmarketed goods. The Tennessee Valley Authority (TVA) was forced to halt construction of the Tellico Dam on the Little Tennessee River because it was thought that the project would cause the extinction of the snail darter, a small fish that had been listed as endangered. No effort was made to estimate the existence value of this fish. The project was simply stopped to protect it from extinction.[10] Similarly, public projects that might disturb Native American burial grounds would be prevented by such laws as the Native American Grave Protection and Repatriation Act (NAGPRA) or laws pertaining to the National Register of Historic Places, which protect sites that have cultural or historic value in the United States. In these cases, U.S. society has concluded that the unmeasured value of these nonmarketed goods outweighs any benefits that might arise from projects that would disturb them.

RISK AND UNCERTAINTY

The costs and benefits of most public projects cannot be predicted with absolute certainty. In general, the initial project costs (e.g., the cost of building a dam or an irrigation system) can be estimated with some confidence, although the history of cost overruns for large government projects such as dams or the development of military hardware is a good reason to use caution in evaluating such cost estimates. More importantly, it is quite likely that estimates of future costs and benefits will be inaccurate because of the inherent risk and uncertainty associated with future events. In situations of *risk*, it is possible to estimate the probabilities that certain events will or will not occur, and this information can be used to condition the results of the benefit-cost analysis. *Uncertainty* exists when probabilities are unknown. In these cases, the analyst may decide to compute several estimates of the project's net present value, using different assumptions about the values of key variables. The idea that a range of values should often be used in benefit-cost analysis has been mentioned several times. Such procedures are sometimes referred to as *sensitivity analysis*. Sensitivity analysis can provide useful information for making decisions when it is impossible to predict impacts with a high degree of certainty. For example, in the hypothetical irrigation project described in Tables 7.1 and 7.2, the probability that the project will actually lead to enough salinity to require the construction of desalination plants may be unknown. In such a case, it would be useful to present results both with and without the plants to explore the sensitivity of the project to this issue.

On the other hand, if the probability that there will be a need for the desalination plants is known, the concept of *expected value* can be used to more accurately assess the social benefits and costs of the project. Suppose that the probability that desalination plants will have to built is 0.6. This means that the probability that the plants will be unnecessary is 0.4. To simplify the discussion, assume for the time being that the figures in Table 7.1 have been corrected for market distortions and government interventions and that the appropriate discount rate is 3 percent. The net present value of the project without the desalination plants is $249.2 million. The present value of the desalination plants in Year 21 discounted at 3 percent is $376.3 million, so the net present value of the project with the plants is -$127.1 million. The expected value of the project is given by the probability that the plants will be needed multiplied by the net present value of the project with the plants, plus the probability that they will not be needed times the net present value without them: (0.6) × (-$127.1) + (0.4) × ($249.2) = $23.4 million.

In this example, the project passes the benefit-cost test. If the probability that the desalination plants would have to be built were 0.7 instead of 0.6, however, the expected net present value would be -$14.2 million. If one has a high degree of confidence in the probabilities, the expected value results provide more useful information than simply using either $249.2 million or -$127.1 million. If one is uncertain about the need for the plants, presenting the results with and without them may be the best that one can do. In general, the best strategy is to provide decision makers with all these results (with and without as well as the probabilities and expected values).

Expected values can be computed for individual years if there is enough information available. Suppose, for example, that the loss of biodiversity predicted for the irrigation project is a risk rather than a certainty. In these circumstances, probabilities could be used to compute the expected value of these losses in each of the years shown in Table 7.2. These expected values would be smaller than the hypothetical costs of lost biodiversity shown in the table, so the aggregate land and environmental costs would be smaller as well. The adjusted figures would be included in the net benefits in the normal way and discounted along with everything else.

Another way to take account of risk in benefit-cost analysis is to use the concept of *certainty equivalence*. In the example of our hypothetical irrigation project, we computed net present values of -$127.1 million if desalination plants are required and $249.2 million if they are not. If the probabilities are 0.6 that the plants will be needed and 0.4 that they will not be required, the expected value of the project is $23.4 million. Policy makers may be uncomfortable with this risky project and be willing to accept a lower net present value if it is certain. Suppose, for example, that there is another project with a net present value equal to a certain $15 million. Suppose that policy makers prefer the project that returns $15 million with certainty to the risky project that generates a higher expected net present value. Suppose further that they would not prefer a project with certain returns lower than $15 million. In these circumstances, the certainty equivalent of the risky project is $15 million. This means that the risky project has to have a higher expected return to be equivalent to a project that generates a return that is lower but certain. The difference between the value of the risky project and its certainty equivalent is known as a *risk premium*. In this case, the risk premium is $8.4 million.

Certainty equivalence can be introduced into benefit-cost analysis by computing the certainty equivalent value of the net benefits in each year. If we believe that the initial costs for the project described in Table 7.1 will be incurred with certainty, no adjustment has to be made. In later years, however, it is possible that the net benefits become increasingly risky. If the probabilities of these risks can be estimated, it would be possible to introduce a risk discount factor that would translate the net present values into certainty equivalents. The values of the certainty equivalents in each year would be lower than the corresponding expected values. The certainty equivalents would then be discounted as usual to compute the net present value of the project in certainty equivalents. The problem with this procedure is that there may be no reference projects that could be used to estimate the risk premium. If enough information on social preferences about risk and the degree of risk associated with a project is available, it may be possible to compute the certainty equivalent. In any case, the basic principle that a risky project needs to generate a higher return than a project that is risk free may be useful when confronted with decisions on risky public investments.

DISTRIBUTION

Suppose that the analysis of a public project shows that it is feasible in that it leads to positive net returns using market prices and that it is also socially efficient, generating a positive net present value when the values are corrected for distortions, market failures, and missing markets. Assume further that there are no important issues of risk or uncertainty that need to be brought into the analysis. Such a project would appear to be socially beneficial, and if there is no alternative project that provides a higher return, one could make a case that this project should be implemented. There is one further issue, however, that should be addressed before concluding that the consequences of the project are positive. The project might impose net costs on a socially disadvantaged group, violating the society's norms with respect to distributional justice. One way to include such concerns is to employ nonconsequentialist rules as inviolable constraints so that the project would be rejected if it violated the rights of the disadvantaged group. Recall that social benefit-cost analysis is nothing more than an input into the decision-making process. Even if the results seem positive, political and ethical problems may override the efficiency criterion of benefit-cost analysis.

Another way to handle distributional issues in benefit-cost analysis would be to attach weights to the monetary values to reflect social values concerning the distribution of benefits and costs. Suppose that the environmental costs associated with the hypothetical irrigation project fall most heavily on low-income families living downstream from the project. This might occur if a socially disadvantaged group, such as Native Americans in the United States, relied on the river for fishing and other uses that are precluded by diversion of the water and the resultant salinity and sedimentation. The figures in Table 7.3 are similar to the figures in the other tables but have been modified to separate out the distributional effects. The labor, materials, and equipment costs shown in the first four columns of Table 7.2 have been aggregated and are shown in the first two columns of Table 7.3. The third column in Table 7.3 includes land and environmental costs adjusted to reflect market failures, government distortions, and the best estimates possible of nonmarket effects. Agricultural returns have also been adjusted in a similar manner. As in the previous example, there is the onetime cost of relocating the cemetery and buildings in the first year (Year 0), followed by growing costs brought on by salinity, sedimentation, and reduced water flows in the main river channel.

The figures in the fourth column reproduce the land and environmental costs with an adjustment based on attaching greater weights to the costs borne by the low-income households. In discussing the time value of money, we noted that the true value of money depends on when it is received or spent. Discounting is used to express monetary costs and benefits in units that have the same value across the time spectrum. One could also argue that the true value of a unit of money depends on who receives or spends it. To this point in the discussion, we have implicitly assumed that a dollar of benefits or costs counts as one dollar regardless of who receives the benefit or bears the cost. The equal weighting of benefits and costs could be thought of as attaching a weight of 1 to all the figures in the tables. This would imply that the social costs of $100 borne by wealthy individuals have the same weight as an identical cost borne by people with limited income. As noted earlier, however, one would expect that a wealthy person would suffer less hardship from the loss of $100 than would a poor person faced with the same loss.

Consider the land and environmental costs totaling $3 million in Year 3 in Table 7.3. Suppose that half these costs fall on the low-income families living downstream from the project. If we could determine that the hardship of these costs caused twice as great a loss of utility for these people as would a similar loss for others, we could weight this loss twice as heavily by simply multiplying it by 2. Thus, the $1.5 million cost to low-income households would be introduced as $3 million and added to the $1.5 million cost to nonpoor households to obtain a total cost of $4.5 million (rounded up to $5 million and shown as -5 in the Distribution Adjustment column for Year 3).

The column of adjusted Land and Environmental Costs is based on the assumption that they fall most heavily on disadvantaged groups that might be expected to suffer disproportionately from these costs. What is being done is to count each dollar of costs to these groups as more than a dollar of cost to society. The results show that although the project has a positive net present value without this adjustment, taking the distributional consequences into account makes the net present value negative. It should be noted, of course, that such a weighting procedure can increase the net benefits as well as reduce them. If many of the unskilled workers employed on the project were drawn from the disadvantaged group living in the area, one might elect to weight their salaries by a factor greater than 1 to include the beneficial impact of these increased incomes. Of course, one could also try to directly measure the beneficial effects of the increased incomes on the local economy, and such a measurement might render the distributional weighting unnecessary.

If the analyst decides to adjust the costs and benefits of a project according to income levels or some other consideration, appropriate weights for the various groups will be needed. It seems intuitively obvious that people with low incomes will suffer more from the loss of $100 than will people who are wealthy. But this intuition does not allow us to determine a specific value for the weight that can be used to reflect the different levels of suffering. This question, of course, returns us to the problem of interpersonal utility comparisons. Many economists would reject the notion that it is either possible or desirable to weight the analysis by some set of factors designed to reflect the impact of benefits and costs on the utilities of different people.

Other economists contend that the formal incorporation of distributional impacts is crucial for meaningful benefit-cost analysis. One way to incorporate these distributional impacts is to estimate the marginal utility of income for different income groups. Blue and Tweeten used statistics on income and a quality-of-life index to estimate equations from which they were able to derive marginal utilities of income for various income groups in the United States.[11] They concluded that an appropriate weight for groups with average incomes equal to 10 percent of the mean income level would be about 1.5. For groups with average incomes equal to the mean, a weight of 1 would be used, while for those with incomes four times as large as the mean, a weight of about 0.3 is recommended. Such weights could be used in the evaluation of a project by multiplying costs and benefits for low-income households by factors greater than 1 at the same time that the impacts on high-income families are reduced in value through multiplication by weights that are less than 1. Such a weighting procedure may change the results quite significantly.

Table 7.3. Benefit-cost analysis for hypothetical irrigation project, weighted for the distributional impacts (in millions of dollars)

Year	All Labor	Materials and Equipment	Land and Environmental Costs	Distribution Adjustment	Operations and Maintenance	Δag Income	Unadjusted Net Benefits	Adjusted Net Benefits	Unadjusted Present Value @10%	Adjusted Present Value @10%
0	-2	-37	-5	-5	-3	0	-47	-47	-47	-47
1	-2	+2	-5	-5	-3	0	-8	-8	-7.3	-7.3
2			-4	-4	-3	7	0	0	0	0
3			-3	-5	-3	23	17	15	12.8	11.3
4			-3	-5	-3	26	20	18	13.7	12.3
5			-3	-5	-3	26	20	18	12.4	11.2
6			-3	-6	-3	25	19	16	10.7	9.0
7			-3	-8	-3	25	19	14	9.7	7.2
8			-6	-9	-3	25	16	13	7.5	6.1
9			-6	-9	-3	25	16	13	6.8	5.5
10	-2	-20	-6	-9	-6	15	-19	-22	-7.3	-8.5
11		+7	-6	-9	-3	25	23	17	8.1	7.0
12			-6	-12	-3	25	16	10	5.1	3.2
13			-10	-14	-3	14	1	-3	0.3	-0.9
14			-10	-14	-3	14	1	-3	0.3	-0.8
15			-10	-14	-3	14	1	-3	-0.2	-0.7
16			-10	-14	-3	14	1	-3	-0.2	-0.7
17			-10	-14	-3	14	1	-3	-0.2	-0.6
18			-15	-21	-3	14	-4	-10	-0.7	-1.8
19			-20	-26	-3	14	-9	-15	-1.5	-2.5
20			-25	-31	-3	14	-14	-20	-2.1	-3.0
Total	-6	-48	-169	-239	-66	345	70	0	22.1	-1.0

CONCLUSION

An important lesson from the preceding discussion is that benefit-cost analysis involves a host of judgments and decisions that must be driven by the particular circumstances of the project under consideration. The basic goal is to modify the values used in the analysis so that they accurately reflect true social values. This may require the use of social discount rates, shadow prices, estimates of values for nonmarketed goods, adjustments for risk and uncertainty, and inclusion of factors to bring distributional considerations into the analysis. In many cases, some formal procedure such as contingent valuation or distributional weights can be used. In others, such procedures may be difficult or impossible, and a better strategy might be to introduce these concerns as side constraints. For example, it may be preferable to simply record the costs to disadvantaged groups, suggesting compensation mechanisms or making an overall recommendation based on explicit arguments about the rights of these individuals rather than attempting to weight the results according to distributional impacts.

Whatever procedures are finally adopted to transform the easily observed impacts of a project into social impacts, it is important to recognize that the results of benefit-cost analysis are simply one set of inputs into the process of making decisions on public projects. The Tellico Dam project in Tennessee passed the initial benefit-cost test, but construction was halted when it appeared that the dam would cause the endangered snail darter to become extinct. It may also be the case that projects that do not pass the benefit-cost test end up being carried out anyway because other considerations outweigh the analysis of economic efficiency. Sometimes a political decision is made that something has to be done regardless of the costs and benefits that may be incurred. In these cases, it is still possible to deploy some of the methods discussed in this chapter to improve the final results. *Cost-effectiveness* analysis (see Chapter 9) uses methods described in the preceding sections to determine the least costly way to achieve a given goal. The least costly alternative might not pass a benefit-cost test, but given that the goal has been judged to be of overriding importance, lower-cost alternatives are usually preferable to costlier ones. Social benefit-cost analysis provides useful information on the consequences of alternative solutions to collective action problems. This information is essential for decisions about the best course of action to pursue. Other consequences (e.g., consequences that are not included in the efficiency criterion of benefit-cost analysis) or nonconsequentialist aspects (e.g., political or ethical effects) are also relevant to this decision and can be brought into the analysis in a variety of ways. The final decision should be based on all these considerations. The case studies in the next two chapters illustrate the use of these methods.

SUMMARY

1. Benefit-cost analysis is a procedure for evaluating investment projects. Private firms use similar procedures to guide their investment decisions. For public projects, it is necessary to correct the values used for any distortions due to externalities, public goods, or government interventions.

2. In evaluating projects or policies, it is always necessary to specify an appropriate base (counterfactual) for comparison.

3. Discounting is the procedure used to take account of the time value of money. Impacts that occur over time have to be discounted to express them in the same units of value. Social discount rates are often used to evaluate public projects that have long-term consequences. The choice of an appropriate discount rate is a complex issue in benefit-cost analysis.

4. Shadow prices are used in place of market prices when there are distortions that cause social values to differ from values reflected in the market.

5. Hedonic pricing and contingent valuation are methods that can be used to measure costs and benefits when markets are missing.

6. Social benefit-cost analysis may include provisions for risk and uncertainty and distributional impacts as well as corrections for price distortions and social time preferences.

7. The method of social benefit-cost analysis is not a recipe. It requires value judgments and ethical decisions.

KEY CONCEPTS

benefit-cost analysis
public projects
planning
counterfactual
discounting
time value of money
net present value
benefit-cost ratio
internal rate of return
social discount rate
social time preferences

price distortions
social benefit-cost analysis
shadow prices
hedonic pricing
contingent valuation
risk and uncertainty
sensitivity analysis
expected value
certainty equivalent
risk premium
cost effectiveness

DISCUSSION QUESTIONS

1. Explain why a dollar of benefits received today is worth more than a dollar of benefits received two years from now.

2. Should market interest rates be used to evaluate policies to combat global warming, or would a social discount rate be preferable? Is this an economic or an ethical question?

3. Give an example of an externality that would distort market values. How might this distortion be corrected in a social benefit-cost analysis?

4. Discuss the problems that might arise in using hedonic pricing or contingent valuation to deal with missing markets.

5. How is sensitivity analysis used in project evaluation?

6. Describe methods for taking account of income distribution in project evaluation. Do you think that it is more reasonable to introduce distributional considerations as side constraints or to include them formally in the analysis?

SUGGESTIONS FOR FURTHER READING

Cline, William. *Estimating the Benefits of Greenhouse Warming Abatement*. Washington, D.C.: Institute for International Economics, 1991.

Ellis, Ralph D. *Just Results: Ethical Foundations for Policy Analysis*. Washington, D.C.: Georgetown University Press, 1998.

Peterson, E. Wesley F. "Time Preference, The Environment and the Interests of Future Generations." *Journal of Agricultural and Environmental Ethics* 6, no. 2 (1993): 107–26.

Schmid, A. Allan. *Property, Power and Public Choice*. New York: Praeger, 1987.

APPENDIX 7.1: THE TIME VALUE OF MONEY

Why is the value of money not constant through time? This is not a simple question. An explanation based on the fact that banks pay interest simply raises the further question of why banks pay interest in the first place. Presumably banks are willing to pay interest on money loaned to them because they wish to raise funds that can be loaned to others at a higher rate of interest. However, the bank could simply charge a onetime fee as compensation for its services in bringing lenders and borrowers together rather than going through the process of paying and charging interest. One might think that the time value of money is connected to inflation. Inflation means that the costs of future goods are greater than the costs of the same goods today, so perhaps interest is paid and demanded as compensation for this decline in purchasing power over time. Yet even when nominal interest rates are corrected for inflation, real interest rates (the nominal interest rate minus the inflation rate) are usually greater than zero, at least in the long run, so there must be something else that would explain the time value of money.

One possibility is simply that people prefer current to future consumption. Consider the case of a farmer who can either consume the entire harvest in the current year or save some of the seeds for planting next year. Insuring that there will be future harvests means that the farmer will have to accept less current consumption. There is a trade-off between consuming now and consuming in the future, and the way in which this trade-off is valued reflects the individual's time preferences. Frequently, the more precarious the current situation, the greater will be the preference for current consumption. Many low-income families save very little because they need to employ all their current resources to survive in the present. In such circumstances, it may not make much sense to further jeopardize an already insecure existence by reducing current consumption to invest for the future. People in such situations are sometimes said to have high rates of time preference, that is, strong preferences for current consumption over future consumption. Such preferences do not mean that no value is attached to future consumption, but they do mean that the present is given more weight than would be the case for people with lower time preferences (see Box 7.1).

Of course, explaining the time value of money on the basis of preferences still does not get to the bottom of the question. After all, why do people attach greater weight to current than to future consumption? The most likely explanation for these preferences is that the future is more risky and uncertain than the present. "A bird in the hand is worth two in the bush" because we can be certain of having the bird that is in our hand, whereas there is a risk that we will not be able to capture the birds in the bush. In other words, the value of something that is held with certainty is greater than the value of something that is less sure. Such an explanation is consistent with the idea that people with few resources or opportunities will have higher time preferences than those with greater security. The wealthy can afford to sacrifice a present gratification in favor of a larger one in the future because their present level of well-being is more than adequate for their needs and desires. Low-income families will prefer benefits today because their needs are high and the risk that they will not survive to enjoy the future benefit, or that the benefit will not materialize, is high.

Box 7.1: Time preferences and the island of Nauru

For many centuries, birds in the South Pacific deposited their droppings (guano) on an island named Nauru. Guano is rich in phosphorous, and the deposits on Nauru were gradually transformed into phosphate rock, a mineral that has great value as fertilizer. The people of Nauru mined the phosphate rock at such a rate that they have largely destroyed the island. It could be said that they had a high rate of time preference in that the future environmental disaster of destroying their native land carried little weight in their decisions about exploitation of the phosphates compared with their desire for current consumption. Perhaps the Naurans simply do not feel the attachment most people do to their homelands. In any case, the Naurans were not irrational. Today, they are extremely wealthy and if they have invested the returns from the mining operations wisely, they will be able to enjoy pleasant and prosperous lives for the foreseeable future. They simply may not be able to do so on Nauru, which now is little more than a barren rock in the Pacific Ocean.

NOTES

1. There is a technical problem with the exercise of market power in the case of hydroelectric generation. Exercising market power usually requires restriction of the amount marketed in order to drive the price received above marginal costs. Reducing the electricity supplied from hydroelectric plants involves restricting the amount of water passing through the generators. But this means that more water is retained behind the dam, making more water available in subsequent periods. Because some of this water has to be moved out of the dam to accommodate water that will arrive after the spring thaw, for example, more electricity will have to be supplied later, driving down the price. In these circumstances, it may not be possible for a firm to exercise market power.

2. Economic planners often use input-output analysis to develop consistent plans. Input-output analysis is described in Chapter 13.

3. The results of the study were presented in a full-page advertisement in the *Lincoln Journal-Star*, 14 December 1998. The ad was paid for by Citizens for Accountable Route Selection.

4. Margaret Behm, "Highway Through History?" *Daily Nebraskan*, August 30, 2000, section 1, volume 100, no. 9, page 1.

5. The Central Arizona Project in the western United States is designed to transfer water from the Colorado River to residential, industrial, and agricultural users in Arizona. Cotton is a major crop in that area. In Mali, a country in the Sahel, the Office du Niger is a project aimed at providing water from the Niger River for the irrigation of cotton and other crops in the southern part of the country.

6. In evaluating projects, it makes no difference whether nominal or inflation-adjusted prices are used as long as they are used consistently. If nominal values are used for the costs and benefits, projected future values will have to include any expected inflation, and a nominal interest rate will need to be used for discounting. The nominal interest rate includes the inflationary expectations of mar-

ket participants. Likewise, if the analysis is done in real terms, all variables—including costs, benefits, and the discount rate—should be corrected for the effects of inflation.

7. See E. Wesley F. Peterson, "Time Preference, The Environment and the Interests of Future Generations," *Journal of Agricultural and Environmental Ethics* 6, no. 2 (1993): 107–26.

8. William Cline, *Estimating the Benefits of Greenhouse Warming Abatement*, Institute for International Economics, Washington, D.C., 1991.

9. The precise meaning of overvalued or undervalued exchange rates can be found in any text on international economics.

10. A subsequent act of Congress exempted the snail darter from this protection, and the dam was completed. Snail darters have been found in other locations, and the populations affected by the dam have been relocated. So far, the predictions that the dam would cause the snail darter's extinction have not been borne out.

11. E. Neal Blue and Luther Tweeten, "The Estimation of Marginal Utility of Income for Application to Agricultural Policy Analysis," *Agricultural Economics* 16 (1997): 155–69.

8

Case Study: Project to Revitalize the Banana Industry in a Low-Income Country in West Africa

INTRODUCTION

This chapter illustrates the practical application of the methods described in the preceding chapter to a specific case, a development project in the West African country of Guinea. About the size of Oregon, Guinea has a population of 7.5 million and per capita income of $570. The annual population growth rate was 2.3 percent in 1999, and 47 percent of the population was less than 15 years old. In 1990, life expectancy at birth was 47 years, and the infant mortality rate was 124 per 1,000 live births (compared with life expectancy of 77 and infant mortality of 7 per 1,000 in the United States). Guinea was part of French West Africa until 1958, when it was granted independence following a referendum organized by the French government. This event was of great significance in Guinea's recent history. The 1958 referendum offered the French colonies in Africa two options. The first was to remain as associates of France until 1960, when full independence would be granted. The second was to become independent immediately. The French government expected the colonies to choose the first option, and all but Guinea did. The Guinean decision to opt for immediate independence soured relations between France and its former colony.[1]

Although independence was granted in 1958, the French government was less supportive of the transition to independence than was the case for its other colonies. The first Guinean president, Ahmed Sékou Touré, turned to the Soviet Union and the United States for assistance but became increasingly radical as Guinea's diplomatic and economic isolation persisted. The socialist development strategy adopted by Touré had disastrous consequences for the country's growth and development. Guinea holds 25 percent of world bauxite reserves, and throughout the years of revolutionary socialism, these resources continued to be exploited by various multinational aluminum companies. However, the mining activities had few multiplier effects on the rest of the economy, which is primarily agricultural. About 80 percent of the workforce is in traditional subsistence agriculture, which has experienced little development since independence. The number of calories available daily per person increased only 2.7 percent between the early 1960s and the early 1980s, from about 2,214 to 2,274. Daily per capita caloric availability fell to 2,036 in 1990, rebounding slightly in subsequent years but reaching only 2,232 in 1997. Annual per capita cereal grain output followed a similar pattern, beginning at 103 kilograms in 1961, rising to 106 in 1980, then falling back to 103 in 1990 and rising again to reach 109 in 1997. These figures are low relative to other countries. Daily per capita caloric availability in 1997 was 2,406 for Africa as a whole, 2,650 for all developing countries, and 3,240 for high-income countries. Per capita grain production in 1997 was 144 kilograms per year for Africa as a whole and 168 kilograms for all developing countries.[2] Overall growth in Guinea was also fairly low, with per capita GNP increasing at an annual rate of about 1.4 percent between 1965 and 1995, compared with an average of 2 to 3 percent for all developing countries.

Touré's death in 1984 was followed by a military coup, and Guinea was run by a military dictatorship until the early 1990s, when a new constitution was adopted. Democratic elections were held in 1993 and 1998, and the country has undertaken economic reforms that have had some impact on economic performance. Per capita GNP grew by 4.6 percent in 1996-97, and while it is still too early to assess the long-term impact of the economic reforms and constitutional changes, it appears that Guinea is making good progress toward the establishment of viable economic and political systems, although this progress is threatened by fallout from the civil war in neighboring Sierra Leone.

In 1997, an international development agency received a proposal from a private firm in Guinea that wished to undertake a project to expand banana production. The Guinean banana industry has declined significantly in the years since independence. Guinea's traditional role as a significant exporter of bananas was eliminated during the predemocracy period as a result of misguided government policies and the closing of French markets to Guinean exporters. As part of the economic reforms in Guinea, the government has made a major commitment to private-sector development and fully supported the request for a development loan to finance the expansion of banana production and exports. For the international agency, the analytical question was whether such a project would be both economically viable and socially beneficial. A simple benefit-cost analysis was done to shed light on these questions.

PROJECT BACKGROUND

Before describing the proposed project, some information on banana cultivation and trade may be helpful. Because banana plants are highly susceptible to disease, it is a common practice to begin the production cycle with disease-free plantlets produced in a sterile environment by in vitro multiplication. The plantlets are available only from laboratories with sophisticated technological capabilities. They are shipped as tiny plants that must be grown out in nurseries before they are transferred to the fields. Once planted, it takes about nine months to produce harvestable bananas. The stem of the banana plant is made up of tightly rolled leaves from which a flower spike with a conical flower head emerges several months after planting. The flower spike bends over so that the tip of the flower cone is pointing toward the ground. The lower part of the flower head contains groups of male flowers that serve no known purpose. Further toward the top of the flower head are "hands" of female flowers that will eventually become the fruit that is consumed. The central stalk of the plant produces one flower head that grows into a bunch of bananas weighing anywhere from 22 to 65 kilograms (48 to 143 pounds).

Once the bunch is harvested, the stalk is cut down and discarded, because it will not produce any more flower heads. At the base of the stalk, five to eight suckers will have begun to grow prior to harvesting the bananas. Healthy suckers are transplanted into new fields, leaving one behind to grow into a new plant. This process can continue for several years, but it is likely that disease problems will eventually require that the plantation be replaced with a new set of in vitro plantlets. Because ripe bananas are highly perishable and easily damaged, commercial bananas are always harvested green. Green bananas can be transported without causing extensive bruising or discoloration and then ripened at the point of sale. Traded bananas are packed in boxes and rigorously controlled to prevent the onset of ripening. Once they have arrived at the final destination, ripening is triggered by administering ethylene gas. Even when bananas are to be consumed locally, they are harvested green and ripened after transportation to the market where they are sold, although the ripening is often done without the gas.

The world banana market is a classic example of north-south trade, with most of the production occurring in tropical developing countries while the major consumption regions are in North America and Europe. Asia is the leading producer (with about 47 percent of world production), followed by South America (26 percent), and Central America (14 percent). Africa produces only 12 percent of the world's bananas, its major producers being Cameroon and Côte d'Ivoire. The world market is dominated by Latin America, with Ecuador, Colombia, Costa Rica, Guatemala, Honduras, and Panama together accounting for 79 percent of world banana exports. The Philippines is the major Asian exporter, with about 11 percent of the world market. On the import side, the United States purchases about 28 percent of world exports, the EU absorbs 27 percent, and Japan and Korea account for 11 percent. Much of world trade is dominated by a small number of multinational firms, most of which have their headquarters in the United States. In 1997, the Food and Agriculture Organization (FAO) of the United Nations predicted modest growth of per capita consumption in the industrialized countries but expected the world market to be oversupplied by about 2000.[3]

The proposed project involves two primary activities. The first concerns the revitalization of the Guinean banana sector, while the second has to do with extracting fibers from banana stems. The relevant markets for the first activity include local, regional, and world markets for bananas. For the second activity, the relevant markets are those for similar fibers such as jute, sisal, or hemp. The project aims to revitalize an important industry in Guinea that has all but disappeared. Bananas are nutritionally beneficial and are part of traditional consumption. In the early years of this project, the increased production will flow primarily to local markets, where it will serve to restrain price increases and contribute to the nutritional needs of local populations. It is expected that the project will have a multiplier effect on development in Guinea, creating employment and new income streams as well as helping to provide the rationale for the development of new infrastructure—including roads, ports, and other facilities to handle bananas. After the initial phases, the project may contribute to the development of markets in West and North Africa as well. The project also aims to discover new uses for by-products of banana production and develop local and international markets for these products.

For the near future, the main market for Guinean bananas will be local. In the 1950s, Guinea exported almost 100,000 tons of bananas, mainly to France. This traditional market is now closed, because the European Union (EU) operates a system of quotas designed to favor producers in former colonies as well as those on certain islands that are part of either France (Martinique and Guadaloupe), Portugal (Madeira), or Spain (the Canary Islands). These quotas are based on historic trade patterns, and since Guinean exports disappeared in the years following independence, Guinea has not been granted a quota to export to Europe. The most efficient banana producers are in Latin America, where large multinational banana companies use modern methods to pack and ship bananas. These firms dominate the North American market and compete vigorously in Europe despite the EU quota system. The United States and some Latin American countries have challenged the EU banana regime at the World Trade Organization (WTO), forcing the EU to modify its politics.

Guinean banana production stagnated following independence but began to increase in the 1980s. The increased production has been consumed almost entirely within the country, although there are limited exports to neighboring countries such as Senegal and Mali. The lack of a quota for export to the EU is a major problem for this industry, because the European market is the largest potential outlet for Guinean bananas. Even if Guinea were able to obtain an EU export quota, however, lack of transportation infrastructure (roads and ports) would render it impossible to take advantage of this market in the short term. The world price for bananas, measured after transportation to U.S. ports, has ranged from $0.37 to $0.49 per kilogram in recent years. Retail prices in Conakry, the capital of Guinea, are between $0.50 and $1.00 per kilogram, while producer prices in rural areas are $0.25 per kilogram. This price structure suggests that Guinean bananas may be uncompetitive even if the structural barriers (long-term contracts, market power of large U.S.-based multinational firms, sophisticated technology required to pack and transport bananas, sanitary regulations, and so on) could be overcome. In the end, it is the internal market in Guinea that has the greatest potential for profitable exploitation.

No market presently exists for banana fibers. The director of the Guinean firm has worked with a German engineer to design a prototype fiber-extraction machine costing almost $100,000. The machine is mobile and can be moved from place to place as different parts of the plantation are harvested. In Africa, the banana stems are generally left in the fields to decompose naturally. While this procedure reintroduces vegetal matter into the soil, it also facilitates the spread of disease and creates other problems associated with the rotting stalks. Processing the stalks to extract the fiber will probably not have serious repercussions for soil fertility and will provide potentially useful fiber products. The fibers extracted from banana stems appear to be quite similar to jute and can be used to make bags, rope, or string, and as filler for mattresses. The world price of jute is about $400 per ton, while that for sisal is about $700 per ton. In recent years, traditional suppliers of jute (India, Bangladesh) have suffered as synthetic fibers have substantially reduced the market for jute. It appears that banana fibers are being exploited in the Philippines, although little information was available on the uses to which these fibers are put or on the technology used for their extraction. The Guinean firm has experimented with the extraction of banana fibers, and there are likely to be local markets for products made from these fibers. Other outlets are less certain, but there may be some scope for development of this activity. The extraction process also produces other by-products (liquid, pulp) that may have value as well. The pulp may be used to feed livestock, although livestock feeding is not highly developed in Guinea. In general, the market prospects for banana fiber in Guinea are not well understood, and this fact adds a substantial element of risk to this component of the project.

The proposed project could have an important impact on local employment and income. In the early years, it will not generate much foreign exchange because there will be limited exports. If Guinea is ever to develop the capacity to export bananas, however, it will be necessary to begin with the careful building of the internal infrastructure that will allow it to confront the exigencies of world markets at a later date. The proposed project has been structured to include three components. The first component is the establishment of a new banana plantation of 100 hectares near Bennah-Moussayah using in vitro plants. This new plantation will greatly expand the production capabilities of the Guinean firm, which currently operates a plantation of only 10 hectares. In addition, the surplus banana suckers from the higher-quality in vitro plants will be sold to other producers in the area, leading to a general upgrading of the region's banana industry. The project will require the construction of several buildings for both the banana and fiber components. Construction of these buildings is the second component of the project. The third component is the establishment of an experimental fiber-extraction project using banana stems from the plantation and from other producers in the region. Detailed descriptions of these components are set out in Table 8.7 in Appendix 8.1.

The duration of the project is eight years. In the first year, the initial investments are made and operations are begun. In the fifth year, the plantation is completely replaced because continued production from the same plant/soil basis after that time would lead to serious disease problems. Initial investments include the establishment of a plantation of 100 hectares (1 hectare = 2.47 acres), the purchase of the in vitro plantlets, the construction of buildings, and the purchase of equipment and vehicles for the production of bananas and the fiber-extraction project. There are also recurring operating and maintenance costs in each year of the project. It is assumed that the project equipment and vehicles will be fully used up during the life of the project, so there is no salvage value for these items. The project will generate revenues from the sale of bananas and banana fiber. In addition, the project will give rise to external effects such as reduced underemployment in the project area, and such impacts will have to be included in the social benefit-cost analysis. The project is to be financed by a loan from the international agency that covers the first three years of the project and totals $1,453,372.

FINANCIAL ANALYSIS OF THE PROJECT

The first step in evaluating this project proposal is to determine whether it is financially viable. The financial analysis is based on the detailed cost information presented in

Appendix 8.2. At the time of this analysis, 1 U.S. dollar was worth 1,000 Guinean francs (FG). In the following analysis, all values are expressed in U.S. dollars. The initial investment totals $813,172 and includes the preparation and planting of the 100-hectare banana plantation ($96,084), construction of a ripening room and several storage buildings ($136,238), purchase of vehicles and equipment ($432,850), and purchase of the in vitro plantlets ($248,000). Operating costs are $180,067, so total costs in the first year are $1,093,238. The Guinean government has agreed to exempt the firm from taxes and import duties, so there is no need to include such charges in the operating costs.

The financial benefits are the first-year sales of bananas and fiber. It is assumed that the fiber operation generates 30 tons of fibers the first year and that these are sold at $400 per ton for total revenue from the fiber operation of $12,000. The banana plantation is established in stages, and only 40 hectares are harvested in the first year (see Appendix 8.1). The expected yield for bananas is 30 tons per hectare, and losses are assumed to be 15 percent of the harvested bananas. Thus, the firm is able to market 1,020 metric tons of bananas (40 hectares × 30 tons/hectare × 0.85). The market price is taken to be 250 FG/kg, or $250 per metric ton. Total revenue from the banana sales are $255,000 (1020 tons × $250/ton). Adding the revenue from the fiber operation gives total revenues of $267,000 in the first year.

In the second year, costs are $180,067 for operations and maintenance. Part of the plantation is harvested twice within the year for a total of 4,800 tons harvested, leaving 4,080 tons to be marketed after losses of 15 percent. The total value of the banana sales is $1,050,000. Fiber production increases to 40 tons for a return of $16,000. In the third and fourth years, fiber output reaches its peak capacity of 50 metric tons for a value of $20,000. Banana production in these years is 3,750 metric tons, leaving 3,187 to be marketed for $797,000 in revenue. In the fifth year, the plantation has to be replaced with new in vitro plantlets from the laboratory in France. Field preparations cost $96,084 while the plantlets cost $248,000. It is anticipated that about one third of the machinery and equipment will have to be replaced as a result of the harsh tropical climate. The expected cost of this replacement is $161,969. Operating costs are still $180,067, so total costs in the fifth year are $686,120. Revenues from the banana operation are only $255,000, as in the first year. Net benefits in the sixth, seventh, and eighth years are essentially the same as in the second, third, and fourth.

These financial costs and benefits are summarized in Table 8.1. The net benefit figures are discounted using a discount rate of 4 percent (the likely interest rate charged by the international agency for the project loan). If the Guinean firm manages to plant all 100 hectares, gets yields of 30 tons per hectare, loses only 15 percent of the harvest during marketing, sells the bananas at $250 per metric ton, and earns the amounts assumed from the fiber operation, the project is essentially paid off at the end of the second year. The net present value of the project (the sum of the figures in the far right column) is $2.495 million. On the basis of this analysis, it would appear that the project is financially viable and that the increased banana sales would easily cover repayment of the loan and the 4 percent interest charges.

Table 8.1. Basic financial analysis of the Guinean banana project (in thousands of U.S. dollars)

Year	Banana	Fiber	Costs	Net	Present Value @ 4%
0	255	12	-1093	-826	-826
1	1020	16	-180	856	823
2	797	20	-180	637	589
3	797	20	-180	637	566
4	255	20	-686	-411	-351
5	1020	20	-180	860	707
6	797	20	-180	637	503
7	797	20	-180	637	484

The fiber operation, however, would not be viable on its own, as shown by the figures in Table 8.2. The fiber-extraction machine has not been field-tested, so the financial analysis of this component is based on a series of assumptions about the processing capacity of the machine and the value of the product. The machine is supposed to be able to process one ton of fresh stems per hour, and each ton yields 50 kilograms of dry fiber. It was assumed that it would be possible to run the machine 8 hours a day, 15 days a month, during 9 months of the year for a total of about 1,000 hours per year. That would mean a processing capacity of 1,000 tons for a total output of 50 tons of dry fiber per year. Using the price of jute ($400 per ton) as an estimate of the value of the banana fiber leads to a total annual value when the machine is fully operational of $20,000.

Table 8.2. Financial analysis of fiber component (in thousands of U.S. dollars)

Year	Machine	Operations and Maintenance	Fiber	Net	Present Value @ 4%
0	-98.9	-11.2	12.0	-98.1	-98.1
1		-11.2	16.0	4.8	4.6
2		-11.2	20.0	8.8	8.1
3		-11.2	20.0	8.8	7.8
4		-11.2	20.0	8.8	7.5
5		-11.2	20.0	8.8	7.2
6		-11.2	20.0	8.8	7.0
7		-11.2	20.0	8.8	6.7

The initial cost of the machine is $98,900, and annual operations costs to cover insurance, maintenance, fuel, and labor are $11,200. It is clear that the net returns from the fiber sales are insufficient to cover the initial investment cost. The net present value of this component is −$49,200. Enough income is generated from the banana sales, however, to finance the fiber component as an experimental project if desired. Because so little is known about the functioning of the machine and the market for banana fiber, it might be interesting to carry out this part of the project simply to find out whether there is some potential for such a market to develop. On the other hand, the operation of complex machines in a tropical developing country is often plagued with mechanical and logistical problems. In this context, the assumptions about the machine's capacity are probably optimistic, and it may turn out that the project would be strengthened by dropping this component altogether.

The financial viability of the project with or without the fiber component depends on the validity of the expectations about yields and the other technical aspects of the project. Given the generally poor infrastructure in Guinea, the difficult climate (annual rainfall in the region of the banana plantation is around 3,000 millimeters, or almost 10 feet), and the susceptibility of the banana plant to disease and other problems, the project is highly risky. Before undertaking the social benefit-cost analysis of the project, one should confront the possibility that the expected yields, production, and prices will not be realized. Suppose that the Guinean firm manages to put only 80 hectares into production (40 the first year and 40 instead of 60 in the second). Suppose further that yields are only 20 tons per hectare and losses during marketing are 20 percent rather than 15 percent. Finally, assume that the price received is the same, that the fiber operation produces no revenue, and that the first-year costs are the same as in the original analysis, although there are increased maintenance costs in years 1 to 3 and 5 to 7. The figures in Table 8.3 reflect these assumptions and show that the project will not generate enough income to pay off the initial investment. The net present value (the sum of the Present Value figures for all eight years) is −$303,000, and the project is not profitable. Note that including some modest revenues from the fiber operation would not change these results.

Table 8.3. Financial analysis with pessimistic assumptions (in thousands of U.S. dollars)

Year	Banana	Fiber	Costs	Net	Present Value @ 4%
0	160	0	-1093	-933	-933
1	480	0	-223	257	247
2	400	0	-223	177	164
3	400	0	-223	177	157
4	160	0	-686	-526	-450
5	480	0	-223	257	211
6	400	0	-223	177	140
7	400	0	-223	177	134

The Guinean firm has not established a lengthy record of production and sales on which to base the analysis. It appears that the firm is generally able to obtain yields that are greater than 20 tons per hectare, although the assumption that yields of 30 tons per hectare will be achieved on the full plantation does seem overly optimistic. Suppose that average yields of 25 tons per hectare can be achieved across the full plantation of 100 hectares, with losses of 20 percent rather than the 15 percent assumed in the initial analysis. In addition, assume that the fiber project generates the same returns as in the initial analysis. Using the cost structure from Table 8.4, the net present value of the project based on the more modest yields and greater losses is $1.2 million (the sum of the Present Value figures for all eight years), as shown in Table 8.4.

Table 8.4. Financial analysis with moderate assumptions (in thousands of U.S. dollars)

Year	Banana	Fiber	Costs	Net	Present Value @ 4%
0	200	12	-1093	-881	-881
1	800	16	-223	593	570
2	625	20	-223	422	390
3	625	20	-223	422	375
4	200	20	-686	-466	-398
5	800	20	-223	593	487
6	625	20	-223	422	333
7	625	20	-223	422	321

The Guinean banana project involves a lot of uncertainty. As noted in the last chapter, about all that can be done when faced with uncertainty is to highlight the problems and carry out a sensitivity analysis using alternative assumptions about the basic parameters of the evaluation. If information on historic banana yields were available, for example, it would be possible to estimate a probability distribution that could be used in computing the expected value of the net returns to the project. Another way to approach the uncertainty associated with this project might be to consult experts in West African banana production in an effort to develop reasonable estimates of their subjective assessment of the likelihood of alternative outcomes. This was not done for this analysis, which was based almost entirely on information provided by the Guinean firm. However, it is possible to reformulate the problem to shed additional light on the likelihood that the project will be financially viable.

Suppose we ask what probabilities would be required for the expected value of the three scenarios analyzed earlier to be negative. The net present values for the alternatives shown in Tables 8.1, 8.3, and 8.4 are, after rounding, $2.5 million, −$0.3 million, and $1.2 million, respectively. It is clear that if the probability that the second scenario will turn out to be true is 100 percent, the project is not viable. Formally, this result is obtained with the following calculation: $(0 \times 2.5) + (1 \times -0.3) + (0 \times 1.2) = -0.3$. Suppose that the probabilities are set at 0, 0.9 and 0.1 for the three cases: $(0 \times 2.5) + (0.9 \times -0.3) + (0.1 \times 1.2) = -0.15$. This combination of probabilities leads to a negative expected value for the project. If the calculation is done with probabilities of 0, 0.80, and 0.20, the expected value of the project is positive. If we believe that the results in Table 8.1 can never be achieved and that the probability that the analysis in Table 8.3 is the correct one is 0.8 or above, we would have to conclude that the project is unlikely to be viable. In evaluating this project, such a conclusion seemed overly pessimistic. While the risk that the project would fail was real, the potential profitability of the banana operation is high enough to leave room for fairly substantial errors in the assumed technical parameters without causing the net present value of the project to become negative.

SOCIAL BENEFIT-COST ANALYSIS

Based on the preceding analysis, it appears likely that the banana project will generate sufficient revenues to pay off the loan, finance the fiber-extraction component as a kind of experimental project, and leave the Guinean firm with fairly substantial profits. These results are sensitive to the assumed technical parameters, and there is a significant risk that the firm will be unable to produce as many bananas as expected. If we accept, however, that the analysis presented in Table 8.4 is broadly accurate, a case can be made for the financial viability of this project. This is an important consideration for the international agency in determining whether to finance the project. In addition, however, the agency is concerned with the social impact of the project. Understanding this issue requires a social benefit-cost analysis, as described in the last chapter.

There are four types of adjustment that are relevant for this project. The first concerns the unskilled labor that is to be employed in setting up and running the banana plantation. These workers will be drawn from the local population, which is made up primarily of subsistence farmers. In general, rural households in Guinea have more family labor than is strictly required to operate their small holdings. There is thus substantial underemployment in the area. Many workers contribute very little to local production, and their withdrawal from these activities would not seriously reduce the total amount being produced.[4] In other words, the opportunity cost of withdrawing these workers from their current activities so that they can work on the plantation is essentially zero. Although the firm will have to pay these workers the going wage rate, the social cost of their employment is negligible. In fact, one could argue that one of the benefits of the project is to increase employment in the region so that from the point of view of society, there is a benefit rather than a cost of hiring these plantation workers.

A second issue is the social opportunity cost of using scarce foreign exchange reserves for the purchase of imported machines, equipment, and other supplies. As noted in the last chapter, an overvalued exchange rate reduces the cost in local currency of imported goods. The foreign suppliers must be paid in their currencies, however, rather than the local currency, and the only way that a country can obtain these foreign currencies is through exporting. The overvalued exchange rate means that the country's exports are more expensive to foreigners, and that may reduce foreign demand for the country's exports, leading to trade imbalances and the eventual exhaustion of foreign exchange reserves. If the Guinean franc was overvalued at the time of this analysis, the true social cost of the imported goods used for the project would be higher than the market prices used in the financial analysis. As part of the economic reforms, the Guinean franc has been left to float so that its value is now determined by market forces, making overvaluation of the currency less likely. Yet because the Guinean franc had been under government control for many years, it is possible that it was overvalued when it was floated. In fact, the Guinean franc has continued to depreciate against the U.S. dollar since the time of the study. In July 1998, there were 1,242 FG per U.S. dollar. It might be inferred that the Guinean franc was over-valued by almost 25 percent at the time of the study. This would mean that the true oppor-tunity cost of all the imported goods is about 25 percent higher than the figures used in the financial analysis.

A third issue is the cost of the land used for the plantation. The Guinean firm owned this land, so no charge was included in the financial analysis for the purchase of the 100 hectares needed for the plantation. From the point of view of society, however, this land could be used for other purposes, so there is an opportunity cost in using it for banana production even though the firm incurred no land acquisition costs. The land market in Guinea is not highly developed, and it is difficult to locate land prices. It is possible, however, to impute a value to land based on the expected net returns that can be earned from it in its normal uses. In the area of the banana plantation, most land is used in sub-sistence farming systems, with maize and manioc (also known as cassava or tapioca) as the main crops. These two crops are frequently intercropped (that is, grown in the same field, with the maize planted first so that it shades the smaller manioc plants, which eventually take over the field when the maize is harvested). When grown separately, yields are one ton per hectare for maize and six tons per hectare for manioc. Prices are about $200 per ton for maize and $120 per ton for manioc. The main production costs are for labor and hand tools. It is estimated that the net returns for an intercropped field are on the order of $50 per hectare. Economists commonly estimate the value of a piece of land as the present value of the stream of future returns that could be earned on the land. Assuming that these low-income households have high rates of time preference, a discount rate of 25 percent was used for this exercise, giving a land value of about $250 per hectare. The imputed land value for the 100-hectare plantation is thus taken to be $25,000.

The final adjustment concerns government taxation. As noted earlier, the Guinean gov-ernment has agreed to exempt the firm from internal taxes and import duties. In the pre-ceding analyses, costs are based on world prices, which are the appropriate measures to use in social benefit-cost analysis as well. One might argue, however, that for Guinea the exemption for local taxes is a cost to the general society. If such is the case, it would be necessary to add local taxes to the firm's operating costs to obtain an accurate measure of the social costs of the project. Local taxes are computed as 5 percent of net returns and average $21,000 per year during the life of the project. They are added into the operations costs for the social benefit-cost analysis. These four adjustments are detailed in Table 8.5, which is based on the cost data shown in Appendix 8.2. The cost of employing unskilled labor is set at zero, the value of imported goods is increased by 25 percent to reflect the overvalued exchange rate, and charges for the opportunity cost of the land and local taxes are added to the total costs.

For the social benefit-cost analysis, the moderate assumptions on yields and losses in Table 8.4 are used. For the time being, it is also assumed that the fiber component is retained as an experimental project that generates a small stream of income. Because

Guinea allows free trade in bananas, the local producer prices would seem to be the appropriate shadow price for the output of the project. Project returns based on these assumptions are shown in Table 8.6, along with adjusted costs from Table 8.5. Table 8.6 also includes present values for alternative discount rates. The Guinean firm will have to pay only 4 percent in interest charges on the loan it receives, but that is clearly less than the rates that would be required for commercial loans, particularly when such loans are for projects with such a high degree of risk and uncertainty. Moreover, it is likely that the social time preference in Guinea is significantly higher than the rate offered by the international agency and perhaps even the commercial rates that would be offered by private banks. The results in Table 8.6 show that the net present value of the project is positive at discount rates less than about 50 percent, even with the four adjustments for social opportunity costs. Net present values (sums of Present Value figures for all eight years) are about $1.44 million, $1.0 million, $235,000, and −$127,000 for discount rates of 4, 10, 25, and 50 percent, respectively.

Table 8.5. Adjustments for social benefit-cost analysis.

Item	Figures Used in Financial Analysis	Adjustment factor	Figures Used in Social Benefit-Cost Analysis
I. Investment Budget			
A. Field Preparation			
Unskilled labor	80,500	0	0
Other	15,584	1	15,584
B. Building/equipment			
Imported equipment	485,906	1.25	607,383
Construction	83,181	1	83,181
C. In vitro plantlets	248,000	1.25	310,000
D. Land acquisition	0		25,000
TOTAL	913,171		1,041,148
II. Operations			
A. Banana plantation			
Unskilled labor	33,950	0	0
Other	10,334	1.25	12,918
B. Operations			
Ethylene gas	8,129	1.25	10,161
Other	48,306	1	48,306
C. Imported fuel	20,108	1.25	25,135
D. Personnel			
Unskilled day labor	13,200	0	0
Other labor	26,040	1	26,040
E. Maintenance	20,000	1	20,000
F. Local taxes	0		21,000
TOTAL	180,067		163,560
TOTAL FIRST-YEAR COSTS	1,093,239		1,204,709

Table 8.6. Social benefit-cost analysis (in thousands of U.S. dollars)

Year	Banana	Fiber	Costs	Net	Present Value			
					@4%	@10%	@25%	@50%
0	200	12	-1,205	-993	-993	-993	-993	-993
1	800	16	-164	652	627	593	502	435
2	625	20	-164	481	445	398	308	214
3	625	20	-164	481	428	361	246	143
4	200	20	-636	-416	-356	-284	-170	-82
5	800	20	-164	656	539	407	215	86
6	625	20	-164	481	380	272	126	42
7	625	20	-164	481	366	247	101	28

Even if the appropriate social rate of time preference is considered to be as high as 25 percent, the project still appears to be socially beneficial, but the benefit-cost ratio at a discount rate of 25 percent is only 1.3. Given the risk and uncertainty associated with this project, some might insist on a higher margin of benefits over costs before concluding that it should be undertaken. On the other hand, there are other factors that might strengthen the case for the project. For example, the people who might gain employment through this project are very poor, and if one weighted their benefits to reflect a concern with distributional issues, the results would be more favorable. In addition, no effort has been made to estimate multiplier effects in local communities as the increased incomes of the plantation workers are spent in the local economy. The project is small enough that it is unlikely to have any significant impact on the regional environment, although further analysis of this issue might be merited. For example, increased fertilizer use (bananas require a great deal of potassium as well as nitrogen and other fertilizers) could have an impact on the local ecosystem. If the project is successful, it could lead to longer-term infrastructural development that would carry with it both costs and benefits. No effort has been made to incorporate these kinds of impacts into the analysis.

Finally, in line with the discussion in chapters 5 and 6, the analyst should ask whether there are any important political or ethical concerns that need to be addressed in assessing the advantages and disadvantages of this project. Politically, the project was well supported. For many members of the Guinean government, this project was particularly attractive because it involved the private sector, and private-sector development is an important goal in Guinea. In the short term, the project is unlikely to generate important infrastructure needs that would require large public expenditures. Even when the effects of using scarce foreign exchange and the local tax exemption are included in the analysis, the project seemed advantageous. Of course, the project is fairly small, so it seems unlikely that it would have significant impacts on the development of Guinea's democracy or the establishment of a functioning and viable political system.

In terms of the ethical considerations that bear on this project, it does not appear that implementation would infringe on the rights of anyone living in the area or that it would have negative effects for future generations, the environment, or people living in other countries. The main ethical concern seems to be the question of distributional justice, and consideration of this issue strengthens the case for the project. Guinea is a poor country in need of economic stimulus to create jobs for extremely poor people as well as to generate incentives and resources for further development of local markets, roads, port facilities, and so on. In addition, bananas are a nutritious food that appeared to be relatively scarce in the capital of Conakry during the field study in 1997. While mangoes, papayas, and pineapples were abundant and inexpensive, bananas were scarce and expensive. Although low-income urban households could probably obtain the same nutrients (with the exception of potassium) from other fruits and vegetables, bananas would add variety to their diets and could make a minor contribution to improving quality of life.

CONCLUSION

For this particular project, the financial and social analyses would seem to point toward a favorable recommendation. The Guinean firm should be able to realize profits even after paying off the loan, and the most likely social impacts of the project appear to be positive. The main concern expressed by the international agency was the risk and uncertainty of achieving the project goals. The fiber project was seen as particularly problematic, but there was also concern that the Guinean firm might be attempting more than it could reasonably expect to achieve. One of my recommendations was that the project be focused on the banana plantation and that, rather than trying to plant the full 100 hectares in the first year, the project be implemented in two phases, each of which would involve installing 50 hectares of banana plants over two years. The international agency also felt that this might be a more appropriate design for the project. In late 1997, it decided to hold off on funding this project until more study of both the fiber and banana operations could be done, with a view toward a rather substantial redesign of the project. All concerned hoped that a way could be found to design a less risky project that could be funded as part of Guinea's development strategy.

DISCUSSION QUESTIONS

1. Is it likely that the results of the analysis of the Guinean banana project would change if the time period were fifteen or twenty years instead of eight? How would you figure this out?

2. One of the people interviewed for the study argued that the production of banana fiber would surely generate demand for this useful item even if there did not appear to be much of a market for it initially. Is this a good argument for including the fiber component in the final project? Explain.

3. Do the results of this analysis appear to be sensitive to the discount rate chosen? What about risk and uncertainty?

4. If you were a manager of a private bank, would you loan the Guinean firm the money for this project? Is this project a public project or an effort by a private firm to get a low-interest loan from a public agency?

APPENDIX 8.1: BANANA PROJECT COMPONENTS

Table 8.7. Banana project components.

Component 1: 100-hectare Banana Plantation

Year/Quarter	Main Activities to be Implemented	Output
1-1	1. Order in vitro plantlets, grow in nursery 2. Clear, prepare, plant 40 hectares 3. Set up irrigation, irrigate as needed	40 hectares of bananas planted
1-2	1. Clear, prepare, plant 60 more hectares 2. Irrigate as needed 3. Fertilize, implement phytosanitary treatments 4. Handle weeding, cultivation	60 more hectares of bananas planted
1-3	1. Maintain, cultivate 100 hectares of growing banana plants	Established 100-hectare plantation
1-4	1. Harvest 40 hectares, planted in Quarter 1-1 2. Remove extra suckers and stems, restart harvested fields 3. Maintain, cultivate other fields 4. Transport harvested bananas to ripening room and other destinations 5. Market ripe bananas	800–1,200 tons of bananas harvested and marketed Excess suckers collected and sold Plantation restarted
2-1	1. Harvest 60 hectares planted in Quarter 1-2 2. Transport, ripen, market new harvest 3. Remove extra suckers and stems, restart harvested fields 4. Maintain, cultivate other fields 5. Irrigate	1,200–1,800 tons of bananas harvested and marketed Suckers sold Plantation restarted
2-2	1. Maintain, cultivate, fertilize, control disease on the 100 hectares; now starting a new round of production	
2-3	1. Begin second harvest, remove suckers and stems, restart harvested fields, etc. 2. Transport, ripen, market bananas 3. Maintain plantation	2,500–3,000 tons of bananas harvested Suckers sold Plantation restarted
2-4	1. Continue second harvest, remove suckers and stems, restart harvested fields, etc. 2. Transport, ripen, market bananas 3. Maintain plantation	Market bananas and other products
3-1	1. Irrigate 2. Maintain plantation	

continued

Table 8.7 *(Continued)*

Component 1: 100-hectare Banana Plantation

Year/Quarter	Main Activities to be Implemented	Output
3-2	1. Harvest new crop, remove suckers and stems, restart harvested fields 2. Transport, ripen, market bananas 3. Maintain plantation	Approximately 3,125–3,750 tons harvested and marketed
3-3	1. Continue harvest and carry out all activities associated with harvest as in previous cells	
3-4	1. Maintain plantation	
Year 4	Fourth year essentially same as third year; at year end, order new set of in vitro plants for next cycle	3,125–3,750 tons harvested and marketed
5-1	1. Clear existing plantation, prepare new fields 2. Begin planting new banana plants at rate of 20 hectares/month 3. Irrigate	
5-2	1. Continue preparing fields, planting new plantation	
5-3	1. Maintain plantation	
5-4	1. Begin harvest, transport, ripening, marketing 2. Remove stems and suckers, etc. 3. Maintain plantation	800–1,200 tons of bananas harvested and marketed
Years 6, 7, 8	Repeat activities indicated for years 2, 3, 4	Average 3,125–3,750 tons harvested per year

Component 2: Construction

Year/Quarter	Main Activities to be Implemented	Output
1-1	1. Order equipment	
1-2	1. Begin construction of ripening room and shed in Conakry 2. Begin construction of shed in Bennah	
1-3	1. Complete construction of ripening room and shed in Conakry; install equipment 2. Complete shed in Bennah; finish floor and install equipment	Functioning ripening room and shed in Conakry Shed in Bennah

continued

Table 8.7 *(Continued)*

Component 2: Construction

Year/Quarter	Main Activities to be Implemented	Output
1-4	1. Begin using buildings to store, process, clean, and handle bananas and fiber from banana stems	Conditioned bananas and other products
Years 2–8	Maintain buildings as necessary	

Component 3: Fiber Extraction

Year/Quarter	Main Activities to be Implemented	Output
1-1	1. Order fiber machine and other needed equipment	
1-2	1. Transport fiber machine to Bennah-M	
1-3	1. Install fiber machine in completed shed 2. Begin tests to determine rate of use	Trials of fiber machine in the field
1-4	1. Begin extracting fiber from stems 2. Experiment with further processing and marketing	Banana fiber and other by-products
Years 2–5	1. Extract fibers from banana trunks 2. Collect other by-products 3. Market fiber and other by-products 4. Experiment with alternative outlets and fiber processing activities 5. Maintain machine and equipment 6. Analyze results of experiment and decide on future steps; design follow-up project	About 50 tons of dry fiber per year plus other products Processed fiber products Decision on feasibility of fiber operation

APPENDIX 8.2: COSTS FOR THE BANANA REVITALIZATION PROJECT

I. Investment Budget ($ = U.S. dollars, FG = Guinean francs)

   A. Preparation and Planting of the 100-hectare Banana Plantation (Component 1)

      1. Labor cost

a. Field clearing:	20 worker-days per hectare
b. Digging of drainage ditches:	140 worker-days per hectare*
c. Field layout:	10 worker-days per hectare
d. Fertilization:	20 worker-days per hectare
e. Digging holes:	20 worker-days per hectare
f. Planting:	20 worker-days per hectare
Total:	230 worker-days per hectare

230 worker-days per hectare for 100 hectares = 23,000 worker-days at daily salary
3,500 FG = 80,500,000 FG
Total Labor Cost: 80,500,000 FG = $80,500

2. Plowing (in FG)	
a. Fuel and oil for tractor:	457,800
b. Driver:	750,000
Total:	1,207,800 FG = $1,207
3. Irrigation (in FG)	
a. Pipes and sprinklers:	620,000
b. Fuel for pumps:	915,600
Total:	1,535,600 FG = $1,536
4. Fertilizer	
a. Urea and 17-17-17: 1,768,000 FG	$1,768
5. Hand tools and miscellaneous (in FG)	
a. 25 sets field clothes at 29,000	725,000
b. 10 wheelbarrows at 52,000	520,000
c. 3 100m ropes at 12,000	36,000
d. 2 50m tapes at 34,000	68,000
e. 5 back sprayers at 60,000	300,000
f. 20 pickaxes at 8,625	172,500
g. 10 pruning shears at 8,625	86,250
h. 20 hoes at 7,500	150,000
i. 20 machetes at 8,625	172,500
j. 10 buckets at 10,350	103,500
k. 20 shovels at 10,350	207,000
l. 10 rakes at 8,625	86,250
m. 5 individual atomizers at 98,000	490,000
n. 4 empty barrels at 14,000	56,000
o. Plastic sacks (sheaths)	2,300,000
p. Infirmary equipment/medicine	3,500,000
q. Garage tools	2,000,000
r. 100 knives at 1,000	100,000
Total	11,073,000 FG = $11,073
TOTAL FOR PREPARATION AND PLANTING:	$96,084
B. Ripening room in Conakry (in FG)	
1. Building and equipment	26,525,507
2. Electrical connections	2,893,200
3. Insulation	8,887,200
4. Electric generator	16,531,250
Total	54,837,157 FG = $54,837

C. Hangar in Conakry (450 square meters; cost in FG)

1. Construction	29,756,250
2. Installation	4,463,438
Total	34,219,688 FG = $34,220

D. Hangar in Bennah-Moussayah
(450 square meters; cost in FG)

1. Construction	29,756,250
2. Scale	892,688
3. Electricity generator	16,531,250
Total	47,180,188 FG = $47,180

E. Equipment (in FG)

1. 2 Tractors at 28,750,000	57,500,000
2. 2 Trailers at 4,025,000	8,050,000
3. 2 Pumps at 7,825,000	15,650,000
4. Fiber extracting machine	98,900,000
Total	180,100,000 FG = $180,100

F. Vehicles (in FG)

1. 1 4 × 4 at 28,750,000	28,750,000
2. 2 Trucks at 92,000,000	184,000,000
Total	212,750,000 = $212,750

G. Spare parts
    1. Assorted spare parts 40,000,000 FG = $40,000

H. Purchase of in vitro plants

1. 200,000 plants (2,000 per hectare for 100 hectares)	
at 1,240 FG per plant:	248,000,000 = $248,000

TOTAL INVESTMENTS	$913,171

II. Operations and recurrent costs
  A. Banana plantation
    1. Labor

a. Mounding:	20 worker-days/hectare
b. Hand weeding:	20 worker-days/hectare
c. Herbicide application:	2 worker-days/hectare
d. Support staking:	15 worker-days/hectare
e. Flower removal:	10 worker-days/hectare
f. Sheathing/sacks:	10 worker-days/hectare
g. Harvest:	10 worker-days/hectare
h. Stem/sucker removal; restart:	10 worker-days/hectare
Total:	97 worker-days/hectare

97 worker-days per hectare × 100 hectares = 9,700 worker-days at daily salary
3,500 FG = 33,950,000 FG

Total Labor Cost:	33,950,000 FG = $33,950

2. Irrigation (in FG)

  a. Fuel for 140 days per year for two pumps: 2,136,400
    Total: 2,136,400 FG = $2,136

3. Fertilization and phytosanitary treatments (in FG)

  a. 4000 kg urea at 460,000/ton 1,840,000
  b. 600 kg 17-17-17 at 322,000/ton 193,200
  c. 5000 kg potassium chloride at 460,000/ton 2,300,000
  d. 80 l. paraquat at 4,600/l 368,000
  e. 36 kg labilite at 2,070/kg 74,520
  f. 400 kg oftanol at 6,025/kg 2,410,000
  g. 40 kg benlate at 25,300/kg. 1,012,000
   Total 8,197,720 FG = $8,198

Total, Bananas: $44,284

B. Operations

 1. Azethyl gas (ripener): 8,129,235 FG = $8,129
 2. Banana Boxes: 3,392,500 FG = $3,393
 3. Insurance (in FG)

  a. Generators 339,250
  b. Tractors 678,500
  c. Trailers 95,000
  d. Pumps 184,500
  e. Fiber machine 1,167,000
  f. 4 × 4 678,500
  g. Trucks 8,684,800
   Total: 11,827,550 = $11,828

 4. Maintenance

  a. Generators 575,000
  b. Tractors 4,600,000
  c. Trailers 241,500
  d. Pumps 161,000
  e. Fiber machine 1,978,000
  f. 4 × 4 3,450,000
  g. Trucks 22,080,000
   Total: 33,085,500 FG = $33,085

Total, Operations 56,434,785 FG = $56,435

C. Fuel (diesel at 700 FG/l and lubrication at 1400 FG/l)

 1. Conakry generator

  180 days at 20 l diesel/day plus 162 l oil: 2,746,800 FG

 2. Bennah generator

  90 days at 20 l diesel/day plus 81 l oil: 1,373,400 FG

3. 4 × 4
    300 days at 10 l diesel per day: 2,100,000 FG
4. Fiber machine
    135 days at 20 l diesel/day plus 122 l oil: 2,060,800 FG
5. Tractors
    135 days at 20 l diesel/day plus 122 l oil: 2,060,800 FG
6. Trucks
    128 round trips Bennah-Conakry at 100 l. per trip
    plus 576 l oil: 9,766,400 FG

Total, Fuel: 20,108,200 FG = $20,108

D. Personnel (annual salaries, in FG)

1. 1 Agronomist/Project head	4,140,000
2. 1 Head of operations	3,000,000
3. 22 Day laborers	13,200,000
4. 2 Vehicle drivers	3,600,000
5. 2 Tractor drivers	3,600,000
6. 1 Fiber machine foreman	2,400,000
7. 4 Fiber machine operators	3,600,000
8. 1 Accountant	3,600,000
9. 1 Secretary-typist	2,100,000

Total, Personnel: 39,240,000 FG = $39,240
  E. Building maintenance: 20,000,000 FG = $20,000

TOTAL OPERATIONS AND MAINTENANCE COSTS $180,067

*Note: Rented equipment could be substituted for the 140 worker-days of labor devoted to digging drainage ditches. The estimated cost for this is 20,000,000 FG instead of the 49,000,000 FG to do it by hand. Thus, using machinery saves about 29,000,000 FG ($29,000) but reduces the amount of employment produced by the project.

APPENDIX 8.3: BUDGET SUMMARY FOR FIRST THREE YEARS (IN U.S. DOLLARS, $)

I. First year
  Investment

Establishment of plantation	96,084
Building construction	136,237
Vehicle and equipment purchase	432,850
In vitro plants	248,000
Operating budget	180,067
Total	1,093,238

II. Second year

Operating budget	180,067
Total	180,067

III.Third year
    Operating budget   180,067
    Total   180,067

Total for First Three Years $1,453,372

APPENDIX 8.4: PROJECT EQUIPMENT AND VEHICLES

1. Ripening room equipment:
   Friga bohu evaporator 8 KW
   Comef condenser group 7.5KW
   Adjustment devices/tubes
   Static press devices
   Ermaflex insulator
   Refrigerant liquid
   Total cost: 19,372,778 FG = $19,373
2. Electric generators: 2 at 16,351,250 FG = 32,702,500 FG
   Total cost: 32,702,500 FG = $32,703
3. 45 hp Agricultural tractors: 2 at 28,750,000 FG = 57,500,000 FG
   Total cost: 57,500,000 FG = $57,500
4. Agricultural trailers: 2 at 4,025,000 FG = 8,050,000 FG
   Total cost: 8,050,000 FG = $8,050
5. Engine pumps: 2 at 7,825,000 FG = 5,650,000 FG
   Total cost: 15,650,000 FG = $15,650
6. Fiber machine: 1 at 98,900,000 FG = $98,900
7. 4 × 4 Vehicle: 1 at 28,750,000 FG = $28,750
8. 10-ton Trucks: 2 at 92,000,000 FG = 184,000,000 FG
   Total cost: 184,000,000 FG = $184,000
9. Irrigation pipes and sprinklers: 620,000 FG = $620
10. Spare parts: 40,000,000 FG = $40,000

Total Cost of Equipment and Vehicles: 491,125,270 FG = $485,546

SUGGESTIONS FOR FURTHER READING

Clapp, J. *Adjustment and Agriculture in Africa: Farmers, the State and the World Bank.* New York: St. Martin's Press, 1997.

Gowan, S. *Bananas and Plantains.* London: Chapman and Hall, 1995.

Kirkpatrick, C. H., and J. Weiss, eds. *Cost-Benefit Analysis and Project Appraisal in Developing Countries.* Northampton, Mass.: Edward Elgar, 1996.

NOTES

This chapter is based on work the author did in Guinea in 1997.

1. Background information on Guinea is from CIA, *World Factbook 1999* (available at Internet

Web site www.odci.gov/cia/publications/Factbook/gov.html); World Bank, *World Development Report 1998-99* (Washington, D.C.: World Bank, 1999); and UNDP, *Human Development Report 1998* (New York: United Nations, 1998). Monetary figures are expressed in U.S. dollars unless indicated otherwise.

2. Figures on caloric availability and grain production were obtained from the United Nations Food and Agriculture Organization's Internet Web site, accessed March 12, 2001, www.fao.org/. Stagnant world prices for bananas in the late 1990s would appear to support the FAO prediction that the market would be oversupplied by 2000.

3. UN Food and Agriculture Organization, Fifteenth Session, *Medium-term Outlook for World Trade in Bananas*, Food and Agriculture Organization, Committee on Commodity Problems, Intergovernmental Group on Bananas, Rome, 7-9 May 1997.

4. The best-known model of the rural sector based on the assumption that there is redundant agricultural labor appears in W. A. Lewis, "Economic Development with Unlimited Supplies of Labor," *Manchester School of Economic and Social Studies* (1954): 22(2): 139-91.

9

Case Study: Agricultural Externalities and Groundwater Contamination

INTRODUCTION

In Chapter 4, externalities were identified as important market failures for which government intervention may be warranted. Externalities occur when property rights are defined in a manner that does not require economic agents to take account of all of the consequences of their actions. Recall that according to the Coase theorem (see Chapter 4), efficient resource allocation can be achieved no matter how property rights are defined, as long as the affected parties can bargain freely and there are no transactions costs. As noted earlier, this bargaining requirement is often violated, and even if transaction costs are low enough to allow the parties to locate the efficient resource allocation, the way in which property rights are assigned has an impact on income distribution. For these reasons, it is often the case that government regulations can increase social welfare by redefining property rights to eliminate persistent externalities.

The case study presented in this chapter concerns one type of agricultural externality and policies that could be put in place to correct this problem. Modern agricultural production depends on the use of a wide variety of chemical inputs, which may not be entirely consumed in the production process. Herbicide and pesticide residues may remain in the soil or on the crops after the harvest, and excess fertilizer applied to a farmer's field may contaminate underground aquifers or surface water. A Coasian solution to this problem is unlikely because of the very high transaction costs that would arise in bargaining between agricultural producers and the general public. As a consequence, government intervention to redefine property rights through regulations or taxation schemes can increase social welfare. This case study focuses on nitrogen contamination of groundwater caused by agricultural production processes. In North America, this problem results primarily from the use of nitrogen fertilizer. High levels of contamination in the European Union (EU) have been caused both by fertilizer applications and by runoff from intensive livestock operations that generate large amounts of nitrogen-rich manure.

Groundwater is an important source of drinking water for many communities. Excess nitrogen fertilizer is converted to nitrates, which can leach into the groundwater. When ingested, the nitrates in the contaminated water are transformed into nitrites, which are harmful to human and animal health. The main benefit of reduced nitrate contamination of groundwater is lowered health risks. Unless it is possible to attach some monetary value to the lowered risk of nitrate-related health effects, it is difficult to compare the value of this benefit with its costs. As noted in Chapter 7, economists have developed ways to measure benefits and costs when markets and market prices are lacking. Some of these methods will be called upon in the following case study.

TECHNICAL BACKGROUND

Nitrogen is an essential nutrient for plants, and the addition of nitrogen fertilizer is of particular importance for crops such as corn. A common practice is to apply nitrogen in a gaseous form (anhydrous ammonia) at planting. The fertilizer is converted by bacteria living in the soil into nitrate, which is the form of nitrogen that can be absorbed by plants. Ideally, farmers would like to apply the exact amount of nitrogen that the crop will need for optimal growth. The amount of nitrates that will actually be absorbed depends on a wide variety of factors, however, so it is difficult to predict how much fertilizer will be needed to achieve the desired yield. For example, if there is substantial rainfall when the plants are young, nitrates may be moved deeper into the soil so that they end up below the root zone where the plants will be unable to absorb them. Because nitrogen fertilizer is relatively inexpensive, farmers often apply more than is theoretically required, thereby insuring that adequate nitrates will be available to the plants when needed. Depending on the weather and other conditions, nitrates may be left in the soil after the crop has been grown and harvested, and the excess can leach into the groundwater.

Groundwater is an important source of water in many parts of the world. In Nebraska, rural households depend on groundwater for their drinking water, and many urban water supply systems are also fed by wells.[1] According to Giraldez and Fox, 25 percent of Canadians rely on groundwater as the primary source of their drinking water.[2] Yadav and Wall suggest that 75 percent of the people living in Minnesota draw their water from underground sources, with most of the state's municipal water systems relying on groundwater for at least part of their supplies.[3] The EU has adopted very restrictive policies on nitrogen discharge because of severe groundwater pollution and the importance of this source of water in supplying both human and livestock needs.[4]

Whether from crop or livestock operations, nitrates that leach into the groundwater can lead to health problems. In humans and animals, nitrates are converted to nitrites, which may affect the ability of the blood to carry oxygen, a condition known as methemoglobinemia, or "blue-baby syndrome." This condition can be fatal in infants and may be harmful to the elderly or those with chronic illnesses. There is some evidence that nitrites may play a role in the development of some forms of cancer. Animal health may also be affected by excess nitrates, and groundwater contamination can lead to broader environmental problems. Groundwater is often linked to surface water so that the nitrates in groundwater may eventually end up in rivers, lakes, and coastal estuaries, where they cause algae blooms and eutrophication. The U.S. Environmental Protection Agency (EPA) has set a standard for nitrates in drinking water of 10 parts per million (ppm). At less than 10 ppm, water is considered safe for all users. Between 10 ppm and 30 ppm, it is safe for most adults and older children, but above 30 ppm, it is not considered safe for anyone to drink.[5] The EU nitrate standard allows a maximum of 50 ppm in groundwater.

Nitrates that originate in nitrogen fertilizer applications are a "nonpoint source" of pollution. Nonpoint sources occur when it is not possible to identify the specific individuals who caused the pollution. In contrast, it may be possible to tie an intensive livestock feed operation to a particular concentration of nitrates, making the feedlot a "point source" of pollution. Pollution from nonpoint sources is difficult to regulate, because it may not be possible to assess liability or monitor compliance with the regulation. Consider, for example, chemical pollution discharged by a specific plant. The pollution may be relatively easy to observe, and there is little question concerning its source. This makes it easy for regulators to establish liability and monitor compliance with any restrictions on effluents that may be put in place. In contrast, nitrates found in groundwater may have their sources in the fields of a large number of farmers scattered across the countryside. In this case, more general policies such as taxes on nitrogen fertilizer or regulations on fertilizer applications may need to be implemented.

Nitrate contamination of groundwater tends to be a local problem. In the United States, the EPA has set general standards for drinking water with which communities are required to comply. The particular policies implemented by local communities to insure safe drinking water can vary widely. In the EU, broad policies are frequently written as unionwide directives that charge the member countries with carrying out the policy using their own resources and institutions. Because the benefits and costs of nitrate policies tend to be site-specific, most of the analyses of nitrate policies focus on particular states or regions. The following case study is based on results reported in several different studies that deal with the site-specific costs and benefits of alternative nitrogen policies.

BENEFIT-COST ANALYSIS OF POLICIES TO PREVENT NITRATE CONTAMINATION OF GROUNDWATER

Many of the studies of this problem focus on methods for measuring either the costs or the benefits of alternative nitrate policies. Two studies that attempt to do complete benefit-cost analyses are discussed first. Yadav and Wall examine the benefits and costs of policies directed at providing incentives for farmers to adopt best management practices (BMP) in a particular watershed in Minnesota.[6] The Rural Clean Water Program (RCWP) is a national initiative funded by the EPA. The RCWP provides subsidies to farmers in certain regions to encourage them to adopt BMP. The BMP adopted in the study region included more precise fertilizer management and the establishment of animal waste management systems, among others. Participating farmers agreed to reduce total nitrogen use by replacing single large applications with several smaller and more timely applications and adjusting the amounts applied to reflect better monitoring of residual nitrogen already in the soil. According to Yadav and Wall, adopting these practices resulted in 21 percent less nitrogen fertilizer used without reducing yields.

This result is curious. Because fertilizer is not costless, one would expect profit-maximizing farmers to use the minimum amount required to achieve a desired yield. Reducing the amount of fertilizer applied reduces costs, and it is difficult to understand why farmers were not already using these cost-reducing practices if they had no impact on yields.

Moreover, the RCWP provided for payments to farmers to give them an incentive to change their farming practices. Such payments would seem to be unnecessary if the new practices actually reduce costs and thereby lead to increased profits. We will return to this puzzle later when we look at some of the studies focusing on costs.

In any case, Yadav and Wall estimate that the RCWP program in the study area involved a government expenditure over three years of $842,409. This is the figure used to represent the costs of the program. To estimate the benefits, they used a method known as the avoidance-cost approach. It turns out that it is fairly simple to take measures to avoid the negative effects of nitrate-contaminated groundwater. The water cannot be purified by boiling or other traditional measures, but there are ways to treat it (e.g., reverse osmosis) to eliminate the nitrates. In addition, if the main concern is methemoglobinemia in infants, parents can solve the problem by substituting bottled water for tap water. Both treatment and the use of alternative water sources require some expenditures. The avoidance-cost approach considers the benefits of reduced nitrate contamination to be equal to the savings resulting from not having to incur costs to avoid the negative effects of the pollution. It has been suggested that this approach establishes a lower-bound benefit estimate because there may be intrinsic values in the resource (options for future use, existence values) that are not measured by this method.

Yadav and Wall imagine three possible future scenarios. The first is an extension of the current situation in which 35 percent of rural wells have water above the EPA standard of 10 ppm and the water supplies in three cities are all within the standard. The second and third scenarios assume that more wells become contaminated in the future and that one or more of the city water supplies becomes contaminated as well. They then compute the costs of correcting the contamination predicted in each of the three scenarios. Their results indicate annual costs ranging from $59,000 to $140,000, depending on the extent of future contamination. In addition to the benefit of avoiding these costs, the adoption of the BMP means lowered expenditures on fertilizer. This benefit amounts to about $103,000 per year. Yadav and Wall add the fertilizer savings to the avoidance-cost savings and compute the number of years required for the present value of this stream of annual benefits to be equal to the initial cost. They find that it will take about six years under the first scenario and four if the most severe level of contamination is realized. Assuming that the annual benefits extend beyond four to six years, the net present value of this policy would be positive.

Another benefit-cost analysis of policies to prevent groundwater contamination is that of Giraldez and Fox.[7] They analyze the costs and benefits of alternative nitrogen policies in the village of Hensall in southwestern Ontario. The authors draw on a number of studies to determine the benefits of reducing nitrate contamination in the village's groundwater. As an initial estimate, they compute the potential mortality from methemoglobinemia and multiply that figure by the value of a human life. The value of a lost life is calculated as the present value of average lifetime earnings. Because the village is small and the probability of death from nitrate contamination is low, their estimates range from $693 to $6,289, depending on the discount rate used and other assumptions.

In addition to these estimates, the authors examined results of some contingent valuation studies in which surveys were done asking people how much they would be willing to pay to prevent nitrate contamination. One of the studies cited by Giraldez and Fox focused on willingness to pay (WTP) by households to be certain that their water meets established standards. Another attempted to measure WTP to insure that an aquifer is protected from potential damage. This second WTP study was aimed at estimating the value of a pristine aquifer, not for immediate human use but for use as a legacy to future generations and to keep open possible options that might be foreclosed if the aquifer were contaminated. Many studies indicate that such values are important for environmental goods, and leaving them out may lead to underestimation of the benefits. Based on estimates of household WTP from these studies and the number of households in the village of Hensall, the authors found benefits of reduced nitrate contamination ranging from $30,000 to $700,000.

These results are fairly discouraging. Depending on method and assumptions, estimated benefits appear to range from $693 to $700,000. Such a wide range is much too imprecise for effective policy analysis. After further reflection, the authors choose values of $2,508 and $11,360 to represent the range of annual benefits from policies to control nitrates. To compute the costs of such policies, the authors first use a complex environmental model to estimate by what amount nitrogen fertilizer applications have to be reduced to meet the nitrate standard of 10 ppm. The environmental model indicates that a reduction in average nitrogen applications from 147 kilograms per hectare (kg/ha) to 140 kg/ha would lower nitrate leaching by about 17 percent, bringing the village's groundwater into compliance with the standard. Such a reduction in fertilizer use will lower costs but will also lower corn yields if farmers are operating at the profit-maximizing level. Giraldez and Fox assume that they are and compute the net impact on farm profits, taking account of the value of the lowered output and the lower production costs. Their estimate places this cost at $1.81 per hectare for the reduction in fertilizer application from 147 kg/ha to 140 kg/ha, or a total of $284.31 for the village's 158 cultivated hectares.

The results of this study show annual benefits of some $2,500 to $11,400, compared with annual costs of only $284. As these costs and benefits are assumed to be repeated without change into the future, they can be directly compared without worrying about converting them into streams and discounting. Because the benefits far exceed the costs, regulations aimed at reducing fertilizer applications would seem to increase social welfare. The authors also estimate the magnitude of a fertilizer tax that would be required to induce producers to reduce nitrogen applications by the required 7 kg/ha. They find that a tax that would have the desired effect would raise the per-hectare cost of the policy to almost $50 for a total of $7,900. The reason for this difference is that a nitrogen tax is applied to all of the fertilizer used. Thus, the price is increased not only for the 7 kg that are not used but also for all 140 kg that are. In contrast, the regulatory ceiling does not raise the price of the fertilizer that is applied and is thus less costly to farmers. The problem with this conclusion, of course, is that the authors are implicitly assuming that the costs of forcing farmers to comply with the regulatory ceiling are zero. It is likely that the state would need to establish a system to monitor and enforce compliance, and the cost of such a system should be added to the estimated farm costs in computing the total costs of the policy. Enforcement costs for the tax policy would probably be small.

THE PROBLEM OF DETERMINING THE COSTS OF NITROGEN POLICIES

If one accepts a particular standard for groundwater quality as given, the analytical problem is to discover the least costly way to reach the standard. In Chapter 7, this type of analysis was referred to as cost-effectiveness. Several authors have studied the costs to farmers of alternative nitrogen policies designed to meet the nitrate standard of 10 ppm as set by the EPA. For example, Hopkins, Schnitkey, and Tweeten consider the impacts on two typical Ohio farms of three policies to control groundwater contamination.[8] The alternative policies include a tax on nitrogen fertilizer, a tax on measured nitrate effluents (runoff), and regulatory control of the amount of nitrates allowed. Their analysis is based on the use of a biological model and a model of decision making on the two farms. Their results suggest that taxes would have to be set at fairly high levels to obtain the adjustment in fertilizer use required to meet the standard. The regulatory standard turns out to be the least costly to the farms because one of them is already meeting the standard and would not need to make any adjustment. This farm would have to pay taxes if a tax policy instrument were chosen.

Wu and associates also integrate biological and economic models in their analysis of four groundwater policy alternatives.[9] The alternatives they consider are nitrogen taxes, taxes on irrigation water, incentives to modernize irrigation technology, and controls on nitrogen applications. The models are calibrated to reflect agroclimatic conditions in the U.S. Southern High Plains, and the four policies are structured to reduce nitrogen leaching and runoff by 5 percent, 15 percent, and 25 percent below a baseline reflecting current conditions. They estimate that a nitrogen tax set high enough to reduce excess nitrogen by 5 percent in the southern part of the study area would reduce farm income by 25 percent compared to the baseline, while the impact of a regulatory policy to achieve the same environmental goal would reduce farm income by only 3 percent. This suggests that restrictions on nitrogen use would be preferable to taxes. The authors point out, however, that a tax on nitrogen fertilizer generates revenue for the government. Thus, the reduction in farm income (decline in producer surplus) is partly offset by the added government revenue. Taking this effect into account makes the taxation policy preferable in terms of total social welfare although farmers would still prefer the regulatory restriction.

Larson, Helfand, and House note that input taxes and restrictions designed to regulate nonpoint source pollution lead to distortions that reduce social welfare at the same time that such policies enhance social welfare by correcting the negative externality.[10] Their approach to the problem of choosing the best policy under these second-best circumstances (see Chapter 5) is to estimate the costs of such distortions associated with various policy alternatives and to pick the policy that generates the least-costly distortions. The model developed to analyze this question is applied to nitrate leaching from lettuce production in California's Salinas Valley. The authors conclude that taxing irrigation water is preferable to taxing nitrogen to achieve the desired reductions in nitrate contamination of the groundwater.

These three studies are examples of policy analyses in which the basic techniques of benefit-cost analysis are used to search for the best way to achieve a given policy objective. No effort is made to estimate benefits, because the standard set by the policy is mandatory regardless of whether the costs of achieving that standard are greater than the presumed benefits. In determining the least-cost intervention for realizing the policy objective, the authors of these studies use models based on the assumption that producers maximize profits. The costs of meeting the groundwater standard are then made up primarily of the loss of profits resulting from compliance. Suppose that farmers are not actually maximizing profits—they are applying more fertilizer than strictly required because that practice saves time, is easier to do, or seems less risky given the uncertainty of fertilizer uptake by the crop. A tax or fertilizer restriction might actually benefit these farmers by moving them closer to the profit-maximizing amount of fertilizer.

This issue arose in the study by Yadav and Wall, who included the value of the fertilizer savings as part of the benefits. In another study, Johnson, Adams, and Perry also found that changing fertilizer management practices can reduce nitrogen use without lowering profits.[11] They use models of hydrological relations, plant response, and economic decision making to estimate the costs of reducing nitrate contamination of groundwater in the Columbia River basin in Oregon. The authors find that some simple changes in application rates and timing can reduce nitrogen use and nitrate contamination with little impact on profits, although further reductions do lower farm profits. Johnson, Adams, and Perry suggest that the apparent overuse of fertilizer is due to unpredictable weather and other technical problems faced by farmers. Of course, these technical factors could be the source of real costs even if the impact of adopting different management practices on short-run farm profits is negligible. For example, adopting different management practices might require additional time to learn and implement the new methods. These are real costs even if they do not show up in the farm accounts.

Another study of the costs of reducing nitrogen use to comply with groundwater standards draws attention to the potential impact of this change on the quality of the final product. Atwood and Helmers point to studies showing that reduced fertilizer applications lower both yields and grain quality.[12] Nitrogen is an essential component of protein, and a high level of protein in such grains as corn is an important quality criterion. Atwood and Helmers find that this quality effect adds 24 percent to 32 percent to the social cost (deadweight loss) of policies to restrict applications of nitrogen fertilizer.

These studies illustrate some of the difficulties in measuring the precise impact of policies to control nitrate contamination of groundwater. Reduced fertilizer applications may lower farm costs, but they may also lead to lower revenues if yield or the quality of the final product also falls. Changing fertilizer management practices could increase exposure to risk or require new equipment or training, and these factors could offset some or all of the cost savings associated with lower fertilizer use. For example, a soluble form of nitrogen can be applied at precisely the time it is needed through center-pivot irrigation systems, reducing the amounts needed at the beginning of the season and insuring that the plants will absorb virtually all that is applied. For farmers who have already invested in center-pivot irrigation systems, the added cost of chemigation is small and entirely offset by lower expenditures for fertilizer. For those who do not own the necessary equipment, however, adopting this technique is prohibitively expensive.

Similar comments apply to modern precision farming using global positioning satellites and detailed field maps to vary fertilizer applications according to the different soil qualities found across the field. Chemigation and precision farming significantly reduce the problem of groundwater contamination through a simple change in technology. Unfortunately, this technological change may be too costly for many farmers. All these considerations come into play in varying degrees in different places. Local variation in climate, crops, and other factors means that the cost of particular policies will not be the same everywhere. Because of this variation, the specific approaches to realizing broad groundwater quality goals have been largely left up to local jurisdictions. This is a sensible strategy, but it may not entirely solve the problem of highly variable compliance costs. Hopkins, Schnitkey, and Tweeten found different costs for two farms in the same region. It may turn out that the most efficient way to achieve the desired level of groundwater purity would be to design a specific policy for each individual farm. But such an approach could prove to be an administrative nightmare.

There is a way to take account of the extremely site-specific nature of the groundwater problem, although to my knowledge this approach has not yet been implemented in any of the regions in which groundwater pollution is a problem. In the Canadian study reported by Giraldez and Fox, it was found that a reduction in nitrogen applications of 7 kg/ha, from 147 kg/ha to 140 kg/ha on each of the 158 hectares in the village, would be enough to ensure that the groundwater in Hensall meets the nitrate standard. Suppose that the village decides to issue 158 permits allowing applications of 140 kg/ha of nitrogen. A farmer with 30 hectares would receive 30 permits, while a farmer with 53 hectares would receive 53. If these permits could be traded among the farmers in the region, those who find it difficult to comply could buy permits from those who are either already in compliance or who have compliance costs lower than the price for which they can sell their excess permits.

Suppose, for example, that the farmer with 53 hectares uses center-pivot irrigation on his land and the cost of installing the valves needed to fertilize through the center-pivot system is $2.00 per hectare. He would be willing to sell some or all of his 53 permits to anyone offering more than $2.00 per permit. The result would be that the average application of 140 kg/ha would be achieved, although some would be applying more than this amount and others less. This would be less costly than forcing each farmer to apply only 140 kg/ha on each hectare owned. The idea of establishing a market in tradable pollution permits is appealing because the overall environmental objective is achieved at the least total cost to producers. Although tradable permits have not been used extensively for groundwater protection, they have been recommended by economists in a variety of contexts. For example, the Kyoto Protocol on global warming includes provisions for tradable permits in greenhouse gases.

Of course, ownership of such permits is a property right, and there may be distributional issues in the way this market is established. For example, it could be argued that a goal of distributive justice would be advanced if all of the greenhouse gas permits were allocated to low-income countries, which could then sell them to the wealthy countries in which greenhouse gas emissions, at least on a per capita basis, are the greatest. Likewise, for groundwater pollution, one might wish to argue for an initial distribution of nitrogen permits favoring low-income, family farms or some other type of farm deemed socially desirable (e.g., organic farms). Political realities make such outcomes unlikely. The most common initial distribution of property rights of this nature is usually the one that causes the least disruption to status quo ante income streams. Still, tradable permits offer a way to realize the environmental objective in the most efficient way, even if the opportunities for greater distributive justice are left unexploited.

The idea of tradable permits is extremely popular among economists, because such a system leads to a more efficient allocation of resources than do the alternatives. Some are uneasy with a system that seems to grant individuals a permit to pollute, however. For example, in discussing solutions for problems of water scarcity in Delaware, several students noted that granting home owners and other water users tradable permits would probably mean that all the available water would be used, whereas regulatory limits could be structured to prevent certain uses and might leave greater quantities of water unused. Even if it is specified that the total amount available under the permit system would be less than what is needed for long-term sustainability, some students still objected to this idea. Their concern seemed to be that certain uses of this scarce and valuable resource might be inappropriate, and these uses would not be prevented by the permit system. Thus, home owners would be able to purchase permits to water their lawns. For consequentialists, the exercise of this choice is Pareto better: no one is harmed, and some people are better off. For the nonconsequentialist students, water should be preserved for important uses, and it is better to leave it in the ground than to use it inappropriately. Of course, it might be thought that the students simply had nosy preferences (Chapter 5). Why should their preferences with respect to water use outweigh the tastes of people who enjoy greener lawns?

THE PROBLEM OF ESTIMATING THE BENEFITS OF GROUNDWATER PROTECTION

Determining the benefits of reducing nitrogen contamination of groundwater also raises complex issues. The main danger of nitrates in drinking water is to infants, who can die from methemoglobinemia. This suggests that an appropriate measure of the benefits of lower levels of nitrates in groundwater would be given by estimating the value of a human life and multiplying that value by the number of infants likely to die from this condition. This approach was used in the Canadian study. It turns out, however, that there have been no reported deaths from methemoglobinemia in North America in recent years. This is because it is a simple matter to prevent these deaths even in areas where the groundwater is contaminated. Parents can simply substitute bottled water for the polluted groundwater, and relatively inexpensive technologies can be used to protect community water supplies. This suggests that the avoidance-cost method described earlier might be a more appropriate measure of the benefits of reduced nitrates in groundwater.

Benefit estimates based on such measures, however, may not capture the uncertainty of the effects of nitrates on health, the possibility that excess nitrates in groundwater may prevent certain options that might otherwise be available in the future, or the intrinsic value that some may attach to knowing that the groundwater resource is being maintained in a pristine state. Some people may believe that maintaining clean groundwater is beneficial from a precautionary standpoint, given that we are uncertain about the long-term health effects of nitrates. Others may feel that clean groundwater is something we should leave to future generations to avoid foreclosing options for these people or simply because there is a value to acting as good stewards of our natural resources. One way to bring these considerations into the measured benefits is to ask people what they would be willing to pay to be certain that groundwater supplies are protected. This approach is known as "contingent valuation."

Abdalla conducted a survey of studies using the avoidance-cost approach to measure the benefits of lowered groundwater contamination from nitrates, pesticides, and other chemicals.[13] The five studies reviewed all found that people do purchase filters and bottled water if they believe that tap water is not entirely safe. The estimated annual expenditures ranged from $32 to $330 per household, depending on the locality. Most studies only measure household avoidance costs, although municipalities and firms may also purchase protective devices to improve water quality. One of the difficulties in evaluating the benefits of nitrogen policies using this approach is that the purchases may be motivated by the presence of a variety of contaminants. Unless nitrates are the only water concern in the area, it may not be possible to separate out the benefits of nitrate reduction from perceptions of the benefits of reductions in groundwater contaminants in general.

The other approach to benefit measurement is contingent valuation. This approach is implemented through scientific surveys asking respondents how much they would be willing to pay to be certain that their water is pure. A great deal of research has been done to determine the kinds of questions that will elicit accurate responses without inadvertently introducing biases. A common approach is to ask if the respondent would be willing to pay X dollars to a nongovernmental fund that would insure that there is no nitrate pollution. Questions are often written to avoid the suggestion that a government agency would be in charge of using the money to correct the problem, because some people do not trust the government. These people might claim to be unwilling to pay anything if the money goes to a government agency, even though they would be willing to contribute to some other organization to insure cleaner water. There are many other sources of potential bias in the way the questions are phrased.

A respondent who indicates a willingness to pay X dollars is directed to a second question asking if she would be willing to pay slightly more than X. A respondent who answers "no" to the first question is directed to a second question asking if she or he would be willing to pay slightly less than X. The value used for X is varied across the sample to obtain information on the full spectrum of household valuation. Statistical techniques are then applied to the sample results to estimate the collective willingness to pay for the environmental change. This latter measure is equivalent to consumer surplus. Recall that "consumer surplus" is measured as the area under the demand schedule and above the relevant price. Consumer surplus is derived by noting that some consumers would be willing to pay more than the equilibrium price. These consumers benefit from being able to purchase the good or service at the lower market price. In contingent valuation studies, the estimated willingness to pay provides the same type of information as the demand schedule.

Boyle, Poe, and Bergstrom examined eight contingent valuation studies that focused on the value of groundwater protection.[14] Estimates ranged from $56 to $1,154 per household per year, depending on the region and particular method used. Powell, Allee, and McClintock estimated annual willingness to pay of $101.84 for an average household that considers its water to be "somewhat safe," while households that believe their water is "very safe" are still willing to pay $36.15 on average for increased water protection.[15] Sukharomana used a contingent value survey to study household willingness to pay to insure that drinking water in Nebraska meets the EPA nitrate standard. She found a mean value of almost $10 per month per household. Sukharomana also estimated avoidance-cost benefits of $6 per household per month and argued that the two measures represent upper and lower bounds on the true benefits of groundwater protection.[16]

CONCLUSION

All of the studies described in the preceding sections concerned costs and benefits in very specific localities. Very few studies have been done to assess national costs and benefits as an input into deciding whether national policies or standards increase social welfare. It would be difficult to find a way to aggregate all these location-specific analyses into one global benefit-cost study for Canada or the United States. Of course, such analyses may not be needed, given the broad consensus that the amount of nitrates in groundwater should be kept within the scientifically established standard.

For the EU, Leuck and associates used a partial equilibrium, multisector model (see Chapters 10 and 13) to predict the effects of alternative nitrogen policies for the countries that make up the EU.[17] This study is also an exercise in cost-effectiveness analysis, because the model is used to predict the impact of compliance with the EU Nitrate Directive on EU agricultural production, nitrogen balances, trade, and world prices. In other words, the EU policy is taken as given, and the costs of achieving the policy objective are measured by lost agricultural production and exports. The authors found that compliance with the EU directive could lower agricultural production significantly, transforming the EU from an exporter to an importer of such livestock products as pork, poultry, and eggs. The provisions of the EU directive include mandatory changes in fertilizer management in the most affected areas to be enforced by member governments. According to this study, these provisions are more effective at reducing nitrate pollution than a nitrogen tax or a more general reform of agricultural policies. The study does not consider the possibility that reduced production and trade within the EU could increase welfare if EU consumers end up spending less for imported food, for example.

While cost-effectiveness studies are informative, it would be interesting to know whether the benefits of establishing and enforcing groundwater nitrate standards are in fact greater than the costs of compliance. The studies by Yadav and Wall and Giraldez and Fox both seem to indicate that the net benefits of reduced nitrate contamination are positive, at least in the regions studied. A rough estimate of the benefits and costs of reduced nitrate pollution in Nebraska can be derived from two of the studies already cited. Sukharomana multiplied the estimated household WTP figure (about $120 per year) by the number of households in Nebraska to obtain an estimate of the total value of lower nitrate pollution of about $60 million. Atwood and Helmers estimated that the costs of a 10 percent reduction in nitrogen fertilizer in Nebraska could reach almost $4 million, far less than Sukharomana's estimate of the benefits.

Of course, the actual reduction in fertilizer use needed to realize the estimated benefits in Nebraska may be greater than the 10 percent reduction assumed by Atwood and Helmers, and it is unlikely that the relationship between farm costs and the size of the reduction is linear. Thus, a 20-percent reduction, for example, may generate costs that are more than twice those estimated for a 10-percent reduction.

All of the estimated values in these and the other studies are dependent on the validity of the methods and data used. While contingent valuation studies may provide a more accurate picture of the nonmonetary values attached to environmental protection, it could overestimate the actual benefits. Suppose, for example, that someone conducted a survey asking people how much they would be willing to pay to prevent the northern spotted owl from becoming extinct. If respondents are unable to place this species in the broader context of species loss, they may overstate the value of preventing this extinction. Imagine how the benefit estimates for preventing the owl's extinction might differ if the question asks about willingness to pay to prevent the extinction not only of northern spotted owls but of ring-tailed lemurs, snail darters, sockeye salmon, white rhinoceroses, and 60,000 other species as well. Despite the technical difficulties of measuring the subjective values that characterize these cases, contingent valuation and the other methods discussed in this case study can provide useful information when markets are lacking, and some indication of relative benefits and costs is better than nothing.

DISCUSSION QUESTIONS

1. Why would farmers apply more nitrogen than needed if reducing the amounts used would have no impact on their profits?

2. Would you favor the use of tradable permits to regulate emissions of air or water pollution? Why or why not?

3. Because it is easy to prevent the negative effects of most levels of nitrate contamination in groundwater, the benefits of regulations to control this problem seem to be small. Is there really a collective action problem in this case? Explain.

4. Technological solutions of the nitrate problem (precision farming and chemigation) can protect groundwater at no added cost to producers who have already invested in the needed equipment. Would it be a good policy for the government to provide subsidies to farmers who do not have the equipment to help them acquire it?

5. Discuss potential inaccuracies that might arise in using contingent valuation methods.

6. Why is the avoidance-cost method often thought to reflect a lower-bound estimate of the benefits of pollution abatement?

SUGGESTIONS FOR FURTHER READING

Hanley, N., and C. L. Spash. *Cost-Benefit Analysis and the Environment*. Brookfield, Vt.: Edward Elgar, 1994.

Hausman, J. A., ed. *Contingent Valuation: A Critical Assessment*. Amsterdam: North Holland, 1993.

Shogren, J. F., ed. *Private Property and the Endangered Species Act*. Austin: University of Texas Press, 1998.

Tietenberg, T. *Environmental Economics and Policy*. New York: Addison-Wesley, 1998.

NOTES

1. Renu Sukharomana, "Willingness to Pay for Water Quality Improvement: Differences between Contingent Valuation and Averting Expenditure Methods" (Ph.D. diss., University of Nebraska, 1998).

2. C. Giraldez and Glenn Fox, "An Economic Analysis of Groundwater Contamination from Agricultural Nitrate Emissions in Southern Ontario," *Canadian Journal of Agricultural Economics* 43 (3)(1995): 387–402.

3. S. N. Yadav and D. B. Wall, "Benefit-Cost Analysis of Best Management Practices Implemented to Control Nitrate Contamination of Groundwater," *Water Resources Research* 34, no. 3 (March 1998): 497–504.

4. European Union, "Discharges of Substances: Pollution Caused by Nitrates from Agricultural Sources," Internet Web site accessed September 8, 2000, http://europa.eu.int/scadplus/leg/en/lvb/128013.htm/.

5. U.S. Environmental Protection Agency, "Nitrates in Groundwater—Frequently Asked Questions," Internet Web site accessed on September 11, 2000, www.epa.gov/region08/...r/dwdrink/dwregs/dwnitrate.html.

6. Yadav and Wall, "Benefit-Cost Analysis," Subsequent statistics relating to the analysis are all from this source.

7. Giraldez and Fox, "Economic Analysis." Subsequent statistics relating to the analysis are all from this source.

8. J. Hopkins, G. Schnitkey, and L. Tweeten. "Impacts of Nitrogen Control Policies on Crop and livestock Farms at Two Ohio Farm Sites," *Review of Agricultural Economics* 18 (3) (1996): 311–24.

9. J. Wu, M. L. Teague, H. P. Mapp, and D. J. Bernardo. "An Empirical Analysis of the Relative Efficiency of Policy Instruments to Reduce Nitrate Water Pollution in the U.S. Southern High Plains," *Canadian Journal of Agricultural Economics* 43 (3)(1995): 403–20.

10. D. M. Larson, G. E. Helfand, and B. W. House. "Second-Best Tax Policies to Reduce Nonpoint Source Pollution," *American Journal of Agricultural Economics* 78 (November 1996): 1108–17.

11. S. L. Johnson, R. M. Adams, and G. M. Perry. "The On-Farm Costs of Reducing Groundwater Pollution," *American Journal of Agricultural Economics* 74, no. 4 (November 1991): 1063–73.

12. J. A. Atwood and G. A. Helmers. "Examining Quantity and Quality Effects of Restricting Nitrogen Applications in Feedgrains," *American Journal of Agricultural Economics* 80, no. 2 (May 1998): 369–81.

13. C. W. Abdalla, "Groundwater Values from Avoidance Cost Studies: Implications for Policy and Future Research," *American Journal of Agricultural Economics* 76, no. 5 (December 1994): 1062–67.

14. K. J. Boyle, G. L. Poe, and J. C. Bergstrom. "What Do We Know about Groundwater Values? Preliminary Implications from a Meta Analysis of Contingent Valuation Studies," *American Journal of Agricultural Economics* 76, no. 5 (December 1994): 1055–61.

15. J. R. Powell, D. J. Allee, and C. McClintock. "Groundwater Protection Benefits and Local Community Planning: Impact of Contingent Valuation Information," *American Journal of Agricultural Economics* 76, no. 5 (December 1994): 1068–75.

16. Sukharomana, "Willingness to Pay."

17. D. Leuck, S. Haley, P. Liapis, and B. MacDonald. "The EU Nitrate Directive and CAP Reform: Effects on Agricultural Production, Trade and Residual Soil Nitrogen," Foreign Agricultural Economic Report No. 255 (Washington, D.C.: USDA, January 1995).

10

Partial Equilibrium Policy Analysis

INTRODUCTION

As noted in Chapter 7, public policy often takes the form of interventions that influence market incentives through the redefinition of property rights or the direct regulation of individual behavior. The purpose of such interventions is to correct market failures that may be reducing efficiency or leading to inequitable outcomes. In the three previous chapters, we have considered policies implemented through government provision of public goods such as irrigation schemes, development projects, environmental protection, or other types of public capital investments. Such projects can be evaluated with the conventional methods of benefit-cost analysis described in Chapter 7. In this chapter, we examine the analysis of policies that are designed to change behavioral incentives through the manipulation of prices or other market conditions.

An example of this type of intervention is the taxation of gasoline in European countries. High gasoline taxes are aimed at correcting externalities associated with the use of petroleum products to fuel private vehicles. These externalities include ground-level air pollution as well as pollution of the upper atmosphere through the release of greenhouse gases (e.g., carbon dioxide) that lead to global warming. In addition, the use of petroleum products may involve an intergenerational externality because petroleum is a nonrenewable resource. Although petroleum supplies will never be completely exhausted, they will be reduced over time, and prices will eventually rise to the point where petroleum will no longer be used. If the interests of future generations are taken into account, the optimum depletion rate, which determines the date at which oil will be too expensive for continued use, may be lower than it would be considering only the interests of the current generation. Higher prices today reduce current petroleum use, relieving future generations of some of the burden of finding alternative energy sources. One could also argue that there are dependency costs associated with the use of petroleum, much of which may have to be imported.

As shown in Chapter 4, negative externalities of this nature are reflected in prices that encourage greater use of petroleum than is socially optimal. In Europe, high taxes are used to raise prices so that the signals to consumers more accurately represent the full opportunity costs of burning fossil fuels. Extraction and processing costs for gasoline are probably quite similar between the United Kingdom (a major petroleum producer from its reserves in the North Sea) and the United States. Yet retail gasoline prices in the United Kingdom averaged $3.97 per gallon in the fourth quarter of 1999, compared with $1.17 per gallon in the United States at about the same time.[1] The difference is largely due to higher taxes levied by the government of the United Kingdom. Higher U.K. prices have resulted in lower per capita gasoline consumption, as expected. Automobiles are smaller and more energy efficient in the

United Kingdom, people walk more and drive less than in the United States, and they have altered their behavior in other ways to conserve energy. Based on data from the U.S. Energy Information Office, British citizens used about 11 barrels of petroleum per person per year compared with about 25 barrels for each U.S. consumer. Total energy consumption per capita in the United States in 1997 was more than twice the per capita consumption levels in the United Kingdom, Switzerland, Spain, or Italy, despite the fact that these countries have similar income levels and standards of living.[2] European energy policies appear to have had the intended effect of lowering consumption and at least partially correcting the negative effects of the associated externalities. The implication of this assessment is that the United States is leaving the negative externality largely uncorrected.

European energy policies are designed to achieve certain objectives by altering the *incentives* that influence the behavior of the economic agents involved. As in the case of the evaluation of government projects, analysis of these kinds of policies requires some estimate of consequences (benefits and costs) as well as attention to nonconsequentialist considerations as appropriate. Unlike the measurement of the benefits and costs of a government project, however, assessment of the likely consequences of policies such as the European energy tax must include some method for predicting individual responses to the incentives that are created by the policy. If we know that the price elasticity of demand (see Chapter 2) for gasoline in the United States is -0.5, for example, we can predict that a tax that raises retail prices by 10 percent will lead to a decrease in consumption of 5 percent.[3] With some additional information, we can estimate the changes in consumer and producer surpluses that will be caused by the tax. Such estimates constitute one measure of the benefits and costs of the policy. Of course, if the purpose of the policy is to internalize a negative externality, it will also be necessary to incorporate the benefits associated with the reduction of the external effects.

To estimate the consequences of policies of this nature, it is often necessary to use a model to predict the way in which individuals and firms will respond to the altered price incentives. In this chapter, we will examine *partial equilibrium analysis* based on models of a single market or subsector of the economy. In Chapter 13, we will introduce more complicated models that attempt to capture a broader range of relationships than is possible in partial equilibrium models. In all cases, the goal is to predict how the economy or some part of it will evolve if a policy is implemented, as compared to its likely evolution if there is no policy change. Partial equilibrium analysis focuses on policy impacts in a single market or a small set of related markets in a particular sector (e.g., the markets for sweeteners, including sugar, sugar substitutes, and noncaloric sweeteners).

As noted in Chapter 7, policy analysis requires a comparison between the expected impact of a policy and an appropriate counterfactual. To evaluate policies designed to alter behavioral patterns, the counterfactual is often determined by using a model to simulate the future without the policy. Comparison of this baseline, or "business-as-usual," simulation with a simulation that includes the policy intervention serves as a measure of the impact of the policy. Thus, in the case of a tax on gasoline, we may want to predict demand next year without the tax and then compare that result with a prediction with the tax. Note that the simple elasticity example given earlier provided comparative information, because it showed that demand would be 5 percent less with the tax than it would have been without. The status quo (without the tax) is the counterfactual in this case.

In carrying out a partial equilibrium analysis of a public policy, the first task is to identify the important relationships that are likely to be affected by the policy. A tax on gasoline will clearly affect gasoline demand. While it is less obvious, the tax will also affect gasoline supply, and it may have other impacts as well. Moreover, these effects will probably not occur instantaneously, because the extent of the responses of consumers, producers, and others will vary with the passage of time. In other words, there will often be a distinction between impacts that are expected to occur in the short run and those that may require a longer-run time frame to become manifest. In identifying the significant relationships as well as the time frame and other dimensions of the problem, it is often essential to draw on the results of economic theory. For example, the theory of demand provides insights on the likely relations between gasoline prices, gasoline consumption, income, and prices of substitutes and complements. Theories dealing with perfect and imperfect competition can provide guidance in determining how supply and demand are generated and how they come together in a market.

Once important relationships have been identified, it is often useful to use a *graphical analysis* to draw out the implications of a policy change. The second part of this chapter is devoted to an extensive graphical analysis of the main elements of U.S. agricultural policy. Such analyses help one understand the nature and direction of the changes that are likely to be induced by a policy. Consider the case of a gasoline tax. For this initial exposition, we will assume that the market is closed so that we do not have to introduce complications related to international trade. The imposition of a tax on gasoline consumption in the United States is illustrated in Figure 10.1. Prior to imposing the tax, market equilibrium is at $1.20 per gallon, with daily consumption of 228 million gallons. It is assumed that the tax raises consumer prices to $1.30 per gallon, while producer prices fall to $1.15 per gallon. The difference (often referred to as a *price wedge*) is $0.15—the tax rate, or amount collected by the government per unit sold. The reason that producer prices fall is that the higher consumer price leads to lower consumption. Consumers will simply not continue to purchase 228 million gallons per day at the higher price although producers still wish to market that quantity. In other words, there is a kind of excess capacity in the market that is eliminated by the producer price decline, bringing gasoline output back into line with the new level of demand.

The advantage of a graphical analysis is that it can help illustrate the full implications of a policy change. Earlier, I noted that while it is clear that a gasoline tax would have an impact on demand, it is less obvious that such a tax would also affect supply. The graphical analysis forces us to recognize that if demand falls in response to a higher price, some other change in the market has to occur if supplies are to be reduced to the new level of demand. That change is the fall in producer prices, which can be determined through reference to the supply schedule shown in Figure 10.1. It is also possible to evaluate the change in producer and consumer surpluses as a result of this tax. At the higher price, consumers lose the areas labeled *A* and *B*. Producers lose the areas labeled *C* and *D*. The government earns tax revenue from this policy equal to areas *A* and *D* (the tax rate, which is the difference between producer and consumer prices multiplied by the quantity on which the tax is levied). It is a simple matter of plain geometry to calculate these areas:

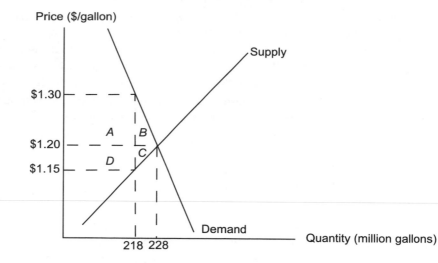

Figure 10.1: A tax on gasoline.

A = 218 × (1.3 − 1.2) = $21.8 million
B = 0.5 × (228 − 218) × (1.3 − 1.2) = $0.5 million
C = 0.5 × (228 − 218) × (1.2 − 1.15) = $0.25 million
D = 218 × (1.2 − 1.15) = $10.9 million

Thus, consumers lose $22.3 million in consumer surplus each day, while producers lose $11.15 million in producer surplus each day, for a total daily loss in producer and consumer surplus of $33.45 million. These losses are partially offset by the government revenue of $32.7 million. Subtracting that revenue of $32.7 million from the total loss of $33.45 million leaves a net loss of $0.75 million. Areas B and C are efficiency losses due to the tax and are just equal to the net social loss of $0.75 million.

Of course, in the preceding example, significant benefits associated with the tax have been ignored. The benefits of reduced pollution are probably greater than the net social loss illustrated in Figure 10.1. An assessment of these impacts would require additional information. In addition, this example is not based on empirical evidence about the location and slopes of supply and demand schedules. While the graph is calibrated to reflect actual U.S. prices and daily consumption of regular unleaded gasoline, the prices and quantities after imposition of the tax are based on purely imaginary parameters. To more fully analyze the impact of this tax, it would be necessary to specify a mathematical model of some nature and then estimate the parameters of the model either through the application of statistical methods to current or historic data or through some other method. Such models can be used to simulate the future with and without the policy change in order to measure its impact. Examples of policy analysis using statistical models will be discussed in the two case studies that follow this chapter. In the remainder of this chapter, graphical analyses will be used in conjunction with some simple elasticity models to illustrate analytical methods that might be applied to agricultural and natural resource policy analysis.[4]

U.S. Agricultural Policy

In all societies, food is an extremely important consumer good that is necessary for survival and that has additional roles associated with ceremony, entertainment, and pleasure. Food production is a source of income for large numbers of people, and even more work on the processing, transporting, and marketing of food products. In low-income developing countries, agricultural production often accounts for more than half of gross domestic product (GDP), and it is not uncommon to find the majority of the labor force engaged in farming. Consumers in low-income countries often spend more than half their disposable income on food. In industrialized nations, production agriculture has less significance in the general economy, and the proportion of disposable income devoted to food is generally less than 20 percent. In the United States, agriculture occupies only about 2 percent of the labor force and accounts for about the same percentage of GDP. Of course, large numbers of workers are employed in industries that process raw agricultural products and distribute the finished food items.

Throughout the twentieth century, the number of farms in the United States declined at the same time that total output and agricultural productivity expanded. From 6.4 million farms in 1910, the number of farms in the United States fell to 5.6 million in 1950 and 2.2 million in 1999. Between 1950 and 1997, the total amount of land in farms fell from 1.2 billion acres to 954 million acres, and the agricultural population declined from 7.2 million to 3.4 million. Despite these changes, total output increased by about 140 percent between 1950 and the late 1990s.[5]

Historically, farm incomes in the United States were lower than the incomes of similar workers in other sectors of the economy. Although this *farm problem* appears to have disappeared in recent years,[6] it has historically provided one of the main rationales for government intervention in the farm economy. Many explanations of the tendency for farm income to lag behind income in other sectors have been offered. One explanation draws attention to the extraordinary gains in agricultural productivity that have occurred in the last one hundred years.

Because farmers are price takers (that is, they are unable to exercise market power as do oligopolistic and monopolistic firms), productivity gains have been largely passed on to consumers in the form of lower prices. These price declines should not pose a problem for farmers if they are completely offset by the productivity increases. New technologies are often diffused relatively slowly, however, so that some farmers end up facing lower prices even though they have not realized the productivity gains that are driving the rest of the industry. The overall effect is to make some farmers redundant, a fact that is reflected in the decline in farm numbers that has characterized the last century. In fact, the exit from farming has often been slower than needed to counter the tendency for prices to fall because many farmers have few alternatives to farming and remain trapped in that occupation even though the returns are very low.

One way that farmers might counter the tendency for prices to fall and for farm workers to be shifted out of agriculture would be to refuse to adopt technological innovations produced through both public and private research. Of course, this would not be desirable for consumers, who benefit from lower prices. Moreover, the resources released from agriculture through this process can be used in other sectors of the economy, thereby contributing to economic growth and more general prosperity. From the farmers' point of view, however, such benefits come at the expense of the farm sector, which would be larger if technological innovation could be prevented. Not surprisingly, there have been occasional efforts to prevent technological change. Some dairy farmers in Wisconsin recognized that the use of recombinant bovine growth hormone (rBGH, also known as bovine somatotropin, BST) could force those who did not adopt the technology out of business as it generated increased production in a market already saturated at the prevailing prices. The problem the farmers face in these circumstances is that they are in a prisoners' dilemma similar to that of the fishing fleets caught up in a process of overfishing. If an individual dairy farmer chooses not to use rBGH, he or she will produce less and the price will still fall unless all other producers also refuse to adopt the new technology. The Wisconsin dairy farmers recognized that they would be unable to overcome defection and free riding in their refusal to use rBGH, so they turned to the government in an unsuccessful bid to have rBGH declared unsafe and banned from use.

In addition to low farm incomes, market instability is often offered as another rationale for government intervention in the farm sector. Because agricultural production is sensitive to highly variable weather conditions, prices and output are inherently unstable. In years of particularly clement weather, bountiful harvests lead to lower prices because demand grows only slightly from year to year. The low prices can cause hardship for producers, some of whom may be forced into bankruptcy. Periods of good weather are inevitably followed by periods of drought, flooding, storms, and other problems that reduce harvests and lead to higher prices. These high prices may cause hardships for consumers, particularly those with low incomes, who often spend a much greater proportion of their disposable income on food than do those with higher incomes. In developing countries, the fear of insufficient food supplies and high food prices is a much greater factor in food and agricultural policy than the problem of low farm incomes. Even in industrialized countries, a secure and stable food supply is often seen as an important policy objective. In the United States, early farm policies were aimed at both stability and low farm incomes.

The Agricultural Adjustment Act (AAA) of 1933 was the first legislation adopted by the U.S. government that specifically targeted market prices for important agricultural commodities.[7] Prior to the AAA, a few government programs had been put in place with the objective of raising living standards in the farm sector. During the Civil War, the Morrill Land Grant Act of 1862 led to the establishment of agricultural colleges where the children of farmers could learn scientific cultivation methods.[8] A few years later, the Hatch Act established agricultural experiment stations in each of the states. These experiment stations are research centers generally located at or near the land-grant colleges and sharing scientific personnel so that there is a direct link between agricultural research and teaching. Finally, around the time of World War I, the Smith-Lever Act established the cooperative federal-state extension service, which is also tied to the land-grant colleges and aimed at transferring research results directly to farmers. Funding for the research and extension activities is shared between the state and federal governments.

Public investment in agricultural teaching, research, and extension has paid off handsomely, as these investments are one of the main reasons for the tremendous increase in agricultural productivity during the twentieth century. Although some types of research lead to products that can be patented, much agricultural research has the characteristics of a public good (see Box 10.1). This type of research will not be supplied by the private sector in socially optimal amounts. Educational programs also have public-good characteristics, and their provision by the public sector has been of benefit to both those who produce food and those who consume it. As previously noted, the benefits of productivity increases in agriculture flow primarily to consumers in the form of lower food prices.

Box 10.1: Basic and applied research

The rationale for public support of agricultural and scientific research often rests on the distinction between *basic* and *applied* research. Research on basic scientific processes is a public good that is both nonexcludable and nonrival. This basic research becomes available to any other interested party through peer review and publication. In general, basic research does not lead to specific products that can be patented but rather to scientific insights or procedures that can be used in a broad range of applications. Because basic research is a public good, the private sector will not generate socially optimal amounts of this type of research, and that justifies public investments to provide more nearly socially optimal amounts. The results of basic research can be used by the private sector to carry out applied research and development (*R&D*) with a view toward discovering innovations that can be patented. Such innovations are excludable, and private firms can recover R&D costs and make a profit on them. This distinction leads to a natural division of labor, with the public sector charged with supporting basic research, leaving applied R&D to private firms. An example of basic research is the work that led to the discovery of genetic engineering methods (gene-splicing). These methods are now widely used by private firms to develop new products that can be patented and sold. The returns to public agricultural research are extremely high. Evenson reports internal rates of return to public agricultural research in developing countries ranging from 19 percent to more than 80 percent, with most results at the upper end of that range (Robert E. Evenson, "IARCs, Aid, and Investment in National Research and Extension Programs," in *Research and Productivity in Asian Agriculture*, Robert E. Evenson and Carl E. Pray [Ithaca, N.Y.: Cornell University Press, 1991] 327). Similar results have been obtained for industrialized countries.

Another early piece of legislation related to agriculture was the Capper-Volstead Act of 1922, which exempted agricultural cooperatives from the antitrust laws that had been established in the early years of the century. Antitrust laws were aimed at preventing collaboration between firms in the same industry, on the grounds that such collaboration or collusion would lead to monopolies that would exploit both workers and consumers. Agricultural cooperatives centralize the marketing of particular agricultural commodities produced by farmers, who could be seen as belonging to the same industry and would thus appear to violate antitrust laws. The Capper-Volstead Act exempts farmer-owned cooperatives from the antitrust rules, allowing farmers to collaborate in the marketing of their products. One reason for this act was the belief that farmers are often at a disadvantage in marketing their goods because they face large processing firms that are able to exercise market power in their dealings with farmers. If farmers can band together, they can bargain with the firms that handle their products on a more equal footing.

In the 1930s, the United States—and eventually the entire world—experienced the most serious economic depression of the modern era. Farming was particularly hard hit by the economic downturn. In addition, a major drought in the Midwest and Plains states (the "Dust Bowl") further devastated the agricultural sector. The AAA was part of the New

Deal legislation initiated by President Franklin D. Roosevelt in an effort to stimulate the economy and end the miseries of the Great Depression. Despite the effects of the drought on agricultural output, it was evident that there was a tendency for farm production to increase more rapidly than demand, leading to low and declining prices. As a result, early farm programs included measures to restrict output in an effort to raise market prices. The initial policy was based on quantitative restrictions (quotas), which were declared unconstitutional because they amounted to allowing the farm sector to act like a monopoly. The final system of *production controls* included quantitative limits on the amounts that could be marketed (marketing quotas) and restrictions on the amount of land farmers could plant (acreage allotments).

Marketing quotas and acreage allotments had to be specified for each farm. Farmers would be assigned an official allotment based on their historical acreage planted to the main program crops (wheat, corn, other grains, cotton, rice, peanuts) and equal to a share of the national acreage that was to be planted to that crop. The secretary of agriculture would then determine the total acreage required to balance supply and demand at a politically acceptable price. The acreage individual farmers could plant would be determined by their share of the total acreage as set out in their allotment. This method of production control is based on restricting the amount of a productive input—land—that can be used in production. In contrast, marketing quotas limit the quantities farmers can sell. Farmers with marketing quotas could produce as much as they wanted, although the amount they could market was restricted. Any production above the amount specified in the marketing quota would have to be consumed on the farm. These quotas were allocated among farmers on the basis of historical marketings. The expected impact of these controls was a shift in the supply schedule, as shown in Figure 10.2, thereby raising the price received by producers.

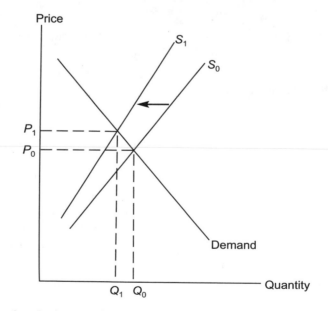

Figure 10.2: Agricultural production controls.

In addition to the production controls, the early farm programs introduced a policy tool that is still in use today, a *price support* known as the *nonrecourse loan*. At harvest, prices are likely to be severely depressed because farmers are all trying to market their crops at the same time. Later in the year, when supplies begin to be used up, prices begin to rise. It would clearly be advantageous for most farmers to put their crop into storage while waiting for prices to increase (assuming that storage costs are not greater than the price differential) rather than selling at the depressed prices that are likely to prevail around the time of the harvest. For many farmers, however, this option is not available. Most farmers in the United States borrow money to purchase the seed, fertilizer, pesticides, herbicides, diesel fuel, and other inputs they need to plant the crop. These loans often come due around the time of the harvest, forcing farmers to market their crops for whatever price they can obtain. Some crops are perishable and have to be marketed as soon as they are harvested. In cases such as these, large amounts of agricultural commodities will be put on the market over a fairly short period of time, with the inevitable result that prices will fall.

The nonrecourse loan program allows farmers to take out a loan from the Commodity Credit Corporation (CCC) using the crop as collateral. Farmers who participate in this program put their crop into storage at harvest rather than marketing it and use the CCC loan to pay off their debts. To be eligible to participate in the loan program, farmers had to comply with the restrictions established by their allotments. Farmers who did not wish to restrict their acreage could plant as much as they wanted, but they would be ineligible for CCC loans if they planted more than their allotment. Clearly this program can be used only for storable commodities such as corn or wheat. As prices rise over the course of the year, farmers can take the crop out of storage, sell it at the higher prices, and pay off the CCC loan. If market prices never rise sufficiently to offset the storage costs and the nominal interest charged by the CCC, the farmer can default on the loan, with the CCC taking possession of the crop. In the case of a default, the farmer does not pay the interest charges. CCC loans are "nonrecourse" in the sense that the government has no recourse to the court system in the case of a default on the loan. Thus, the fact that a farmer has defaulted on the loan is not counted against his credit rating and the government is not allowed to claim any compensation other than taking possession of the crop. The amount of the loan is determined by the amount the farmer wishes to store and the predetermined government loan rate. Loan rates are set as a price per bushel or other unit of output on the basis of rules included in the farm legislation. Thus, if the loan rate has been set at $2.00 per bushel for corn and a farmer wishes to put 150,000 bushels into the loan program, he would be eligible for a CCC loan of $300,000.

The possibility of defaulting on the loans without penalty means that the loan rate acts as a *price floor*. Assuming that all or most farmers are participating in the program, the lowest price they will receive is the loan rate. If market prices are at or below the loan rate (suppose that there is still enough corn left over from the previous harvest so that the price is low), farmers will keep their corn off the market, taking CCC loans and storing it instead. If prices never rise above $2.00, they simply default on the loan and take the $2.00 loan rate as their price. These relationships are illustrated in Figure 10.3, which shows the expected evolution of market prices during the course of the year. At harvest, market prices would fall. The loan rate serves as a price floor, however, so prices do not fall below the level of the loan rate. Later in the year, as stocks are used up, prices begin to rise and farmers begin to pay off the CCC loans and sell their harvest at the higher market prices. The effect of the loan program is to prevent the steep fall in prices that would normally occur at harvest.

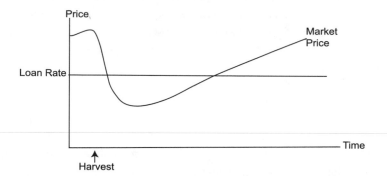

Figure 10.3: Market prices and the loan rate.

When the original farm programs were established, it was expected that the loan rate would be below the average market price for the year and that farmers would generally redeem their crops and sell them on the market. The purpose of the loan program was to smooth out the seasonal downturn in prices and to act as a safety net in years when prices remained depressed throughout the marketing year. The crops accumulated by the CCC during such years were available to be put on the market in years when drought or other natural disasters led to shortages and price increases. Such events were expected to be fairly unusual.

The anticipated relation between the loan rate and market prices is illustrated in the two-part Figure 10.4. In Figure 10.4a, the average market price (MP) for the year is established where supply equals demand. LR_1 represents a loan rate set below MP. At this level, the loan rate has no effect on the market.[9] However, if the loan rate is set at LR_2, the program will interfere with the market. Because the loan rate acts as a floor, the market price will not fall below LR_2. In Figure 10.4a, consumers will only buy Q^D at LR_2, while producers will wish to sell Q^S at that price. In this case, the CCC will accumulate stocks equal to $Q^S - Q^D$. Similar effects arise when an abundant harvest shifts the supply curve out, as shown in Figure 10.4b. In this case, the original loan rate (LR_1) ends up above the market-clearing price as a result of the supply shift from S_0 to S_1. As in Figure 10.4a, the CCC accumulates stocks as producers default on their loans.

The idea that the loan rate should be set at levels that would not normally interfere with the market raises an important question: When are times normal, so that the observed market price represents the correct benchmark below which the loan rate should be set? In fact, the decision on where to set the loan rate has always been controversial. Farmers may argue that the loan rate should be set at a level that will allow them to cover their costs or earn a fair return on their investment, for example. But costs of production vary dramatically across the United States, and it is not clear what should be included in calculating a fair return. The price needed to generate a fair return on investments that include an expensive, high-powered, air-conditioned tractor is higher than would be needed for similar investments with a more modest tractor. During the early years of the program, loan rates were often set in relation to *parity*, a measure comparing farm receipts to farm expenditures (see Box 10.2). Farm receipts depend on the prices received, while expenditures represent costs of production. A parity price, therefore, might be thought of as the price that would allow farmers to cover costs of production.

Box 10.2: The concept of parity

Agricultural price parity is based on the idea that the purchasing power of the products sold by farmers should remain constant. Stated differently, parity means that the amount of goods a farmer can buy with the proceeds from the sale of his or her products should be maintained at a constant level. Thus, if a bushel of wheat would earn enough money to allow the farmer to buy a shirt at some time in the past, it should still be the case today that a bushel of wheat will be worth a shirt.

To operationalize this concept, the USDA has kept track of the prices paid by farmers and the prices received for their commodities. Because farmers buy and sell many things, the prices have to be aggregated through an indexing procedure. This is the same type of procedure used to determine the consumer price index (CPI) used as an indicator of the cost of living. Price indices of this nature begin with a base-year set at 100. For the next year, a weighted average of the relevant prices is computed and compared with the base year average. Suppose that the average in the second year is 3 percent higher than in the previous year. Then, 3 percent of 100 is added to the base-year index so that it becomes 103 in Year 2. The parity index is based on an index of prices received by farmers divided by an index of prices paid. In the base year, both are at 100, so the parity index is 100/100 = 1 (usually the result is multiplied by 100, so the base year of the index is 100). Suppose that in the second year, the prices paid have increased 5 percent while the prices received have only increased 3 percent. The parity index would then be: (103/105) × 100 = 98.1.

Whenever the parity index is below 100, prices paid have increased more rapidly than prices received. In some farm legislation, the law was written to require that loan rates be set at "90 percent of parity," which means that they should be set at 90 percent of the price that would make the parity index equal to 100. One problem with this idea is that the parity index is highly sensitive to the base year chosen. In fact, the original parity index took the period 1910–1914, a period of very high farm prices known as the golden age of agriculture, as the base. In subsequent years, the ratio of prices received to prices paid declined dramatically, and loan rates set in relation to the parity index began to rise well above market clearing prices. Other problems with indices such as the parity index include the effects of changes in quality, productivity, and other factors over time. Such changes undermine the idea that a bushel of wheat should always "buy" a shirt.

It is interesting to note that Third World countries made an identical argument in the 1960s and 1970s. In this case, the index was referred to as the "terms-of-trade" index, computed as the ratio of an index of prices that developing countries received for their exports divided by an index of prices paid for imports. As with parity in U.S. agriculture, the developing countries based the index on a period of high prices (during the Korean War, when prices for commodities thought to be exported mainly by low-income countries were high). As part of their call for a New International Economic Order (NIEO), developing countries pointed to the deterioration of the terms of trade and called for commodity prices that would insure that a ton of cocoa would procure the same quantity of manufactured goods in 1975 as it would have during the 1950s.

In fact, the construction of the parity indices was such that setting loan rates at levels that would maintain parity over time would have meant that the loan rates would be above market equilibrium prices, so that the CCC would continually accumulate stocks. Consequently, farm legislation often specified that loan rates be set at some percentage of the price that would maintain parity between the indices of prices received and prices paid. During World War II and the Korean War, there was much less concern about agricultural prices, as wartime demand for agricultural commodities led to market prices that were well above the loan rates no matter how they were established. After these wars, however, agricultural output expanded rapidly as demand increases slowed and loan rates set in relation to parity began to interfere with markets for the main agricultural commodities. Throughout these years, marketing quotas and acreage allotments had been maintained, and farmers were required to comply with these restrictions to be eligible to participate in the loan program. During the 1950s, an additional program, the Soil Bank, was introduced in an effort to reduce output and raise prices. Farmers could enroll land in the Soil Bank by agreeing to take it out of production and put it into long-term conservation uses. In return, they received payments from the government to compensate for the income lost from the idled land.

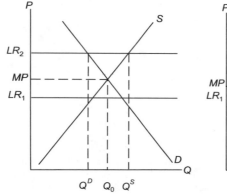

Figure 10.4a: The relation of loan rates to market equilibrium.

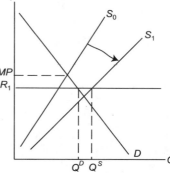

Figure 10.4b: Loan rates in the presence of a supply shift.

Despite reductions in the amounts of land planted through the allotment and Soil Bank programs (the Soil Bank was phased out during subsequent years, although a similar program, the Conservation Reserve Program or CRP, was reintroduced in the 1980s), output continued to increase and the CCC continued to accumulate stocks as loan rates were set at congressionally mandated levels that were above market clearing prices. In 1954, another type of policy tool was introduced. Public Law 480 (P.L. 480) represented an attempt to increase demand by providing subsidized food aid to low-income, developing countries. The intent of P.L. 480, as well as other programs aimed at promoting the sale of U.S. commodities overseas or providing export credits for foreign sales, is to shift the demand schedule to the right, as illustrated in Figure 10.5. This shift in the demand schedule reduces the stocks accumulated by the CCC at the loan rate (LR in Figure 10.5) from $(Q_S - Q_D^0)$ to $(Q_S - Q_D^1)$. If demand is shifted far enough, it may even raise the market price above the loan rate so that the CCC would cease accumulating stocks. Of course, P.L. 480 also has humanitarian objectives. Subsidized food exports directed at low-income countries benefit consumers in those countries, although such sales depress prices and act as a disincentive for farmers in developing countries to modernize and expand production. Part of P.L 480 is directed at disaster and famine relief, and these programs are more clearly humanitarian than the other parts of P.L. 480, which, arguably, could be characterized primarily as a surplus disposal program that benefits U.S. producers by reducing market-depressing stocks.

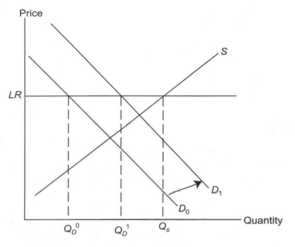

Figure 10.5: The loan rate in the presence of a demand shift.

Surplus disposal and demand enhancement both have a long history in agriculture. Programs to store commodities for later release onto the market in times of shortage have been a common response to surplus production. Another approach is to divert surplus commodities from their primary market to an alternative outlet. The European Union (EU) has historically supported farm prices at very high levels, with the result that the intervention agencies in the member countries have accumulated large surpluses of beef, wine, butter, and grain. To reduce these surpluses, the EU has implemented policies to channel the surpluses into alternative markets. For example, surplus wheat is denatured to prevent it from being used as bread wheat, diverting it into the livestock feed industry. Surplus wine was distilled into ethyl alcohol.

Both the EU and the United States have used *food aid* and other export subsidies to divert surplus commodities from the domestic to foreign markets. The U.S. government subsidizes market promotion programs in foreign countries and uses revenues from the P.L. 480 program to introduce foreign buyers to U.S. exporters. A specific objective of P.L. 480 is to increase the likelihood that importers in developing countries will turn to U.S. suppliers as the countries begin to enter commercial markets. This is precisely what has occurred in countries such as Taiwan and Korea. All of these efforts can be thought of as ways to increase demand and dispose of surpluses generated through the effects of government price support programs.

By the 1970s, loan rates in the United States had become increasingly inconsistent with market conditions. U.S. farmers were not able to take advantage of expanding foreign demand because the loan rate was raising prices above world market levels. In 1972, a new policy was implemented to help with this problem. The concept underlying this new program is a well-known theory in macroeconomic policy, usually attributed to the 1969

Nobel laureate in economics, Jan Tinbergen, which shows that the number of policy tools needs to be equal to the number of policy objectives if they are to be effective. Loan rates were being called upon to accomplish two objectives: to support prices and to increase farm incomes.[10] The objective of increasing farm incomes meant that there was pressure to raise loan rates, and this undermined their effectiveness as price supports. The policy added in 1972 was to support incomes with the expectation that loan rates could be reduced so that they would return to their original role of smoothing out seasonal price variation and acting as a safety net in years of unusual circumstances. The new policy tool was to offer farmers direct payments as income supplements. In this way, prices would not have to be raised to allow farmers to realize incomes that were comparable to incomes in other sectors. These *income supports* were known as *deficiency payments* and were based on the difference between a politically determined target price and market prices.

The deficiency payment program is illustrated in Figure 10.6a. The target price (TP) is set at a level that is expected to cover the costs of production and provide a return to the farm family labor. At TP, farmers will deliver Q_0 for sale on the market. For consumers to purchase this quantity, the market price will have to settle at MP. MP is the market-clearing price given the quantity, Q_0, offered by producers. The deficiency-payment rate is the difference between TP and MP. In effect, farmers receive MP from consumers and ($TP - MP$) from the government, for a total price of TP. The total cost to the government is the deficiency payment rate ($TP - MP$) multiplied by the quantity marketed, Q_0. Note that the loan rate (LR) has been set below the market-clearing price so that it can serve as a price support that is normally not expected to interfere with the market.

An important aspect of this program is that it represents a subsidy to both consumers and producers. The consumer price, MP, is below the free-market price, PE, while producers receive a price that is higher than the equilibruim price. Consumer surplus is increased by the areas marked D and E, while producer surplus is greater by the areas marked A and B. The cost of the program is greater than these surplus gains, however. The government pays ($TP - MP$) multiplied by the total quantity, Q_0—that is, the areas marked A, B, C, D, and E. Area C is an efficiency loss generated by the deficiency payments. In Figure 10.6b, the supply schedule has shifted from S to S_1—due, perhaps, to particularly good weather, for example—with the result that the market-clearing price is below LR. In these circumstances, the CCC will wind up with stocks equal to $Q_1 - Q_2$.

The actual functioning of the deficiency payment program was quite complex. Farmers were required to report the acreage they had planted to various crops and their average yields to the local Agricultural Stabilization and Conservation Service (ASCS) office, generally located in the same county as the farm. These records were used to establish each farmer's base acreage and program yield. The deficiency payment the farmer received was calculated as his or her base acreage multiplied by his or her program yield multiplied by the deficiency-payment rate. The deficiency-payment rate was set as the difference between the preannounced target price and an average market price as determined by the USDA. A farmer's actual yield, acreage, and price received could all differ from the variables used in these calculations.

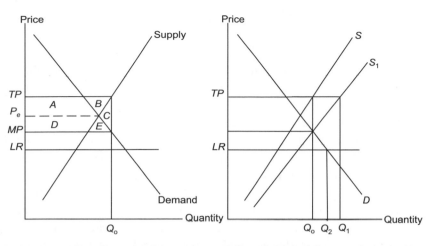

Figure 10.6a: The welfare effects of deficiency payments.

Figure 10.6b: Deficiency payments with supply increase.

During the 1970s, there were other changes in the basic farm program. Allotments were replaced with programs requiring that some portion of a farmer's land be set-aside and, later, that a percentage of the farmer's base acreage be left uncultivated. Under the Acreage Reduction Program (ARP), farmers could be required to put some of their base acreage for corn, wheat, cotton, and other program crops into an approved conservation use. As with the earlier production control programs, eligibility to receive loan support and deficiency payments was contingent on compliance with the requirements (if any) of the ARP.

During the last two decades of the twentieth century, this program continued to evolve, becoming increasingly complex. In periods of extremely low prices, special programs to increase farm income were added. During periods when world demand was low, explicit export subsidies became more common. In times of favorable prices or when the CCC began to accumulate stocks, the programs were made less generous.[11] The policy objectives expanded beyond the traditional price and income supports to include food safety issues, environmental concerns, and the maintenance of U.S. export volumes. The result was a program of great complexity. Some program modifications made participation in the basic program less attractive. Many of these changes were driven by beliefs that policies should be *market-oriented*, that is, that farmers should rely more on markets and less on the government.

In 1983, 65.5 percent of net farm income came from direct government payments to farmers. Between 1983 and 1995, almost a third of net farm income, on average, was the result of these payments.[12] In addition, some agricultural prices (e.g., dairy and peanuts) were supported at levels that transferred substantial amounts of money from consumers to producers. Farmers were able to use the intricacies of the farm programs to extract extra benefits. For example, deficiency-payment rates of $1.00 per bushel were not uncommon for wheat during the 1980s. At that deficiency-payment rate, a farmer with 5,000 acres of wheat and average yields of 45 bushels per acre might be eligible for a direct payment of roughly $225,000. To avoid such extravagant payments, the government capped direct transfers at a maximum of $50,000 per farm. Farmers responded by dividing their holdings into separate enterprises. In the preceding example, the 5,000-acre farm might be divided into five 1,000-acre farms incorporated under the names of the farmer's wife, children, and cousins. Each of these entities would be eligible for $45,000, assuming the same yields, and the family would receive the full $225,000 rather than being limited to the $50,000 maximum payment.

Developments of this nature gave rise to the charge that farmers were "farming the farm programs" and added support to efforts to replace these programs with a more market-oriented approach. Support for government intervention in agriculture has always been based on the perception that farmers were less well off than others in society and the belief that farming is a particularly virtuous way of life in need of protection and nurturing for the general benefit of society. Costly government programs gave rise to the perception that farmers are just like any other rent-seeking, special-interest group feeding at the public trough. Moreover, as noted earlier, the old farm problem of low incomes seems to have disappeared. About 75 percent of the 2.2 million farms in the United States today are best viewed as part-time or hobby farms, in which most household income is derived from nonfarm activities. The remaining 25 percent are full-time commercial farms that earn almost all their income from agriculture and produce the majority of the food in the United States. In fact, the largest 8 percent of U.S. farms produce two-thirds of agricultural output. The largest farms are generally fairly prosperous, and the hobby farms are not dependent on agricultural prices for their income. While there are many commercial farms that are vulnerable to market fluctuations, one of the primary justifications for government intervention in agriculture seems to have disappeared as farm incomes have caught up with the levels observed in other economic sectors.

The culmination of the drive toward more market-oriented policies was the Federal Agriculture Improvement and Reform Act (FAIR) of 1996, also known as "Freedom to Farm." FAIR eliminated the acreage reduction and the deficiency-payment programs along with all the paraphernalia that had gone along with them (e.g., acreage bases, program yields). The loan rate was retained but was to be set at levels expected to be below market-clearing prices. The rationale for this policy change was that farmers should be free to farm as they wish without government constraints, obtaining their income from the market rather than government programs. An example of a problem the FAIR act aimed to correct concerns the range of crops planted. Because farmers wanted to maintain their historic acreage bases in the main program crops, they were reluctant to experiment with alternative crops. Acreage bases were the average amount planted to given program crops over the past, with adjustments for ARP requirements. If a farmer elected to plant a nonprogram crop (e.g., popcorn) one year, this average would be lowered. To maintain base, farmers tended to plant only program crops. With the changes initiated by the FAIR act, maintenance of acreage base was no longer necessary, and farmers became "free to farm" whatever they wished.

To ease the transition to greater reliance on the market, the government offered farmers transition payments, which were designed to decrease over time and disappear in 2002. Initially, FAIR was reasonably well accepted by farmers; market prices were unusually good, and generous transition payments added a substantial windfall gain. By 1998, however, increased output in response to the favorable prices had begun to come onto the market, and prices began to fall. Drought and flood disasters that caused great distress among some farmers but did not reduce overall production also served as an incentive for the government to seek ways to increase transfers to farmers. In 1999, $22.7 billion was transferred to farmers as direct payments to compensate for low prices and lost crops.[13] This sum amounts to more than $10,000 for every farm in the United States. Of course, almost all the payments went to the largest farms rather than the small part-time and hobby farms. Assuming that these payments were distributed exclusively among the top 25 percent, the average transfer to each farm would have been about $41,300.

In the preceding discussion, an important consideration has been left out of the model of U.S. farm programs. About 40 percent of the total revenue in agriculture is generated through export sales. About half of all U.S. wheat production is exported, and the United States is a major supplier of many agricultural products on world markets. Agricultural trade and trade policy have an important influence on the policy options the government faces. As noted earlier, if loan rates are set too high, the United States runs the risk of losing substantial export sales. A full treatment of international trade is beyond the scope of this chapter, but it is important that the model of a typical U.S. agricultural market that is being developed take account of this important effect. Appendix 10.1 presents the derivation of excess supply and excess demand curves on the world market and uses the model developed there to examine the U.S. Export Enhancement Program (EEP). National differences give rise to comparative advantages that show up in differences in equilibrium prices when countries do not trade, a state referred to as autarky. It is the intersection of excess supply and excess demands on the world market that determines the world price (see Appendix 10.1).

For our purposes, a simple graphical model can be developed that incorporates foreign demand. In Figure 10.7, the excess demand derived from the aggregated supplies and demands of other countries is simply added to domestic U.S. demand to represent the total demand facing U.S. producers. In the absence of trade, demand would be defined by the segment a-b, with the equilibrium price/quantity at b. Foreign demand derived from the excess demand schedule of other countries (see Appendix 10.1) is added from point c on the domestic demand schedule. Total demand is thus represented by a-c-d, with an equilibrium (world) price of P_w. At P_w, U.S. consumers will purchase Q_{US}, and the quantity $(Q_T - Q_{US})$ will be exported. Total demand is obviously Q_T, where domestic supply intersects total demand.

With the apparatus set out in Figure 10.7, it is possible to carry out a simple graphical analysis of FAIR. In this case, the counterfactual is the basic farm program just prior to the adoption of FAIR. The main elements of that program included the loan program, acreage reduction, and deficiency payments. These program elements are shown in Figure 10.8. The ARP shifts supply in from S_0 to S_1. Total demand (D_T) is derived by adding foreign demand to U.S. demand (D_{US}). The target price (TP) is set above the equilibrium where S_1 equals D_T. At TP, producers supply the quantity Q_{DP} to the market. The market will clear at price MP, the price at which consumers are willing to purchase the quantity Q_{DP}. Note that at MP, U.S. consumers will purchase Q_{DPUS}, with the difference between that quantity and Q_{DP} being exported. The loan rate has been set at a level that does not interfere with any of these relationships. The government incurs a cost equal to $(TP - MP) \times Q_{DP}$, or the areas marked *a*, *b*, *c*, *d*, and *e*. This is a simplified representation of the policy setting prior to the implementation of FAIR.

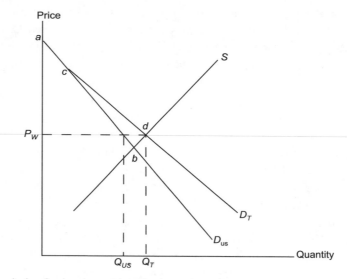

Figure 10.7: Introducing foreign demand into the domestic market.

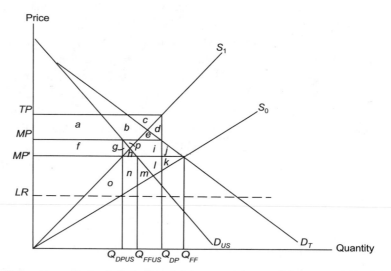

Figure 10.8: The welfare effects of eliminating deficiency payments and the acreage reduction program (FAIR).

FAIR eliminated deficiency payments and the ARP. The transition payments were not tied to production in any way, so the expectation was that they would not affect decisions on the quantities that are supplied. We continue to assume that the loan rate is set at a level that does not affect the market. Thus, the impact of FAIR was to give rise to a new equilibrium at the intersection of D_T and S_0, the supply schedule that would prevail in the absence of the ARP. At this equilibrium, the market price is MP' (for both consumers and producers), and the quantity exchanged is Q_{FF}. U.S consumers will purchase Q_{FFUS}, and the difference between that quantity and Q_{FF} is exported.

To analyze this policy change, we can begin by examining the changes in producer and consumer surplus and the cost to the government. Recall that producer surplus (PS) is the area below the relevant price and above the relevant supply schedule, which were TP and S_1 prior to FAIR and MP' and S_0 after. Likewise, consumer surplus (CS) was given by the area below total demand and above MP prior to FAIR, which changes to the area below D_T and above MP' under FAIR. Finally, the government (GOV) no longer has to support the cost of the deficiency payment. Using the labels in Figure 10.8, these relationships can be represented as:

$$\Delta PS = -a - b - c - p - g - f + o + n + m + l + k$$
$$\Delta CS = f + g + p + h + i + j$$
$$\Delta GOV = a + b + c + d + e$$

When these areas are added up, the net effect of the change is positive ($o + n + m + l + k + j + i + h + d + e$), the area between S_0 and S_1 plus the deadweight loss from the deficiency payment program, area d.

Producers have lost some of the surplus from the earlier program but have gained the surplus associated with the output that was forgone under the ARP. Consumers benefit from the lower price, and the government saves the cost of the deficiency payments. Note, however, that this graph does not include the transition payments. These payments are supposed to be phased out, so the analysis more accurately reflects the effects of FAIR once there are no more transition payments. In the early years of the program, the costs to the government were about equal to what had previously been spent for deficiency payments offsetting the savings shown in Figure 10.8.

It appears that FAIR, at least after the transition payments cease, would be socially beneficial. The net effect on producers depends on whether area ($o + n + m + l + k$) is bigger or smaller than the loss of ($a + b + c + p + f + g$). Note that this loss is a transfer from producers to consumers and taxpayers. It would be possible to compensate producers by transferring ($a + b + c$) from the government savings and ($p + f + g$) from the consumer savings back to farmers. Consumers would still gain ($j + i + h$) while taxpayers would gain ($d + e$). Of course, compensation for the loss is overly generous, because most of the producer loss is already offset by the gain of area ($o + n + m + l + k$).

These general results depend on several conditions. First, the slopes of the supply and demand relationships are simply assumed in Figure 10.8, and the empirical reality would certainly be quite different. It would be possible to redraw this figure to make the results appear somewhat different. Imagine, for example, what it might look like if a much higher target price had been assumed. Second, there are several assumptions implicit in this representation that might be questionable. For example, how will farmers actually respond to the elimination of the ARP? They may not return corn or wheat acreage taken out of production under the ARP to corn or wheat production. They may elect to produce something else, or they may decide to plant even more corn or wheat than their original acreage bases. Thus, the elimination of the ARP may not result in a simple movement from S_1 back to S_0. In fact, ARP requirements just prior to the adoption of FAIR were not in force because prices were high and there seemed no need to reduce acreage. Under these circumstances, it is difficult to predict how farmers will respond to the changes.

In addition, the policy change depicted in Figure 10.8 covers only a part of the range of policy measures that were in place before and after the adoption of FAIR. For example, most producers were not receiving the full deficiency payment prior to FAIR, and other conditions had made the program less attractive. As a result, many farmers had elected not to participate. These elements are not taken into account in Figure 10.8. Programs such as the CRP, crop insurance, subsidized loans, and many others were in use both before and after the change. A precise assessment of the likely impact of FAIR would require a much more complex model to capture the overall impacts of these different and sometimes contradictory influences on producer and consumer behavior. Nevertheless, the graphical analysis can help identify some of the fundamental relationships and suggest how the farm sector might be affected by the new policy. Depending on the purposes for which a policy analysis is being done, this may be enough to draw useful conclusions about the potential benefits and costs of a change, or it may serve as a useful first step in specifying a more complete and accurate model that can be used to obtain more precise estimates of the impacts.

The policies described in the graphical exercise apply only to grains and cotton. There are many other types of programs that have been used to regulate agriculture in the United States. The U.S. peanut program is described in Appendix 10.2 as an example of another policy that has been used to support producer income. In Appendix 10.3, the main elements in the history of the EU's *Common Agricultural Policy* (CAP) are described as another example of the kinds of programs often employed in industrialized countries to support farmers. In the next section, we will extend the graphical analyses through the use of some simple models based on elasticities. More complex simulation models will be presented in the two case studies that follow this chapter.

SOME SIMPLE ANALYTICAL MODELS

In the preceding discussion, graphical representations were used to determine the overall direction of changes that might be expected to occur as a result of a policy change. While this type of analysis can be quite useful, it does not provide information on the magnitude of the changes. In evaluating a policy change such as the imposition of a tax on gasoline or the elimination of deficiency payments and acreage restrictions, estimates of the nature and size of the likely costs and benefits are needed, and deriving such estimates requires more than just the likely direction of the changes. As noted earlier, a common approach to this type of problem is to build a mathematical model of the relationships and use the model to simulate the likely effects of the policy change relative to an appropriate counterfactual.

Consider, for example, an analysis of the impact of the North American Free Trade Agreement (NAFTA) on Mexican corn imports from the United States. Imports are roughly equal to the difference between domestic supply and demand. A simple model of Mexican supply and demand could be estimated using standard statistical methods and used to predict the impact of the free trade agreement on imports. Economic theory would suggest that demand for corn in Mexico depends on its price and the income of Mexican consumers. Mexican supply depends on the producer price for corn, weather, technological changes that affect yields, and the prices of substitute crops. Using historical data, demand and supply equations could be estimated. These equations could then be used to predict the future evolution of Mexican supply and demand if prices continue to evolve as in the past and these results contrasted with the likely future supply and demand if prices fall to U.S. levels, as would be expected to happen under NAFTA. Imports in the two cases would simply be the difference between the projected supply and demand.

This type of analytical procedure is known as *simulation*. The basic idea is to build a model that captures the important relationships and that can then be used to predict what will happen under alternative policy scenarios. In essence, the model is used to perform experiments that compare the effects of the policy change being evaluated with the likely evolution of the economic variables if the policy change is not implemented. In the case of our NAFTA analysis, the problem is to predict what is likely to happen if NAFTA is adopted and compare those results with the likely outcome if NAFTA is not adopted. The difference between these two sets of predictions can be taken to represent the impact of NAFTA on Mexican corn imports. Note that the comparison in this case involves a counterfactual that is based on the model predictions, not on some observed state of affairs. For example, it is inappropriate to note that U.S. corn exports to Mexico in 2000 were 20 percent higher than they were in 1994 and attribute this change to NAFTA. Many things besides NAFTA changed between 1994 and 2000. Mexican corn imports might have increased even if NAFTA had not been implemented. The model aims to answer two questions: (1) What would have happened if NAFTA had not been put in place? and (2) What did happen given that NAFTA was adopted? The answer to the first question is the appropriate counterfactual in assessing the impact of NAFTA. As noted frequently, such consequentialist results are only one input into the decision-making process in which nonconsequentialist considerations related to rights, justice, politics, and so on often override the simple enumeration of benefits and costs.

In Chapters 11 and 12, the use of models and simulation are illustrated with case studies of partial equilibrium policy analysis. In this chapter, we develop a fairly simple approach to the analysis of policies that affect market variables. *Log-linear models* are based on the use of elasticities and basic supply and demand relationships.[14] Throughout this book, supply and demand have been represented as functions of prices. The graphs all show price on the vertical axis and quantities on the horizontal axis, however, suggesting that prices are functions of the quantities demanded and supplied. In fact, of course, either representation can be used. In this example, supply and demand will be represented as inverse functions, with price as the dependent variable.

The following model is based on Figure 10.1.

Let: P^S = the price received by suppliers
         P^D = the price paid by consumers
         Q = equilibrium quantity = equilibrium quantity supplied
         (Q^S) = equilibrium quantity demanded (Q^D)
         t = tax (subsidy if negative) = $(P^D - P^S)/P^S$
and $D^{-1}(Q)$, $S^{-1}(Q)$ = the inverse demand and supply functions, respectively.

The equilibrium equations are:

(10.1)   Supply equation: $P^S = S^{-1}(Q)$
(10.2)   Demand equation: $P^D = D^{-1}(Q)$
(10.3)   Price wedge: $P^D/P^S = 1 + t$

Equation 10.3 sets out the relationship between producer and consumer prices in the presence of a tax. Recall from Figure 10.1 that in the case of a gasoline tax, the price consumers pay is higher than the price producers receive; the consumer price is not equal to the producer price. The tax rate, t, is defined as the percentage difference between these two prices. This definition can be reduced to the expression in equation (10.3). Note that a positive value for t represents a tax, while negative values are used for subsidies, and that if $t = 0$, then $P^D = P^S$.

To analyze the impact of a tax, this system can be expressed in logarithms and differentiated. See Box 10.3 for the rules of *logarithmic differentiation* used in the following discussion. If equations (10.1) and (10.2) were written as direct rather than inverse relationships, log differentiation would lead to: $dlnQ^S = \alpha dlnP$, where Q^S is the quantity supplied and α is the price elasticity of supply; and $dlnQ^D = \eta dlnP$, where Q^D is the quantity demanded and η is the price elasticity of demand. Because the equations are written in the inverse, the relationships are:

(10.4) $dlnP^S = 1/\alpha \, dlnQ$
(10.5) $dlnP^D = 1/\eta \, dlnQ$
(10.6) $dlnP^D - dlnP^S = dln\,(1 + t) \approx dt$

This system of equations can be solved as follows. Multiply equation (10.4) by α to obtain $\alpha dlnP^S = dlnQ$. Do the same for equation (10.5) to obtain $\eta dlnP^D = dlnQ$. Since both of these expressions are equal to $dlnQ$, they must be equal to each other: $\alpha dlnP^S = \eta dlnP^D$. From equation (10.6), we know that $dlnP^D - dt = dlnP^S$, so we can substitute $(dlnP^D - dt)$ for $dlnP^S$ in the preceding expression to obtain:

(10.7) $\alpha(dlnP^D - dt) = \eta dlnP^D ==> \alpha dlnP^D - \eta dlnP^D = \alpha dt$
                   $==> (\alpha - \eta)dlnP^D = \alpha dt$, or $dlnP^D = \alpha/(\alpha - \eta)dt$

The same procedure can be followed to obtain an expression for $dlnP^S$:

(10.8) $dlnP^D = dlnP^S + dt ==> \alpha dlnP^S = \eta(dlnP^S + dt) ==>$
$\alpha dlnP^S - \eta dlnP^S = \eta dt ==> (\alpha - \eta)dlnP^S = \eta dt$, or
$dlnP^S = \eta/(\alpha - \eta)dt$

From equation (10.4), we know that $dlnQ = \alpha dlnP^S$. But we also have $dlnP^S = \eta/(\alpha - \eta)dt$, and the right-hand side of this expression can be substituted for $dlnP^S$ in the expression for $dlnQ$ to obtain:

(10.9) $dlnQ = \alpha\eta/(\alpha - \eta)dt$

Box 10.3: Rules of logarithmic differentiation

Natural logarithms have properties that are particularly convenient for the type of analysis being described in this chapter. Recall the basic rules of logarithmic manipulation:

$ln(xy) = ln(x) + ln(y)$ and $ln(x/y) = ln(x) - ln(y)$

Logarithmic differentiation $[dln(x)]$ of relationships uses these rules to obtain the following:

if $z = xy$, then $dln(z) = dln(x) + dln(y)$
if $z = x/y$, then $dln(z) = dln(x) - dln(y)$

Two other rules related to logarithmic differentiation are:

if $z = x + y$, then $dln(z) = [x/(x + y)]dln(x) + [y/(x + y)]dln(y)$
and
$dln(z) = (1/z)dz$

This final relationship is based on the rule for derivatives of logarithmic functions. It shows that the expression $dln(z)$ is equivalent to the expression for the percentage change in z. The percentage change in a variable is given by $(z_1 - z_0)/z_0$, or $\Delta z/z$, which is approximately equal to dz/z. Thus, $dln(z)$ represents the percentage change in the variable z. Recall that an elasticity was defined as the percentage change in one variable brought about by a 1 percent change in some other variable: $(\Delta z/z)/(\Delta y/y)$ or $dln(z)/dln(y)$.

Now suppose that we represent a demand equation as $Q = f(P)$. Totally differentiate this expression to obtain: $dQ = (dQ/dP)dP$. Now divide both sides of the equation by QP:

$dQ(1/QP) = (dQ/dP)dP(1/QP)$.
Write $(dQ/dP)dP(1/QP)$ as $[(dQ/dP)/Q](dP/P)$ to obtain
$dQ(1/QP) = [(dQ/dP)/Q](dP/P) = [(dQ/dP)/Q]dlnP$

Now, multiply both sides of this expression by P:

$dQ/Q = dlnQ = (dQ/dP)(P/Q)dlnP$.

$(dQ/dP)(P/Q)$ can be rewritten as $(dQ/Q)/(dP/P)$, which is the expression for an elasticity.

This is the result that is of particular use in analyzing market displacements. Let η represent the price elasticity of demand. Then the demand equation $Q = f(P)$ can be log differentiated to obtain: $dlnQ = \eta(dlnP)$.

Equations (10.7) through (10.9) show the percentage changes in prices and quantities as functions of dt (the change in the tax) and the elasticities of supply and demand. Suppose that we wish to analyze the impact of a gasoline tax as illustrated in Figure 10.1. It has been suggested that an increase in gasoline taxes of 50 percent might be required to comply with the Kyoto Protocol on global warming. In terms of the simple model presented in this section, such a tax would be represented by a value for dt of 0.5. To calculate the impact of this increase on consumer and producer prices and the equilibrium quantity, we need to have estimates of the elasticities, α and η. The values for these parameters could be obtained by estimating a statistical model of the gasoline market. It might also be possible to use estimates from other studies of petroleum markets. For purposes of exposition, I will simply assume some reasonable values for these elasticities.

In the short term, consumers and producers may be unable to make large adjustments in their patterns of consumption or production. If this is true, then their response to the gasoline tax may be modest. Thus, appropriate short-run elasticities are likely to be fairly small in absolute values. Let us assume that the price elasticity of supply (α) is +0.4, while the price elasticity of demand (η) is -0.3. Elasticities of this magnitude indicate that a price change of 10 percent will lead to changes in consumption or production of only 3 or 4 percent. If these values are inserted into the equations derived earlier, we obtain the following results:

$$dlnP^S = \eta/(\alpha - \eta)dt = [-0.3/(0.4 + 0.3)](0.5) = -0.214$$
$$dlnP^D = \alpha/(\alpha - \eta)dt = [0.4/(0.4 + 0.3)](0.5) = +0.286$$
$$dlnQ = \alpha\eta/(\alpha - \eta)dt = [-0.3(0.4)/(0.4 + 0.3)](0.5) = -0.086$$

The 50 percent tax is split between producers and consumers, with a decrease in producer prices of 21.4 percent and an increase in consumer prices of 28.6 percent. The quantity exchanged falls by 8.6 percent. In Figure 10.1, the initial equilibrium was found at a price of $1.20 per gallon, with 228 million gallons exchanged. The tax would lead to consumer prices of $1.54 ($1.20 × 1.286), producer prices of $0.94 ($1.20 × 0.786), and a quantity exchanged of 208 million gallons (228 × 0.914). Note that the difference between consumer and producer prices is $0.60, which is precisely 50 percent of the original price of $1.20. This result follows naturally from the value for dt of 0.5.

These results can be used to calculate changes in consumer and producer surplus and government revenue as indicated in Figure 10.1. Areas A and D are rectangles, which can be computed as the vertical distance on the price axis multiplied by the horizontal distance (208) on the quantity axis. Areas B and D are triangles, the value of which can be calculated using the formula for the area of a triangle (one half base multiplied by height). The change in consumer surplus is [(1.54 − 1.20) × 208] + [(1.54 − 1.20) × (228 − 208) × 0.5], which equals $74.12 million. Using the same procedures, the change in producer surplus is $56.68 million. The government earns $0.60 on each gallon sold (208 million), so government revenue is $124.8 million, which is $6 million less than the combined losses of producer and consumer surplus (74.12 + 56.68 = 130.8). Thus, the gasoline tax generates a cost to society of $6 million. The benefits of the tax would include the savings realized due to reduced global warming as well as any other positive effects that might be associated with reducing the negative externalities that go along with gasoline use.

In the long run, consumers and producers will be able to make greater adjustments to the price changes. Consumers will find it advantageous to replace their older vehicles with more energy-efficient automobiles, as opposed to purchasing gas-guzzling SUVs. Producers may close down unprofitable wells as the lower prices persist. It is therefore likely that in the long run, the responses will be more substantial, and this implies elasticities with higher absolute values. Suppose that the long-run elasticities are -1.2 on the demand side and +0.8 for supplies. With these values, the 50 percent tax increase will lead to a fall in long-run producer prices of 30 percent, an increase in long-run consumer prices of 20 percent, and a long-run decline in the quantity exchanged of 24 percent. This would imply producer prices of $0.84, consumer prices of $1.44, and a total quantity exchanged of 173 million gallons. Government revenues would be $103.8 million, set against producer losses of $72.2 million and consumer losses of $48.1 million. The efficiency loss from the tax is $16.5 million, compared to the short-run estimate of $6 million. Government revenues are lower in the long run, as the quantity to which the tax is applied has fallen quite a lot as a result of the tax. Of course, the long-run benefits are also greater because of the lower consumption.

A rough idea of the environmental benefits of the gasoline tax can be obtained by making an adjustment to the supply schedule in Figure 10.1 to reflect the presence of a negative externality. Recall that supply schedules are derived from the horizontal summation of individual firms' marginal cost curves. Adding the costs associated with the negative externality to the private costs reflected in the private supply schedule, S_p, will lead to a new supply schedule, S_S, that includes both the private and the social costs of gasoline use. In Figure 10.9, S_S is located above S_p, and the area between these two schedules up to the point established by the equilibrium quantity, Q_e, is the total social cost of the externality ($b + c + d + e + f + g$). Because the externality is not internalized by producers and consumers, the market equilibrium is established at P_e and Q_e. The sum of producer and consumer surplus at this equilibrium is ($a + b + c + d + e + g + j + h + k$). Subtracting the social cost from this area leaves ($a + j + h + k - f$) as the net social welfare in the presence of the externality.

Suppose that a gasoline tax is introduced that is just sufficient to raise private costs to the level of the social costs. This is represented in Figure 10.9 by the consumer price, P^D, set at the intersection of S_S with the demand schedule. The tax forces the producers and consumers to internalize the externality, shifting the equilibrium from that established at the intersection of D and S_p to the intersection of D and S_S. This eliminates the social cost, leaving total social welfare equal to ($a + j + h + k$). Note that welfare has increased by the amount of area f. By inspection, it appears that the net gain resulting from the tax, area f, is about equal to the two triangles, d and e. If the two supply schedules were parallel and linear, f would be exactly equal to ($d + e$). This suggests that the net benefit of the tax, based on the assumptions used previously, is around $16.5 million. Although the exact magnitudes of the other areas shown in Figure 10.9 cannot be estimated with the model described earlier, we can conclude with some confidence that the tax will lead to an increase in social welfare as the negative externality is eliminated.

The method used for this exercise can be extended to provide a preliminary analysis of the impact of the FAIR act. As in the graphical analysis sketched previously, we wish to examine the simultaneous effects of the ARP and deficiency-payment programs, and this will require some modifications in the basic model set out in equations (10.4) through (10.6). Let $d\beta$ represent elements that shift the supply curve vertically and $d\gamma$ elements that generate vertical shifts in the demand schedule. Add these terms to the original model:

(10.10) $dlnP^S = 1/\alpha\ dlnQ + d\beta$
(10.11) $dlnP^D = 1/\eta\ dlnQ + d\gamma$
(10.12) $dlnP^D - dlnP^S = dt$

This system can be solved using the same procedures as in the simpler example to obtain:

(10.13) $dlnP^S = \eta/(\alpha - \eta)dt + \alpha/(\alpha - \eta)d\beta - \eta/(\alpha - \eta)d\gamma$
(10.14) $dlnP^D = \alpha/(\alpha - \eta)dt + \alpha/(\alpha - \eta)d\beta - \eta/(\alpha - \eta)d\gamma$
(10.15) $dlnQ = \alpha\eta/(\alpha - \eta)dt + \alpha\eta/(\alpha - \eta)d\beta - \alpha\eta/(\alpha - \eta)d\gamma$

In this model, changes in producer and consumer prices and the quantity exchanged are functions of elasticities, changes in the tax or subsidy, and changes in the supply and demand shifters.

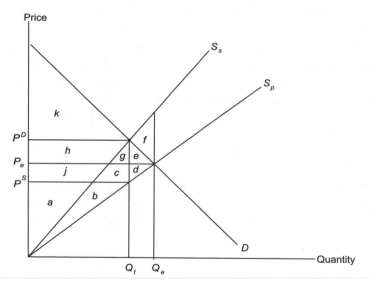

Figure 10.9: The welfare effects of a negative externality.

To analyze the impact of FAIR, we can measure the effects of the ARP and deficiency-payment programs relative to the market equilibrium without these programs. In the following exercise, we will actually be analyzing the impact of the ARP and deficiency-payment programs on the free-market equilibrium rather than their removal. These results will still provide information on the impact of FAIR, however. We simply have to reverse the signs so that the producer price increase due to deficiency payments becomes a reduction due to their removal. Recall that the deficiency payment is actually a subsidy, so that dt is negative rather than positive, as in the case of a tax. The ARP can be represented as a supply shifter, $d\beta$. Suppose that the effect of the ARP is to shift the supply schedule up by 15 percent and that the deficiency payment amounts to a subsidy of 20 percent. Then, $d\beta = 0.15$ and $dt = -0.20$. Assume that $\alpha = 0.6$, $\eta = -0.5$, and that there are no demand shifters for this analysis ($d\gamma = 0$). Using these values in equations (10.13) through (10.15) results in predictions that producer prices will increase 17.3 percent, consumer prices will decrease by 2.7 percent, and the quantity exchanged will increase 1.4 percent. All of these changes are relative to the equilibrium prices and quantities.

Suppose that these results are obtained in analyzing the impact of FAIR on the U.S. corn market. If we know that prior to adoption of this new legislation, producers were receiving $2.35 per bushel and that the market-clearing price was about $1.90, then the removal of the deficiency payment and 15 percent ARP would lead to an equilibrium price of $1.94 per bushel. In 1995, the quantity of corn produced and consumed was about 9.0 billion bushels, which would increase to about 9.13 billion bushels (9.0×1.014) following the policy change.[15] Results such as these could be used to measure the changes in economic surplus and government costs. Most of the values for the parameters used in the calculations have simply been assumed, however, and the results should be seen as illustrative rather than exact.

The limitations of the approach outlined in this section can be clarified by considering some of the events that affected corn markets around the time FAIR was adopted. Just prior to the act, worldwide shortages and rapidly growing demand in several Asian countries had resulted in extremely high prices. As a result, ARP limitations had been eliminated. As noted earlier, a wide range of other programs limited the amount of acreage planted to corn and other field crops, subsidized exports, and so on. None of these factors is included in the simple analysis presented here.

In the years following the adoption of FAIR, many of these elements changed. Good weather and increased planting stimulated by the high prices of the middle of the decade led to large increases in supplies. At the same time, the Asian financial crisis led to some fall in demand, and prices for corn and other temperate crops fell dramatically. The U.S. government put through various emergency measures to assist farmers during this crisis, and, as we have seen, these measures led to a transfer to farmers of almost $23 billion in 1999. Meanwhile, export subsidies have largely been eliminated as the United States complied with the Uruguay Round trade agreement of the GATT/WTO, which entered into force in 1996. Under this agreement, countries must lower the quantities of subsidized exports by 21 percent and reduce budgetary outlays for export subsidies by 36 percent. Of course, other countries also had to adjust their trade policies to comply with the agreement. Given the high dependency of U.S. agriculture on foreign markets, these policy adjustments, as well as other events outside the borders of the United States, had significant impacts on the U.S. corn sector. Clearly, none of these many factors is taken into account in the simple analysis described in this chapter.

CONCLUSION

The graphical and elasticity models described in this chapter allow analysts to obtain initial assessments of the likely economic impacts of policy changes that are under consideration. Despite the fact that such analyses cannot capture all the complexities associated with a particular policy change, they can provide helpful insights into the direction and possible magnitudes of some of the effects of the change. In addition, they can often be used to provide important clarifications of public policy issues. Consider, for example, the growing opposition to globalization that was manifested in a series of protests directed at the World Trade Organization (WTO), the International Monetary Fund (IMF), and the World Bank in late 1999 and 2000. One of the groups leading these demonstrations made the claim that World Bank programs harm poor people in low-income countries, offering the case of cashew production in Mozambique as an example of the bank's negative impact on the poor. According to the protesters, the World Bank forced Mozambique to eliminate an export tax on raw cashew nuts. This led to higher prices for raw cashew nuts, causing the demise of the Mozambique cashew-processing industry. Without some model of these markets, this claim seems incoherent. How could eliminating a tax on cashews lead to an *increase* in their price?

Consider Figure 10.10, which illustrates the impact of an export tax in the case of a small exporting country. In the absence of the tax, the price in Mozambique would be the world price, P_w, and the quantity ($Q_{SF} - Q_{DF}$) would be exported. An export tax reduces the internal price in Mozambique to P_t and reduces the quantity exported to ($Q_{ST} - Q_{DT}$). Consumer surplus increases by areas a and b, while producer surplus is reduced by areas a, b, c, d, and e. The revenues from the tax earn the government area d, which offsets part of the producer loss, leaving deadweight-loss triangles equal to areas c and e as the net welfare effect of the tax. Note that if the tax were set high enough (e.g., at levels equal to the difference between P_w and P_a), all exports of raw nuts could be prevented. Professor Paul Krugman, an eminent economist who writes a regular column for the *New York Times* Op-Ed page, looked into the claim that the elimination of this export tax had harmed Mozambique.[16] Although Krugman did not present the graphical analysis shown here, it is almost certain that something along the lines of Figure 10.10 was in his mind as he thought about this issue.

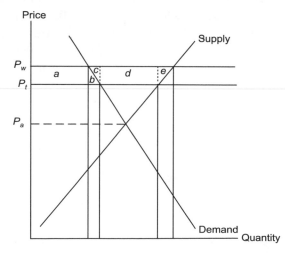

Figure 10.10: Export tax.

Krugman noted that cashews are produced by peasant farmers in Mozambique and sold to a local processing industry. It is not clear whether any raw cashew nuts are exported, but it is likely that they are not suggesting that the tax is at least equal to $(P_w - P_a)$. Krugman pointed out that the export tax was introduced when the state-run monopoly that purchased raw cashews and commercialized processed nuts was privatized. In many developing countries, state-run enterprises have been used to extract producer surplus from rural areas. As legal monopsonies, these enterprises set low prices for the raw products they bought from rural producers. The difference between the low producer price and the world price ended up either in government coffers or in the pockets of those running the inefficient state enterprise.

For the Mozambique cashew industry, raw cashews are a major part of the cost of producing processed nuts. The low cost of this input allows firms that would otherwise be too inefficient to compete on the world market to survive. Privatization, of course, subjected the new firms to competitive pressures, among which would have been the pressure of having to pay market prices for the raw nuts if the export tax had not been implemented. In other words, the tax served to maintain the low prices for the raw nuts. If the tax were set at a level that would prevent exports $(P_w - P_a)$, the government would raise no revenue from it, as growers would be forced to sell their entire crop to the newly privatized processors at the autarky price. Lower taxes would allow some exports of raw cashews, but growers would still receive only the after-tax price, P_t, If the tax were set high enough (greater than $P_w - P_a$), Mozambique could end up importing raw cashews for its processing industry, although it would have to use an import subsidy to bridge the gap between the world price and the very low internal price.

According to Krugman, the Mozambique processing industry employed about 10,000 people, while raw nuts were produced as part of the output of some 5 million peasant households. Elimination of the export tax did raise the price the processing industry had to pay for the raw nuts and may have resulted in its destruction. Cashew growers, however, whose income was certainly lower than the income of the workers and owners in the processing industry, could sell their products at the world price either to whatever remained of the Mozambique processing industry or to importers in other countries. Thus, the policy change benefited low-income people in a very poor country at the expense of a small number of relatively well-to-do individuals associated with the cashew nut processing industry. Krugman's simple analysis does not prove that globalization is good or that the World Bank always acts in the interests of the poor. It does show that the Mozambique cashew nut case does not support the argument of those who wish to claim that the World Bank routinely causes harm to the poor. The antiglobalization forces will need to find a better argument for their position if they are to be taken seriously in the public debate over globalization and development policy. The insights provided by Krugman's simple analysis clarify these issues, and such clarification can contribute to better public decisions.

Of course, it is often necessary to back up such insights with more precise estimates of the benefits and costs of public policies, and that is likely to require more sophisticated models that can be used to run policy experiments. Such models will be explored in subsequent chapters. The tools discussed in this and the following chapters focus on the consequences of alternative policies as measured by such economic values as producer and consumer surpluses or government costs. In many cases, the analysis will be incomplete if the distributional consequences are left out, and there may be other kinds of considerations that need to be specifically addressed in reaching a conclusion on the desirability of a particular policy change. It is no accident that much of the literature on agricultural policy concerns the desirability of protecting and encouraging an agricultural system based on family farms, even if measurable economic benefits and costs seem to indicate that modern, industrialized agricultural systems are more efficient. For many, the demise of the family farm would severely damage fundamental rights and values, causing irreparable harm to both rural areas and the broader society. Even if one does not agree with this assessment, it would be inappropriate to rule out these concerns a priori in analyzing a major policy shift as occurred with FAIR.

SUMMARY

1. Many public policies aim to influence behavior through changes in incentives. Evaluating the costs and benefits of such policies requires the use of some kind of model to predict key economic variables with and without the policy.

2. Partial equilibrium analysis focuses on single markets or a small set of related markets. A useful first step in designing a partial equilibrium analysis is to do a graphical analysis to identify important relations and the likely directions of change.

3. A complex set of policies has been put in place to deal with distributional and efficiency problems in U.S. agriculture. These policies have included production controls, price and income supports, trade barriers, and many others. The effects of these policies are illustrated with graphical analysis.

4. While graphs can help identify the direction of change, they offer little information on the size of changes induced by public policies. Such information can be obtained from statistical models that allow the analyst to simulate policy impacts.

5. Rough indications of the size and direction of policy effects can often be obtained using log-linear models based on elasticities and data on supply and demand.

6. In Europe and North America, a large array of policy tools has been used to raise farm income, support and stabilize prices, and achieve other objectives related to food safety, the environment, economic development, and famine relief. Graphical analysis and log-linear models can help foster an understanding of the welfare effects of these policies.

KEY CONCEPTS

incentives	deficiency payments
partial equilibrium analysis	market-oriented policies
graphical analysis	simulation
price wedge	log-linear models
farm problem	logarithmic differentiation
production controls	import tariffs
price supports	large-country cases
nonrecourse loan	export subsidies
price floor	marketing quota
parity	Common Agricultural Policy
food aid	variable levy
income supports	export restitutions

DISCUSSION QUESTIONS

1. Contrast the evolution of agricultural policy in the United States and the European Union.

2. Explain the impact of a gasoline tax on producers, consumers, prices, and the quantity of gasoline exchanged in the market.

3. Suppose the government wished to subsidize gasoline consumption. Draw a graph showing the impacts of the gasoline subsidy on consumers, producers, prices, and quantities.

4. Use the "large-country" graphs from Appendix 10.1 to analyze the welfare effects of an import tariff levied by the rest of the world (ROW).

5. Suppose the government places a 20 percent tax on cigarettes. Use the log-linear model described in equations (10.7) through (10.9) to compute the changes in prices and quantities, assuming that the supply elasticity is 0.8 and the demand elasticity is -0.4. Do you think cigarettes should be taxed to reduce consumption of this harmful product?

6. Discuss the rationale for agricultural support in the United States. Do you think that farm policies are still needed, or is the farm sector just another rent seeker? Explain.

SUGGESTIONS FOR FURTHER READING

Fennel, R. *The Common Agricultural Policy: Continuity and Change*. New York: Oxford University Press, Clarendon Press, 1997.

Gardner, B. *The Economics of Agricultural Policies*. New York: McGraw-Hill, 1990.

Orden, D., R. Paarlberg, and T. Roe. *Policy Reform in American Agriculture*. Chicago: University of Chicago Press, 1999.

Sadoulet, E., and A. de Janvry. *Quantitative Development Policy Analysis*. Baltimore: Johns Hopkins University Press, 1995.

Thompson, Paul B. *The Ethics of Aid and Trade*. New York: Cambridge University Press, 1992.

Timmer, C. P. *Getting Prices Right*. Ithaca, N.Y.: Cornell University Press, 1986.

Tsakok, Isabelle. *Agricultural Price Policy: A Practitioner's Guide to Partial Equilibrium Analysis*. Ithaca, N.Y.: Cornell University Press, 1990.

Tweeten, L. *Agricultural Trade: Principles and Policies*. Boulder, Colo.: Westview Press, 1992.

APPENDIX 10.1: LARGE-COUNTRY TRADE EFFECTS AND THE U.S. EXPORT ENHANCEMENT PROGRAM

In Chapter 2, the welfare effects of an *import tariff* were illustrated for a small country facing perfectly elastic international supplies. In this appendix, we will examine the derivation of excess supply and excess demand schedules for cases where the introduction of trade barriers can be expected to have an impact on world prices. Such situations are often referred to as *large-country cases* in recognition of the fact that changes in imports or exports by countries having a large share of the world market will have an impact on world prices. In Figure 10.11, the excess supply and excess demand schedules for a large exporting country (the United States) and a large importing region (the rest of the world) are derived. Abundant land and capital, a favorable climate, and a skilled workforce mean that the United States has a comparative advantage in the agricultural good (corn) represented in Figure 10.11. This fact is reflected in the equilibrium price in the United States when the country is in autarky, that is, when it is not trading, as shown in the figure's left-side graph. The autarky price in the United States is P_a, which is lower than the autarky price in the rest of the world (ROW), P'_a, as shown in the right-side graph.

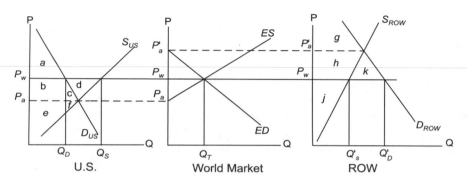

Figure 10.11: Large-country trade model.

Define excess supply (ES) as the difference between U.S. supply (S_{US}) and U.S. demand (D_{US}). At the autarky price, P_a, ES is obviously zero ($S_{US} = D_{US}$). The center graph in Figure 10.11 represents the world market. If the world price were equal to P_a, there would be no reason for the United States to export corn. ES would be zero, and this locates the intercept on the price axis in the center graph representing the world market. The rest of the ES schedule is derived as the horizontal distance between S_{US} and D_{US}. In a similar manner, the excess demand (ED) schedule is derived as the horizontal distance between D_{ROW} and S_{ROW} beginning at the intercept given by the autarky price in the rest of the world. The intersection of ES and ED establishes the world price, P_w, and the quantity exchanged in the world market, Q_T. Q_T is equal to ($Q_S - Q_D$) and also to ($Q'_D - Q'_S$) by construction. In Figure 10.7 earlier in the chapter, ED from Figure 10.11 was added to the demand schedule in the United States to derive a total demand curve in the U.S. market that included both U.S. and foreign demand.

Figure 10.11 can be used to illustrate the gains from international trade. In autarky, domestic supply will equal domestic demand at the relevant autarky prices in both regions. Consumer surplus in the United States is equal to the areas marked a, b, and c, while producer surplus is given by areas e and f. With trade, the price increases from P_a to P_w. Consumer surplus is reduced to area a, while producer surplus expands to areas b, c, d, e, and f. The consumer loss (b and c) is transferred to producers, who gain area d in addition. Thus, for the United States as a whole, there are net gains from trade equal to area d. It would be possible to compensate consumers for their losses by taking away areas b and c from producers and transferring them back to consumers. Under these circumstances, producers are still better off by area d, while consumers after the transfer are no worse off. A similar analysis in ROW would show that the importing region experiences a net gain of area k: consumer surplus increases by h and k, producer surplus falls by h, so the net change is the addition of area k to social surplus.

In Figure 10.12, this framework is used to analyze the welfare effects of an *export subsidy* such as the U.S. Export Enhancement Program (EEP) or the export refunds (also referred to as export restitutions) employed by the EU. The free-market equilibrium is at the point where Q_T is exchanged in the world market at a price of P_w. The export subsidy raises the price in the U.S. or EU market to P_S. At this price, the quantity supplied increases to Q'_{SU}, while the quantity demanded falls to Q'_{DU}, and exports expand to $Q'_T = (Q'_{SU} - Q'_{DU})$. *ROW* will only purchase this quantity if the world price falls to P'_w. At that lower price, demand in *ROW* increases to Q'_{DR}, while supply falls to Q'_{SR}, and imports increase to $(Q'_{DR} - Q'_{SR})$, which is equal to Q'_T. The government in the exporting region provides a per-unit subsidy, which is equal to the difference between P_S and P'_w. The changes in producer (*PS*) and consumer (*CS*) surplus as well as the cost to the government (*GOV*) of the subsidy can be represented using the letters marking the areas in the graph. For the U.S./EU,

$$\Delta CS = -a - b$$
$$\Delta PS = +a + b + c + e + f$$
$$\Delta GOV = -b - c - d - e - f - g - h - i - j - k$$

The government cost is equal to total exports $(Q'_{SU} - Q'_{DU})$ multiplied by the subsidy $(P_S - P'_w)$. The net effect of the subsidy in the exporting countries is $(-d - g - h - i - j - k - b)$, which is unquestionably negative (the costs to the government are greater than the net of producer gains and consumers losses). In *ROW*, consumers gain $(w + u + x + y + z)$, while producers lose areas w and u. The net effect in *ROW* is positive, as the consumer gains due to the lower world price outweigh the producer losses.

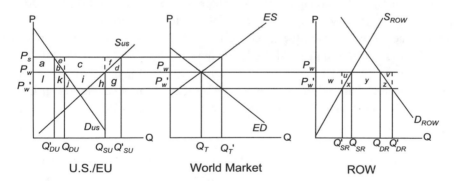

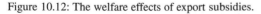

Figure 10.12: The welfare effects of export subsidies.

For the world as a whole, the effects of the subsidy are: $x + y + z - d - g - h - i - j - k - b$. Note that the area $(u + x + y + z + v)$ in ROW is equal to area $(g + h + i + j + k)$ in the U.S./EU. Thus,

$$(x + y + z) = (g + h + i + j + k - u - v)$$

If we substitute the right-hand side of this equation for $(x + y + z)$ in the expression for the net effects of the subsidy, everything cancels out except for four small deadweight loss triangles $(-b - d - u - v)$. The export subsidy benefits U.S./EU producers and ROW consumers. It harms U.S./EU consumers and ROW producers. Welfare in the exporting countries declines as a result of the high cost of the subsidy, while welfare in the importing countries increases as they take advantage of the lower world price. The net costs in the exporting countries are greater than the net gains in the importing countries, as indicated by the four small triangles, which reflect efficiency losses stemming from the distortion introduced by the subsidy.

The EEP has had some rather interesting side effects. The formation of NAFTA meant that trade barriers between Canada, the United States, and Mexico were reduced or eliminated. Canada and the United States are both major wheat producers and exporters. When the United States used the EEP to promote U.S. wheat exports, domestic prices increased, as expected given the results in Figure 10.12. Because the United States is a large country in the wheat market, the EEP also had the expected side effect of lowering world wheat prices. Faced with a choice of exporting to a depressed world market or taking advantage of NAFTA provisions to sell in the U.S. market for the more favorable U.S. price, it was inevitable that Canadian wheat would begin to flow into the United States, and this is precisely what happened.

As noted in the text, food aid (P.L. 480) amounts to an export subsidy as surplus food is offered at favorable terms to buyers in low-income countries. Figure 10.12 can also be used to analyze the effects of programs such as P.L. 480. Consumers in low-income countries benefit from the lower price, while producers find their returns to agricultural production lowered. In the United States and the European Union, in contrast, food producers benefit from higher prices, while consumer surplus is reduced. It is no accident that U.S. farm groups have always favored P.L. 480, which reduces surpluses and increases demand for farm products, leading to higher prices. The price-depressing effect of food aid in the recipient countries, however, has led to widespread criticism of these programs. Farmers in low-income countries have little incentive to modernize or expand their output when market prices are depressed by subsidized food imports.

From the point of view of developing countries, it would generally be preferable to receive technical assistance aimed at enhancing agricultural productivity rather than this subsidized food. Most people in low-income countries earn their living from agriculture, and one of the surest ways for these countries to begin solving the problems of poverty and low economic growth is to enhance the productivity of their farmers. Increased productivity could make these countries less dependent on U.S. food exports, however, and might give rise to agricultural industries that could compete with U.S. producers on world markets. Farm groups in the United States have tended to oppose technical assistance for farmers in low-income countries while favoring subsidized food aid, for obvious reasons. The fears of U.S. farmers were realized when the Brazilian and Argentinean soybean industries entered world markets as major exporters in the 1970s and 1980s. A small part of the success of these industries stemmed from the technical assistance provided by scientists at U.S. universities working with the U.S. Agency for International Development (USAID). U.S. soybean producers lobbied successfully for legislation making it illegal for USAID to provide foreign aid that would lead to increased output of agricultural products that might compete with U.S. agricultural exports. This lobbying effort and the resultant legislation were direct responses to the development of soybean production in South America.[17]

APPENDIX 10.2: THE U.S. PEANUT PROGRAM

The agricultural policies described earlier in this chapter were applied to grains and cotton. A wide variety of different policy measures have been used for other crops and livestock products. The U.S. peanut program is an example. The central element of this program is a *marketing quota* that restricts the amounts producers can market in the United States. When the program was originally set up, farmers who had established a record of producing and marketing peanuts were granted marketing quotas set at levels that reduced the total amount of peanuts put on the market. These quotas are property rights that have economic value. They can be sold or rented to other producers, inherited by one's heirs, ceded to the bank if the farm goes bankrupt, and so on. For most of the history of the program, quotas could be sold or rented only to producers residing in the same county. This provision tended to freeze production in traditional peanut-growing areas. In recent years there has been some easing of this constraint.

In Figure 10.13, the world price, P_w, is below the U.S. autarky price. Suppose that the marketing quotas are set so that the quantity marketed is exactly equal to Q_1. If there is no barrier to trade, the quota would have no effect on the market. The United States would import $(Q_2 - Q_1)$ at the world price, which would be the price U.S. peanut growers would receive. Because the program aims to raise producer prices, a trade barrier is needed to keep cheaper imports off the market. In this case, the trade barrier is set high enough to prevent imports altogether. Under these circumstances, the quantity supplied to the market is Q_1, raising the price paid by consumers to P_1. Relative to the free trade situation, producer surplus has increased by area b, while consumer surplus has been reduced by areas b, c, and i. There is a net social loss of areas c and i, but producers' welfare has increased substantially, and that is the goal of the program.

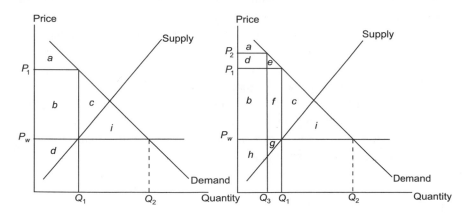

Figure 10.13: Marketing quotas.   Figure 10.14: Quotas and the U.S. peanut program.

During the 1980s, another element was added to the peanut program. In Figure 10.14, the marketing quota has been set to the left of Q_1 at Q_3, raising the price for quota peanuts to P_2. The new element is the opportunity for producers to market peanuts beyond their quota on the world market at P_w. Note that the world price is high enough to give rise to supplies beyond the quota in the segment of the supply curve between Q_1 and Q_3. Producers who owned peanut quota rights could market their quota at the quota price (P_2) and sell over-quota peanuts on the world market. Producers who did not have any quota rights could now produce peanuts but only for sale on the world market. Peanuts destined for the world market are referred to as *additional* peanuts and must be crushed and exported as oil to insure that they do not reenter the U.S. market for whole nuts. Note that this program further reduces consumer surplus compared to the pure quota program by areas d and e. Producers gain areas d and g but lose area f. The social loss expands to $(c + f + i + e)$.

It is interesting to note that the peanut program was left largely unchanged by FAIR. As previously noted, a peanut quota has a value to its owners equal to the difference between P_2 and the world price, multiplied by the quantity of peanuts the holder is allowed to market. Over time, the ownership of these quotas has changed substantially. Many producers now have to rent the right to market their crops from retired farmers or their spouses who own the quota but no longer farm. For these people, the quota is a source of income in the same way that ownership of a corporate stock provides income in the form of dividends. Because the quota is part of the assets of a farm, many quotas have wound up in the hands of banks following foreclosure on farms that have been unable to repay their loans. Some of the rules about sale and rental of quotas have been relaxed, and it is now possible for new producers in areas with no quotas to obtain quota rights after producing additionals for some period of time. This modification means that peanut production can be started in areas outside the traditional peanut zones in the deep South. However, all the people who own a peanut quota have a strong interest in seeing this program continue. If the government were to eliminate the peanut program, quota holders would lose a valuable asset.

Note, however, that this asset was created out of thin air by a government program, which has conferred substantial economic benefits on quota holders. While it would be politically difficult to eliminate the program, it is hard to argue that doing away with it would violate the canons of distributive or social justice. After all, most people do not have government-generated assets handed to them for virtually nothing. Creation of quota rights generally leads to this kind of outcome. Under the multifiber agreement and U.S. textile import policy, several textile-producing countries (mainly in Asia) were granted quotas to sell on the U.S. market. These quota rights became so valuable that some textile producers retired and lived off the proceeds from renting their quotas to others. The multifiber agreement is to be phased out under the Uruguay Round trade agreement.

The European Union has operated a program for sugar similar to the U.S. peanut program. Sugar refiners are granted two types of quota. Sugar counted under the A quota is supported at a high price. A certain amount of sugar beyond the A quota (the B quota) can be marketed to allow for variations in market demand. Any sugar beyond the A and B quotas can be sold only on the world market. In effect, EU sugar producers benefit from high price supports, with the possibility of dumping onto the world market any sugar they may have in excess of their quotas. The U.S. sugar program, described in the case study in Chapter 12, has the effect of restricting imports, thereby lowering world demand. The EU sugar program results in additional supplies of sugar to the world market. The effect of these two programs is to lower world sugar prices dramatically, to the detriment of countries such as Australia and Brazil, which are major sugar exporters.

APPENDIX 10.3: AGRICULTURAL POLICY IN THE EUROPEAN UNION

This appendix discusses the formation of the European Union (EU) and its Common Agricultural Policy (CAP).[18]

Origin and History of the CAP

The European Economic Community (EEC) was established in 1957 with the signing of the Treaty of Rome by the governments of Germany, Italy, France, the Netherlands, Belgium, and Luxembourg. Previous regional agreements among the countries of Western Europe tended to be organized around relatively narrow concerns, as in the case of the European Coal and Steel Community (ECSC) established in 1951. In contrast, the EEC treaty was a comprehensive agreement to remove trade barriers between the members of the community and to coordinate external trade policy. The European Community (EC) included the EEC, the ECSC, the European Defense Community (established in 1952), and other regional organizations. In 1972, the United Kingdom, Denmark, and Ireland joined the EC, which was further enlarged with the accession of Greece in 1981 and Spain and Portugal in 1986. In 1992, a new treaty—the Maastricht Treaty—was adopted, and the EC became the European Union (EU). Subsequently, Austria, Finland, and Sweden joined the EU, which as of this writing has a total of fifteen members. Several Eastern European countries are candidates for membership, and it is likely that the EU will continue to expand. In 1998, the EU became a monetary union with the adoption of a single currency, the euro.

Agriculture in the six original members of the EC was characterized by great diversity.[19] Producers in France and the Netherlands were relatively efficient, and exports were an important component of their agricultural economies. In contrast, producers in Germany, Italy, Belgium, and Luxembourg had small holdings and were less efficient than their counterparts in France and the Netherlands. For French agricultural interests, the advantages of economic union consisted of access to German agricultural markets. The Germans expected to benefit from increased industrial sales in France. In general, farm structure, prices, and the organization of food and agricultural marketing chains differed greatly among these six countries.

Although there had been a long tradition of government intervention in agriculture, the specific policy mechanisms employed by the original members of the EC were fairly diverse. Tariffs were widely used to raise European farm prices during the economic depression of the 1930s. Following the Great Depression, European trade barriers in agriculture were largely left in place because of the perception that agriculture is subject to market instability and the belief that family farming is an important way of life. It should be noted that agricultural protectionism was as extensive in the United States as it was in Europe at this time. Widespread protectionism of both industry and agriculture contributed to the length and severity of the Great Depression.

The World War II left the European countryside in ruins, and there were widespread food shortages. Memories of these difficult times reinforced the desire of European governments to insure stable domestic supplies through the use of trade barriers and price supports. Policies established to raise producer prices above world prices and to dispose of surpluses through export subsidies and storage programs were well entrenched prior to the formation of the EC. Because of the great diversity in the different agricultural sectors of the original EC members, however, prices were supported at different levels, and widely divergent agricultural regulations were found in each country. Fennell estimates that 30,000 different regulations had to be dealt with in setting up a common agricultural policy.[20]

Why was a common policy for agriculture needed? The creation of the EEC required the elimination of tariffs and other barriers to trade among the members of the community. As previously noted, the original members of the EC had pursued different agricultural policies, with the result that agricultural prices differed from one country to the next. Different prices cannot be sustained unless a tariff or some other measure is used to adjust the prices of traded goods. Without such trade barriers, it would be possible to purchase agricultural commodities in countries with low prices for sale in countries where policies to maintain high prices were in place. The resultant flow of goods would undermine the high-price policies. Thus, the elimination of internal barriers to trade made it impossible for the member states to continue their separate policies.

The main economic advantage of a customs union such as the EEC, of course, is that production of particular goods is shifted from regions where it is costly to produce them to areas where production is carried out more efficiently. This rationalization of production in Europe had been prevented by the trade barriers that were needed to isolate national agricultural markets. Such isolation was necessary if the European countries were to maintain different agricultural policies. The decision to form the EEC meant that national markets could no longer be isolated and the divergent national policies had to be harmonized to form a single community-wide policy. Thus, the *Common Agricultural Policy* (CAP) was made necessary by the formation of the EEC, a move that was expected to generate widespread benefits in the form of lower prices due to more efficient production.

The original CAP was based on several principles. The first, the principle of common prices, was unavoidable because of the elimination of trade barriers within the EC. A second principle is common financing. Each member of the EC contributes a portion of its tax revenues to the EC budget, which includes a central fund, known as the European Agricultural Guidance and Guarantee Fund (EAGGF), from which the CAP is financed. This fund replaces the financing of agricultural policy by the individual member states. Revenues from tariffs and levies applied to agricultural goods imported from countries outside the EC are also held in the central fund rather than by the treasuries of the countries that collected them. A final principle is the principle of community preference. Community preference means that no agricultural goods imported from countries outside the community can be sold at prices lower than the prices of other members of the EC.

Memories of difficult times during and after the World War II have also influenced the structure of the CAP. Price stability is seen as a major policy objective, and the CAP is designed to isolate EU food markets from world market fluctuations. Fear of food shortages has led to a desire to increase the degree of food self-sufficiency, although that is not explicitly listed in the objectives of the CAP. Price stability and food security can be attained in many different ways. The particular approach to these issues followed by the EC involves extensive government intervention, reflecting historical patterns of distrust for free markets in Europe. Agricultural policy has also been influenced by prevailing beliefs that rural life and family farms have special value and that it is important to maintain a relatively large rural population. These beliefs make it easier for EU consumers to accept the high food prices that have been induced by the CAP.

Basic Mechanisms of the CAP

The CAP has two sets of programs corresponding to the two parts of the EAGGF: guidance and guarantee. Structural programs fall under the smaller guidance fund, while conventional price and income support programs are covered by the guarantee fund. When the CAP was established, the EC was a net importer of many temperate-zone agricultural commodities. In addition, the level of the common prices agreed upon by the founders of the EC was determined through a compromise that prevented individual members from experiencing price declines as the CAP was put in place. As a result, internal prices were set above world prices, making some form of trade barrier essential. The CAP included a wide range of market regulations aimed at maintaining high internal prices and preventing imported commodities from interfering with the evolution of internal markets.

The programs for cereal grains can serve as an example of the types of market interventions practiced in the EU. Until the recent reforms, these programs were based on a politically determined target price representing the desired level of wholesale prices for a given commodity. Note that the EU target price differs from the U.S. target price used to determine deficiency payments. The EU target price was a true target, something to be aimed at, and was used to calculate an internal support price, known as the intervention price, and a minimum import price, known as the threshold price. The intervention price was supported by government agencies that were required to purchase whatever producers wished to sell at the intervention price. This price, thus, served as a floor for market prices. Unlike the early farm programs in the United States, producers were not initially required to reduce production to receive the benefits of market support.

The threshold price was enforced through the application of a trade barrier known as the *variable levy*. The variable levy is set as the difference between the world price and the threshold price. Because the world price changes from day to day, the variable levy also varies on a daily basis to insure that imports never enter the EU at prices lower than the threshold price. If the EU imports grain, internal prices would tend to rise to the level of the threshold price, and no grain would be sold into intervention at the lower intervention price. In the early days of the CAP, the EC was a net importer of grain, prices were near the threshold price, and little grain was accumulated by the intervention agencies. As EC farmers began to respond to the high market prices, production expanded, imports fell, and surplus grain began to pile up in the intervention agencies. This led to the introduction of a new mechanism, the *export restitution*.

Export restitutions are variable export subsidies. They are calculated as the difference between the world price and the higher market prices in the EU. The increased production in the EU meant that market prices fell below the level of the threshold price, reaching the level of the intervention price as surplus stocks began to accumulate. The export restitution was designed as a mechanism for diverting the surplus production from the internal market to the world market. Because world market prices are lower than the internal prices in the EU, grain cannot be sold on the world market without an export subsidy. The effect of these policies was to transform the EC from a major importing region to one of the world's leading exporters of agricultural products. In addition, these mechanisms provide almost perfect insulation from world market variations for internal markets, thus insuring internal price stability.

Evolution of Agricultural Policy in the EU

Not surprisingly, these agricultural policies generated a great deal of conflict with other exporting nations. For the United States, increased production in the EC meant the loss of an important foreign market and the rise of a major competitor on the world market. For the EC, the CAP led to price stability and increased self-sufficiency but at a high cost. By the late 1970s, external pressure and budget constraints had forced the EC to modify its policies. In 1977, the coresponsibility levy was introduced to help manage the growing surplus dairy production. The coresponsibility levy is a tax on total receipts paid by producers to help defray the costs of the dairy program.

A major policy shift was made in 1984 with the introduction of measures to limit the amounts of grains and dairy products that would be supported by the CAP. Prior to this decision, the EC provided price supports for whatever amounts were produced within the community (i.e., there were no production controls like the acreage set-asides in the United States). The 1984 agreement introduced a system of production quotas for the dairy sector and expanded the use of guarantee thresholds for cereal grains. The milk quotas are a form of mandatory production control that forces individual producers to limit the amount of milk they produce and market. Guarantee thresholds are a less restrictive production control. They set a limit on the total amount of a commodity that will receive CAP support but do not set a rigid limit on production.

Another important issue arose in 1986 with the accession of Spain and Portugal to the EC. Spain was a major market for U.S. corn and soybean exports. Membership in the EC meant that Spain would have to adopt the CAP in its entirety, including variable levies on grains. The United States feared that its corn and sorghum exports to Spain would fall following implementation of the CAP and threatened to take action under the provisions of the GATT/WTO that regulate the formation of customs unions. A trade war was averted in early 1987 when the EC agreed to allow Spanish importers to make annual purchases of 2 million metric tons of corn and 300,000 metric tons of sorghum from outside the EC.

Despite the 1984 policy reforms, the spending on the CAP continued to expand more rapidly than the resources available through the EC budget. In 1988, another major reform effort was undertaken. To ease the budget pressures, EC policy makers increased the amount member states were required to contribute to the EC budget. To control expenditures, a land set-aside program was introduced, and coresponsibility levies were extended to grains. In addition, the maximum quantity of grain for which CAP supports would be available was set at 160 million metric tons. Annual production greater than this amount would trigger an automatic 3 percent reduction in support prices for the next year.

The main innovation of the 1984 and 1988 CAP reforms was the elimination of open-ended support for whatever quantities are produced. The system of target, threshold, and intervention prices, variable levies, and export restitutions remained intact. Although there has been an effort to restrain increases in price supports, the fundamental problems of agricultural policy in the EC were not resolved by the changes. The EC continued to produce large surpluses because the incentives for producers to expand production remained strong. The surpluses were disposed of at great expense through subsidized exports (which annoy other exporting countries) and a variety of measures to shift surplus production to secondary markets. For example, surplus wine was distilled into alcohol, and surplus wheat was dyed purple to prevent its use for anything other than livestock feed.

Agricultural policy reforms in 1992 represented a significant departure from earlier practices, instituting policies that were almost the same as those of the United States prior to the adoption of the Federal Agriculture Improvement and Reform Act (FAIR). Support prices for many commodities were lowered, with the loss in producer income made up by direct payments similar to deficiency payments in the United States. The set-aside program was expanded, eliminating entirely the open-ended support of agricultural production that had characterized the CAP in the past. This shift in policy took place in anticipation of potential international requirements that were being negotiated under the Uruguay Round (1986–1994) trade negotiations. The agricultural agreement of the Uruguay Round was largely the product of debate between the United States and the EU, the two main antagonists on agricultural trade. It requires reductions in tariffs and nontariff trade barriers as well as other changes. One of the most important provisions of this agreement for the EU is the requirement that export subsidies be reduced. Without export subsidies to remove surplus production in the EU, prices for the main commodities will fall.

The long history of conflict over agricultural trade between the EU and the United States stems from fundamentally different conditions in the two regions. In general, the EU does not have a comparative advantage in temperate agricultural products and, in a world of free agricultural trade, would end up as a major importer of these goods, with an agricultural sector much reduced in size. The United States has a comparative advantage in grains and oilseeds so that free agricultural trade would benefit most U.S. farmers. The result of these differences is that EU policy makers generally favor protectionist policies, while their U.S. counterparts are more inclined to promote free trade (unless the subject is peanuts or sugar, for example, two commodities in which the United States does not have a comparative advantage). There are other differences that lead to significant conflicts over food and agricultural trade. Europeans tend to be more suspicious of scientific pronouncements, particularly those that have to do with the safety of foods produced by modern industrial systems. This is particularly true of genetically engineered foods, which are widespread in the United States but generally perceived negatively in Europe (see Chapter 12). One implication of globalization is that neither the United States nor the EU will find it as easy to protect sensitive domestic industries such as textiles in the United States or agriculture in Europe.

Other pressures were shaping the evolution of the CAP at the end of the twentieth century. In the early days of the EC, the CAP was the main common program requiring funding by the community and took the major share of the community budget. As recently as 1978, 73 percent of the EC budget was spent on agriculture. By 1998, this figure had fallen to 55 percent, but that amount was still considered fairly extravagant by some, leading to pressures to reduce agricultural spending. The other issue facing the EU was the entry of several Eastern European countries. Countries joining the EU take on all the obligations and receive all the benefits of membership, including the generous agricultural supports that make up the CAP. Without some reform of the CAP, enlargement seemed likely to be extremely costly.

In 1999, the EU adopted a set of policy reforms referred to as Agenda 2000. A principal objective of this new policy was to ease current budget costs as well as those anticipated to arise with the entry of the Eastern European countries by freezing the CAP budget at the 1999 level. In addition, support prices for grains were reduced and other policy changes were made in an effort to comply with the Uruguay Round requirements and in light of expected further liberalization of the agricultural trade regime. It is interesting to note that the basic elements of U.S. agricultural policy have been modified every four or five years with the approval of a new Farm Bill. In the EU, this periodic modification of farm policy is referred to as policy reform, which also seems to occur every four or five years (e.g., 1977, 1984, 1988, 1992, 1999). It seems reasonable to expect that gradual modification of agricultural policy will continue on both sides of the Atlantic and that these modifications will increasingly result in the harmonization of the two sets of policies.

NOTES

1. Energy Information Agency, U.S. Department of Energy, Internet Web site accessed April 4, 2000, www.eia.doe.gov. U.K. price in U.S. dollars, converted from British pounds.

2. World Bank, *World Development Report 2000/2001*, Internet Web site accessed March 10, 2001, www.worldbank.org.

3. The price elasticity of demand is defined as the percentage change in demand for a given percentage change in price. This can be written: %Δ demand/%Δ price. If the value of this elasticity is −0.5, then %Δ demand/%Δ price = −0.5, which implies that %Δ demand = −0.5 × (%Δ price). If the change in price is 10 percent (0.10), we have the change in demand equal to (−0.5) × (0.10) = −0.05 (minus 5 percent). The effect of a gasoline tax also influences supply, and incorporating these effects will change the prediction somewhat.

4. There are many types of models that can be used to carry out a partial equilibrium analysis. Some of these are quite complicated and may require advanced statistical methods to obtain estimates of the model parameters. Expertise in these methods can be developed only through advanced

study in statistics and economics. The purpose here is to provide some basic information on partial equilibrium analysis so that students can better understand complex studies written by professional analysts. There are many books on statistical methods and economic modeling. A good reference on policy modeling is Isabelle Tsakok, *Agricultural Price Policy: A Practitioner's Guide to Partial Equilibrium Analysis* (Ithaca, N.Y.: Cornell University Press, 1990). See also E. Sadoulet and A. de Janvry, *Quantitative Development Policy Analysis* (Baltimore: Johns Hopkins University Press, 1995) for a somewhat more advanced treatment focusing on developing countries. Bruce Gardner, *The Economics of Agricultural Policies* (New York: McGraw-Hill, 1990) provides an analytical treatment of U.S. agricultural policies.

5. National Agricultural Statistics Service, *Farms and Land in Farms* (Washington, D.C.: USDA, February 2000). See also the Internet Web site www.nass.usda.gov:81/iped/report.htm/ and the *Economic Report of the President, 2000*.

6. See "Farm Operator Household Income and Wealth Compare Favorably with All U.S. Households," *Rural Conditions and Trends* 8, no. 2 (1997): 79-85.

7. There are many descriptions of the development and evolution of U.S. agricultural policy. A succinct history of U.S. programs is provided by J. Harwood and C. Jagger, "Agriculture's Safety Net: Looking Back to Look Ahead," *Choices* 4 (1999): 54–60. More complete histories are provided in D. Orden, R. Paarlberg, and T. Roe, *Policy Reform in American Agriculture* (Chicago: University of Chicago Press, 1999); and M. C. Hallberg, *Policy for American Agriculture: Choices and Consequences* (Ames: Iowa State University Press, 1992).

8. The Morrill Act granted federally owned land to state governments, which were to use the proceeds from the sale of the land to establish colleges for working-class people. These colleges were to focus on the "agricultural and mechanic arts, not neglecting the military arts."

9. In fact, by removing the seasonal price fall, the loan rate results in an average annual price higher than the average would be if the low seasonal prices were included. Thus, the loan rate may lead to slightly higher average prices. The impact of this program on prices and quantities will be quite limited, however, as long as the loan rate is set below expected market prices.

10. Note that production control programs (allotments and marketing quotas) could be thought of as a second policy aimed at raising farm income. Although the controls contributed to higher prices, these prices applied to reduced total output, so the impact on incomes was diminished. In addition, there is the problem of "slippage." Slippage occurs as rational farmers, in complying with the acreage controls, retire their least productive land and apply more inputs on the remaining land under cultivation, thereby obtaining a higher yield. A requirement that 10 percent of farmers' corn acreage be left uncultivated may lead to only a 3 percent fall in total output as a result of this slippage.

11. Examples include the Payment In Kind (PIK) program of 1983; the Farmer-Owned Reserve (FOR); the Export Enhancement Program (EEP); marketing loans (currently referred to as loan deficiency payments) that allowed farmers to sell their stocks on the market at prices below the loan rate, keeping the difference between the loan rate and these lower market prices; the 50-92 program, which provided 92 percent of the potential deficiency payment on up to 50 percent of a farmer's land taken out of cultivation; subsidized crop insurance; the Conservation Reserve Program (CRP); and many more. The details of these programs are quite complex, and their interaction with each other and the ARP and deficiency payment system meant that farmers were faced with a bewildering problem in deciding whether it would be more profitable to participate in these government programs or not.

12. These figures are taken from Orden, Paarlberg, and Roe, *Policy Reform*, 62 (Table 9).

13. "Congress Agrees to $7.1 Billion in Farm Aid," *New York Times*, 14 April 2000, sec. A , p. 18.

14. The approach described in this section is based on a set of lecture notes: Richard K. Perrin,

"The Log-linear Approach to Comparative Statics" (University of Nebraska, February 1998, photocopy). It requires some understanding of calculus.

15. U.S. Statistical Abstract. 1997.

16. P. Krugman, "A Real Nut Case, "*New York Times*, 19 April 2000, sec. A,Op-Ed page.

17. See Paul B. Thompson, *The Ethics of Aid and Trade* (New York: Cambridge University Press, 1992) for a full account of this issue.

18. This Appendix is based on E. Wesley F. Peterson and Samarendu Mohanty, *Agricultural Policy Reform in the European Community: Implications for Markets for Livestock Feed Ingredients*, University of Nebraska Research Bulletin 320 (September 1993).

19. There are many books on the CAP and EU agriculture: S. Harris, A. Swinbank, and G. Wilkinson, *The Food and Farm Policies of the European Community* (New York: John Wiley, 1983); R. Fennell, *The Common Agricultural Policy: Continuity and Change* (New York: Oxford University Press, Clarendon Press, 1997); Michael Tracy, *Agriculture in Western Europe* (London: Granada, 1982); and USDA Economic Research Service, *The European Union's Common Agricultural Policy: Pressures for Change*, Economic Research Service report WRS-99-2 (Washington, D.C.: GPO, October 1999).

20. R. Fennell, *The Common Agricultural Policy of the European Community*, (London: Granada, 1979).

11

Case Study: The Welfare Effects of the U.S. Sugar Program

INTRODUCTION

In this chapter, a case study of the U.S. sugar program is used to illustrate the nature of partial equilibrium analysis. This case study is based on a report written by Dr. Stephen Haley of the Economic Research Service (ERS) of the U.S. Department of Agriculture (USDA).[1] The model used to analyze this program incorporates numerous elements to take account of links to other markets in the sweetener sector. Because of the complex interrelationships between sugar and substitute caloric and noncaloric sweeteners as well as the significance of international trade in sugar, the sweetener model developed for this study is somewhat more elaborate than many partial equilibrium models. It was developed by analysts working in the Specialty Crops Branch of the Market and Trade Economics Division of the ERS for use in responding to requests for analysis of potential policy changes at a time when agricultural policy reform was on the political agenda.

BACKGROUND ON THE U.S. SWEETENER SECTOR

Human beings have always desired sweet-tasting foods, and various forms of sugar have been extracted from different plants since the invention of agriculture in the Neolithic revolution. In Western Europe, honey was widely used to satisfy these desires prior to the development of widespread sugarcane cultivation. For most of human history, sugar has been a luxury good available in small quantities and consumed primarily by the well-to-do. In the 1400s, the great age of European exploration began, and the cultivation of sugarcane was spread to newly discovered tropical lands. In the 1700s, sugar was an important part of the triangular trade that began with the shipment of sugar (and such derivatives of sugar production as molasses and rum) and other raw materials produced in the New World to Europe. The second leg of the triangle was the export of manufactured goods from Europe to Africa, where they were traded for slaves. The triangle was completed by the transportation of slaves to the Caribbean and the American colonies to work on the plantations producing sugar, tobacco, cotton, and other raw materials for export to Europe. The development of trade in sugar meant that prices in Europe fell, and sugar became a staple in the diets of workers participating in the industrial revolution.[2]

Sugar has been produced from cane and beets in the United States since the 1800s. According to Polopolus and Alvarez, per capita sugar consumption in the United States rose from about 10 pounds per year in the early 1800s to 33 pounds at the time of the Civil War to 86 pounds during World War I.[3] Until the 1970s, sugar consumption continued to grow, but the development of caloric and noncaloric substitutes led to a rapid decline in per capita consumption beginning in about 1972. From a peak of 102 pounds that year, annual per capita consumption fell to less than 70 pounds by the late 1990s (see Table 11.1). The decline in per capita sugar consumption was due primarily to the development of the corn wet-milling industry, which produces starch used to make high-fructose corn syrup (HFCS). HFCS has completely replaced sugar in soft drinks and many other beverages, in part because it is liquid and therefore easier to mix with the other ingredients. The desirable properties of HFCS are not the only reason that it has been widely substituted for sugar in food manufacturing industries, however. The U.S. sugar program supports U.S. sugar prices at levels well above the world price for sugar through the use of price supports and import barriers. High domestic prices for sugar make HFCS an attractive substitute.

Table 11.1. Annual per capita caloric sweetener and sugar consumption (pounds), United States

Year	Total Caloric Sweeteners	Refined Sugar	Corn Sweeteners*	Corn Sweeteners as Percent of Total
1970	122.3	101.8	20.5	16.8
1972	124.9	102.3	22.6	18.1
1975	118.0	89.2	28.8	24.4
1980	123.0	83.6	39.4	32.0
1985	128.8	63.6	65.2	50.6
1991	137.9	65.1	72.8	52.8
1994	147.4	65.5	81.9	55.6
1997	154.1	67.2	86.9	56.4

Sources: Polopolus and Alvarez, *Marketing Sugar*, data for 1970–1985; *Sugar and Sweetener*, SSS-226 (Washington, D.C.: ERS/USDA,September 1999):p. 346, data for 1991–1997; and author's calculations.

*Includes minor caloric sweeteners.

Because the use of noncaloric sweeteners has expanded dramatically in recent years, the growth in demand for caloric sweeteners may not have been as rapid as it would have been in the absence of these noncaloric substitutes. Still, as shown in Table 11.1, per capita consumption of total caloric sweeteners did grow between 1970 and the end of the last century. In the case of sugar, there is another important factor in this market. About half the sugar delivered for use in the United States was imported prior to the 1980s.[4] The figures in Table 11.2 show that the proportion of total deliveries accounted for by imports has declined quite substantially. The cause of this decline is the U.S. sugar program, which restricts imports to protect the domestic industry from foreign competition. Raw sugar prices in the United States have been supported at around 21 cents per pound, compared with world prices that have ranged from as low as 5 cents per pound to 15 cents per pound in more normal years. The U.S. sugar program will be described in more detail shortly.

Table 11.2. Total deliveries and imports of raw sugar
(in 1,000s short tons), United States

Year	Total Sugar Deliveries	Sugar Imports	Imports as a Percentage of Total Deliveries
1980–81	9,825	4,881	49.7%
1984–85	8,097	2,677	33.1%
1987–88	8,193	874	10.7%
1991–92	9,006	2,194	24.4%
1994–95	9,337	1,853	19.8%
1997–98	9,815	2,163	22.0%

Source: Polopolus and Alvarez, *Marketing Sugar*; *Sugar and Sweetener*, SSS-226 (Washington, D.C.: ERS/USDA, September 1999); and author's calculations.

Sugar can be produced from both sugar beets and sugarcane. The juice from sugarcane is perishable and must be rapidly processed. The initial processing of the cane juice produces raw sugar, which is further refined to produce the sugar used by food manufacturing industries and individual consumers. Sugar beets are processed directly into refined sugar. Historically, raw sugar has been the most widely traded commodity, although the recent emergence of the European Union (EU) as a major exporter has led to increased quantities of refined sugar traded on world markets. The EU relies on sugar beets almost exclusively for its sugar. In the United States, sugarcane is produced in Louisiana, Florida, Hawaii, and Texas. Sugar beets are grown throughout the United States, with most production concentrated in the Northern Plains, Idaho, and California. The amount of sugar derived from beets in the United States is slightly larger (about 53 percent of total sugar production in 1998–99) than the amount of U.S. sugar coming from sugarcane.[5] Consumers purchase refined sugar for direct use in home cooking, in sweetening tea or coffee, and in sprinkling onto foods. In restaurants, the price of a meal typically includes granulated sugar in packets or cubes for sweetening beverages and foods to individual tastes. People also consume sugar indirectly through their purchases of manufactured foods such as baked goods, candies, and sweetened dairy products. Direct consumption (referred to as nonindustrial use) accounted for about 40 percent of total refined sugar consumption in 1999.[6]

THE U.S. SUGAR PROGRAM

Polopolus and Alvarez provide a full account of the evolution of U.S. sugar policy prior to the 1990s.[7] As with other agricultural commodities, there was little government intervention in sugar markets before the 1930s, although Polopolus and Alvarez document the early use of tariffs on imported sugar both to raise revenue and to protect domestic producers. The first Sugar Act was adopted in 1934. This legislation provided for quantitative restrictions on both domestic producers and foreign suppliers through quotas allocated to sugar refiners. It is interesting to note that the quotas were applied at the processing level rather than at the farm level. This is a characteristic of sugar policies and policies for other products (e.g., dairy) where the active markets are for the processed good rather than the farm-level commodity. There is virtually no world trade, for example, in the primary commodities (cane, beets, or milk) used to produce the raw and refined sugar and the dairy products (butter, cheese, etc.) that are traded.

To comply with the quotas, refiners reduced their purchases from cane and beet producers, and the law required that they do so equitably. The purpose of the quota restrictions was to stabilize domestic sugar prices at levels advantageous to both producers and consumers. The act also provided for a tax on processors, the proceeds of which were to be rebated to cane and beet producers to compensate them for having to reduce their deliveries of the raw product. Polopolus and Alvarez note that this tax was deemed unconstitutional and was subsequently replaced by an excise tax not related to the producer payments, which continued to be made but from general appropriated funds.

This basic act underwent much refinement and modification during the years it was in force. Most of the changes had to do with details on the way in which quotas were established and allocated. The most significant event in the development of U.S. sugar policy was the Cuban revolution of 1959, which brought Fidel Castro to power. Cuba had benefited from a large quota and was a major supplier of sugar to the U.S. market. Soon after the revolution, however, U.S. trade with Cuba was cut off and the Cuban sugar quota allocated to other countries.[8] In the early 1970s, world commodity prices, including those for raw sugar, rose substantially. The original Sugar Act expired in 1974 and was not replaced. Problems began to arise in 1977 when the commodity price boom slowed and declining sugar prices led to pressures to reactivate government support programs. In 1977, a nonrecourse loan program, similar to the one described in Chapter 10, was established for refined sugar. Later acts specified that the loan program should generate no government outlay. Such outlays would occur if low market prices induced refiners to default on their loans, forcing the Commodity Credit Corporation (CCC) to take possession of large amounts of sugar. To prevent this, the USDA was given the authority to limit imports through the allocation of quotas to foreign suppliers. Since the early 1980s, import restrictions have been used to maintain domestic prices at or above the loan rate, which has been set well above world market prices.

The sugar quota system is a voluntary export restraint (VER), as described in Chapter 5. VERs set limits on the amounts that particular countries can export, with the sum of these national quotas set to maintain U.S. prices at a predetermined level. The governments of the individual exporting countries distribute the export quotas among their domestic refiners, which are then able to sell their allocation to U.S. importers at the higher U.S. price. Technically, VERs allow the exporting country to capture the quota rents associated with the trade barrier. Throughout the 1980s, the total quantities of imported sugar were severely restricted. Average annual imports fell from around 4.6 million short tons in the 1970s to 2.2 million tons for the period from 1982 to 1990. It was during this period that the divergence between U.S. market prices and world prices increased dramatically. From 1983 to 1990, U.S. import prices averaged 21.89 cents per pound, compared with world prices (Caribbean prices plus transportation to New York) of 8.26 cents per pound.[9] It is worth noting that the world market for sugar is somewhat peculiar. The EU operates a sugar program similar to the U.S. program for peanuts (see Appendix 10.2), which results in the dumping of large amounts of sugar on the world market for whatever price it will fetch. Much trade in sugar is conducted under long-term agreements so that the sugar never actually enters the world market. The result of these interventions and special arrangements is that the world market for sugar is quite thin and subject to great instability.

In 1989, the U.S. system of trade restrictions for sugar was found to be in violation of U.S. commitments by a dispute resolution panel set up by the General Agreement on Tariffs and Trade (GATT). The U.S. agreed to replace its VERs with a tariff-rate quota (TRQ), which allows a certain quantity of sugar to be imported duty-free or with a very low tariff. Any sugar above the duty-free quantity is subject to a high import tariff (16 to 18 cents per pound). This tariff is high enough to severely restrict the amounts imported or to prevent imports altogether, protecting the high U.S. prices. In the 1990s, the U.S. sugar program underwent further modifications following the Uruguay Round trade agreement that created the World Trade Organization (WTO) and the Federal Agriculture Improvement and Reform Act (FAIR) of 1996. The requirement that the sugar program generate no government cost is no longer in force, and the United States has agreed to reduce the tariff (from 17.62 cents to 15.82 cents per pound in 2000) applied to sugar imports above the low-duty quota set at about 1.26 million short tons.[10] The nonrecourse loan is still in force, although some details of its administration have been modified. This is the program that is analyzed in the study by Haley.

MODEL FOR THE WELFARE ANALYSIS OF THE U.S. SUGAR PROGRAM

Analysis of the U.S. sugar program will have to take account of a number of critical features of the sweetener sector. The production and consumption of sugar in the United States is influenced by domestic sugar prices, the prices and availability of HFCS and other substitutes, and sugar exports from other countries. In addition, it is possible that beet producers and cane producers will react differently to changes in market conditions, and this possibility will have to be explicitly recognized in the analytical model. The ERS model described by Haley takes account of these factors. In addition, it is based on the use of some fairly advanced statistical methods expected to yield more precise estimates of the model parameters. Before describing the model in detail, it may be helpful to reexamine the basic procedures followed in developing and using analytical models of this nature, as described in the last chapter.

Consider the problem of predicting how much sugar will be produced in the United States if the current sugar program is modified or eliminated. Economic theory as well as practical experience suggest that sugar refiners will adjust their output in light of the costs of raw sugar (in cane-producing regions) or sugar beets as well as the price of refined sugar. All of these variables depend on other prices and quantities in the sweetener sector, and this makes the problem of predicting changes in refined sugar production more difficult. If we ignore these complications for the time being, however, we can examine how an analyst might go about making such predictions. The first step would be to specify a model of sugar production that identifies the important factors influencing this activity. In a sense, we have already done so by noting that production depends on the output price and the costs of either sugar beets or raw sugar from cane. Suppose we decide to specify two models, one for the production of refined sugar from sugar beets and the other for production from raw sugar. In both cases there are other factors, such as the cost of other inputs, that need to be taken into account. The general form of these models would be to relate production to the expected output price, the cost of the relevant source of sugar (beets or cane), and the cost of other inputs.

Recall that supply of a good is represented as a function of its own price, with other factors such as input prices acting to shift the supply schedule. Such a relationship can be written as an equation showing that supply is a function of these prices. If we know that such an equation has a particular form and we have specific values for the different parameters, then we can use the equation to predict the impact of altering one of the variables. Suppose we know that refined sugar is related to its own price and the prices of inputs in the following manner:

$$(11.1) \qquad S = 9,000 + 0.36P - 0.14C$$

where S = sugar supply in thousands of tons, P = price, and C = input cost. This equation can be used to predict supply with and without the policy change. Suppose that the only impact of the policy change will be to lower sugar prices (P). Then we could simulate sugar supply at the prices expected to prevail if the policy is not changed and compare the results with those obtained by estimating supply under the new policy with lower sugar prices. Let C equal $250 per short ton and the price in the absence of the policy change be set at $400 per ton. Then equation (11.1) indicates that supply would be about 9,109,000. If the implementation of the new policy causes the price to fall to $200 per ton, then the expected supply would be 9,037,000, and the impact of the policy would be to reduce supply by 72,000 tons.

There are two obvious procedural questions that arise in considering this exercise. The first concerns the functional form of the equation: Why would one expect equation (11.1) to be linear? The second question is where the parameters (9,000, 0.36, and 0.14) come from. These questions are beyond the scope of this book. We are concerned here only with providing an overview of policy modeling and cannot hope to address all the complexities associated with the specification and estimation of economic models. Although many alternatives to linear equations are available for economic modeling, it is often more convenient to use some sort of linear representation to facilitate the statistical estimation of the parameters. An equation such as equation (11.1) can be parameterized by using linear regression methods and data on supplies, prices, and costs. Estimation of equation parameters requires attention to a host of problems related to the nature of the data, the kinds of equations being estimated, restrictions that hold across equations (e.g., that the predictions from the equations for cane supply and beet supply have to add up to be equal to the predicted total supply), and so on. These complications cannot be dealt with here and will be left to later courses in statistics and econometrics.

What is important in the preceding discussion is to understand that the various relationships in the sweetener sector described earlier can be represented as a series of mathematical equations and that the parameters in these equations can be estimated on the basis of historical data and various statistical methods. In some cases, a critical model parameter does not need to be estimated statistically, because it is known from the technical relationships involved. For example, the ERS sugar model employs a technical parameter referred to as the recovery rate. The recovery rate for sugarcane shows the amount of refined sugar that can be obtained from a ton of sugarcane, and this value is known precisely by the technicians who operate the processing plants. Once a model is specified and its parameters determined, it can be used to predict the impact of a policy change by altering some of the explanatory variables in the model. Note that the procedure followed in considering the impact on supply of a change in policy was to use the model to predict what would happen without the policy change and compare that with what is expected to happen with the policy change. This should come as no surprise, as we have repeatedly emphasized the importance of locating an appropriate counterfactual for comparison with the predicted effects of a policy change.

The actual model used to analyze the U.S. sugar program includes models of primary sugar production, industrial and nonindustrial sugar demand, equations to represent the movement of sugar through the various processing stages, and an international component used to close the system. The equations in this model are described in some detail in Appendix 11.1. The production model includes two systems, one for beet sugar and one for cane sugar. In the case of sugar from beets, the system is further broken down into regional supplies, while for cane sugar, separate supply relationships are modeled for each of the four states in which sugarcane is grown. In studies of agricultural production, it is common to exploit the fact that production is by definition equal to the number of acres planted multiplied by the yield per acre. Faced with changing price incentives, farmers are likely to make adjustments in the number of acres planted to a given crop. High prices may induce them to expand acreage, while low prices may lead them to plant more of some other crop. Once planted, the total amount produced will depend on the yield, which, in turn, depends on such unpredictable matters as the weather. Average yields are well known, and it is often more accurate to estimate an acreage equation and use that to predict the acreage planted, with total production determined by multiplying average yields times the predicted acreage. This is what is done in the analytical model developed by the ERS team.

There are acreage equations for each of five beet-producing regions as well as for the four states in which cane is produced. Farmers are assumed to adjust the acreage they plant to sugar-bearing crops in light of sugar prices, the costs of production, and a regional processing capacity constraint. Monetary values (prices, costs, expenditures, etc.) included in the model are expressed in real terms, that is, corrected for inflation. The amount of sugar-processing capacity in a region is assumed to influence farmer planting decisions. Initial processing of sugar is generally done in the region where the crop is grown, and the potential for processing sugar in a region is likely to have an impact on planting decisions. Regional processing capacity is not static; new plants can be built or existing plants can be run more intensively if economic conditions are favorable. Changes in regional processing capacity (measured as the maximum amount of sugar produced in a region in recent years) are likely to be made on the basis of net returns derived from sugar prices and the costs of processing the raw product. In the actual model used for this study, these functional relationships are statistically estimated using prices and costs.

Consider the model for Florida sugar production and its use in analyzing a policy change. Reform of the sugar program will have an impact on sugar prices. The Florida production model can be used to simulate production in that state, with the price change and without it. The first step is to use the capacity equation to predict processing capacity in the area. Predicted processing capacity is then introduced into the acreage equation along with the two sets of prices (with the policy reform and without) and costs (assumed not to change as a result of the policy reform) to predict acreage. The predicted acreage is then multiplied by the average per-acre yield of sugarcane and by the recovery rate described earlier to obtain the amount of refined sugar represented by the cane produced in Florida. The same procedure is followed for the other regions and states, and all the predicted amounts of sugar produced are added up to obtain total U.S. sugar production.

The second component of the model is a representation of sugar demand. This model has three stages. The first stage is made up of six equations representing retail demand for bakery, beverage, dairy, confectionery, canned/frozen/bottled, and other food products that contain sweeteners (sugar, HFCS). Demand for each sector's output depends on real consumer prices and a measure of real food expenditures. The food manufacturing sectors demand both sweetener and nonsweetener inputs. The demand for sweetener inputs is modeled as a function of a weighted average of all sweetener prices and the output of the various industrial sectors. For example, demand for bakery goods from the first stage is taken as that industry's output and used, along with the sweetener price, to explain the bakery industry demand for sweeteners. In the third stage, demand for sugar by each of these industries is determined by its price, the price of HFCS, and the industry sweetener demand. Nonindustrial demand is estimated in a parallel system, with final household demand depending on sugar prices, the prices of sugar-containing products, and overall sweetener demand. All these sources of demand are then added up to obtain total retail demand for refined sugar.

Between the production and demand components of the model are all the activities associated with processing, wholesaling, and retailing. To link the production component, which provides estimates of the quantities of sugar equivalent to the amounts of the raw products produced, with the demand component, which models demand for refined sugar at the retail level, this model employs a series of estimated margin equations, with prices at the different levels in the system shown as functions of marketing costs, sugar demand, and variables to represent other factors that might affect these prices. To determine the prices within the system being modeled, it is necessary to introduce world trade. Figure 11.1 is similar to the trade model illustrated in Appendix 10.1. If the United States imported no sugar, then the equilibrium price would be established where the demand schedule (derived from all the industrial and nonindustrial demands and linked to the supply side by the estimated margin equations) intersects supply at P_a.

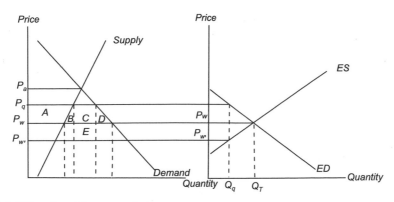

Figure 11.1: U.S. and world sugar markets.

If the United States imported sugar freely, excess demand from the United States would intersect excess supply from the rest of the world at a price of P_w in the right-side graph of Figure 11.1. Sugar imports are limited by the application of a TRQ, however. Instead of importing Q_T at P_w, the United States limits imports to Q_q, raising the domestic price to P_q. The U.S. policy also has the effect of depressing the world price from P_w to P'_w. A model of excess supplies from the rest of the world is not constructed for this analysis. Instead, the authors use an independently estimated elasticity of excess supply to predict the impact of U.S. trade liberalization on world prices. This is necessary because when the model is used to simulate the elimination of the U.S. trade barrier, the world price will return to P_w, which will then be the effective price in the United States.

But, of course, P_w will depend not only on excess supply from the rest of the world but on U.S. excess demand determined by all of the supply and demand components included in the rest of the model. In a sense, one might think of this model as using a large number of equations to figure out the initial change in U.S. prices following trade liberalization and then measuring the adjustments in supply and demand to this initial change as well as the impact of those adjustments on the price change itself and so on until the system converges to a new equilibrium. It should be emphasized that the way this model is structured is just one of many possible specifications. Appropriate model structures depend on the use to which the analysis is to be put as well as technical and theoretical considerations. As in the case of benefit-cost analysis, the design of an analytical model is more an art than the simple application of a set of rules.

RESULTS OF THE ANALYSIS

The preceding description of the ERS sugar model may be somewhat confusing. The important concept to retain from this discussion is that the ERS analysts have used their understanding of the relationships in the U.S. sweetener sector to structure a set of equations that can be used to conduct an experiment. The experiment is to simulate supply, demand, and imports under current circumstances (the baseline scenario) and compare that with a simulation of what is likely to happen if the TRQ is eliminated and sugar is allowed to freely enter the U.S. market (the trade-liberalization scenario). Each simulation will generate predictions of prices and quantities that can be used to compute changes in producer and consumer surplus. Because of the rich detail in the model, changes in producer surplus, for example, can be computed for Florida cane growers, northwestern beet growers, and so on.

The results of these simulations indicate that elimination of the TRQ will lower U.S. retail, wholesale, and raw sugar prices by 13 to 23 percent, compared to the baseline scenario in which the TRQ is not eliminated. The largest impact is at the unprocessed level. World prices are predicted to increase by 98 percent as U.S. excess demand rises following elimination of the TRQ. Note that such changes would be consistent with a world price of 8 or 9 cents per pound, compared to a U.S. price of 22 cents per pound (the world price would almost double to 16 to 18 cents, while the U.S. price would decline by 23 percent to about 17 cents). The policy change will also lead to a fall in HFCS prices as sweetener consumers substitute the lower-cost sugar for corn sweeteners. These price changes cause and are caused by adjustments in production and demand. Compared to the baseline, imports increase by 453 percent as domestic production falls and demand increases. Cane production is predicted to be 38 percent lower without the TRQ, while beet production declines by 19 percent. The production changes vary from a fall of about 15 percent in Great Plains beet production to a 56 percent decline in Louisiana cane production. Total U.S. sugar production is expected to fall by 28 percent, from around 9 million tons to 6.5 million tons. Total supplies of raw sugar from domestic sources and imports is 27 percent higher after elimination of the TRQ, while HFCS deliveries decline by almost 8 percent. Industrial demand rises 20 percent, while nonindustrial demand increases around 12 percent and total demand is 17 percent higher. The greatest increase in sugar demand is in the beverage industry, which, according to the model, will substitute some sugar for the HFCS that is virtually the only type of sweetener used in that industry in the baseline simulation.[11]

Haley uses these results to estimate the changes in consumer and producer surpluses for sugar users and producers and for HFCS producers. Figure 11.1 can be used to illustrate the impact of trade liberalization in the U.S. sugar sector. The effect of eliminating the TRQ is to lower internal prices in the United States from P_q to P_w. Consumers gain the areas labeled A, B, C, and D. Producers lose area A. Under current policies, exporters with quota rights are able to purchase sugar at the depressed world price, P'_w, and sell their allotments at P_q, thereby capturing the quota rents labeled C and E. Elimination of the TRQ means that area C is transferred back to U.S. consumers, while area E disappears as the world price returns to P_w. Areas B and D are efficiency gains. Haley does not present separate estimates of areas B, C, D, or E, although areas B, C, and D can be determined by calculating the difference between the increase in consumer surplus and the loss of producer surplus in the sugar sector. Because Haley is not concerned with the welfare effects in other countries, no effort is made to estimate the quota rents. He does estimate the change in producer welfare in the HFCS industry. The net effect of trade liberalization is the gain in consumer surplus minus the losses to producers of sugar and HFCS.

Note that eliminating the TRQ does not affect government revenues. The reason is that the program was structured so that there would be no government costs, although some of the recent modifications to the program may result in the potential for the program to generate government expenditures. For the period analyzed in this study, however, the tariff on imports beyond the duty-free quota was set high enough to prevent imports of sugar greater than the basic quota. The program has been based on an implicit tax on consumers resulting from the import restrictions, with very little tariff revenue or costs associated with defaults on the CCC loans. The welfare implications of eliminating the TRQ as estimated by Haley are presented in Table 11.3, along with earlier estimates published by the U.S. General Accounting Office (GAO).[12] The figures in the table represent annual costs and benefits of eliminating the program. One of the objectives of Haley's study was to contrast the results of the GAO study with those based on the ERS model. As can be seen in Table 11.3, the GAO found that the sugar program represented a much larger annual cost to society, as shown by the net benefits from elimination of the program (Net Impact column) of $291 million. In contrast, Haley's estimate of these net benefits is only $34 million.

Table 11.3. Welfare effects of eliminating the U.S. sugar program, as measured by Haley and GAO (in millions of dollars)

	Gain in Consumer Surplus (C)	Loss of Sugar Producer Surplus (S)	Loss of HFCS Producer Surplus (H)	Net Impact ($C - S - H$)
Haley/ERS	$674	$437	$203	$34
GAO	$1,400	$561	$548	$291

Haley and the GAO actually use their results to answer the slightly different question of how much the sugar program costs the United States. The consumer benefits of eliminating the program are just equal to the costs to consumers of not eliminating the program. From this perspective, the program benefits sugar and HFCS producers by $640 million (437 + 203 = 640) for Haley or $1,109 million (561 + 548 = 1,109) for the GAO while costing consumers $674 million (Haley) or $1,400 million (GAO). The reason that Haley's estimate of the net effect of the program is lower than that of the GAO is because the ERS model includes more details, which lead to smaller decreases in consumer prices than in the GAO study. Although the net effect of the sugar program is still negative (costs greater than benefits), Haley's results suggest that it is less costly than indicated by the GAO study. This result could be interpreted to mean that the cost of delivering program benefits to sugar and HFCS producers is less than suggested by the GAO result.

Suppose that we accept this conclusion. Would that mean that the program is justified because it is relatively efficient in delivering benefits to producers? In the context of U.S. consumer expenditures of about $5.5 trillion in 1998, $674 million is insignificant. However, consumers would still be better off simply writing a check to sugar and HFCS producers for an amount equal to the benefits transferred by the sugar program ($640 million, according to Haley) rather than paying an implicit tax on the goods they buy, a tax that is not collected by the government to provide public goods but rather simply transferred to producers. Moreover, it is important to emphasize the study's limitations. It does not include impacts on sugar consumers, producers, and refiners in other countries; environmental effects of sugar production in the United States and in other countries; the impact of changing sugar consumption on health; or the full distributional implications of the program.

Consider the international repercussions of the sugar program. In Figure 11.1, area *A* ($437 million) is the benefit of the program to *sugar* (Figure 11.1 does not include HFCS supply, demand, or economic surplus) producers. The difference between this benefit and the consumer cost ($674 million) is $237 million, most of which is transferred to foreign suppliers as quota rents. The elimination of the TRQ would put an end to these quota rents at the same time that it would almost double the world price. With the exception of Cuba and the EU, most of the major sugar exporters appear to have sugar quotas for the U.S. market. These countries would lose the quota rents currently being transferred from the United States if the program is eliminated but would benefit from the general rise in world prices and the ability to sell more to the United States. Of course, sugar consumers in the exporting countries would face higher prices and a loss in consumer surplus.

A recent study published by the GAO sheds some light on this issue.[13] The study was done by a research team at Iowa State University using a large multisector model (see Chapter 13). The study analyzes sweetener markets in 1996 and 1998 with and without the U.S. sugar program. The authors of the report find that U.S. sugar prices would fall substantially if the program were eliminated, while world prices would increase modestly. Producer benefits from the current U.S. program were $788 million in 1996 and $1,045 million in 1998. Total U.S. consumer losses were $1,471 and $1,938 million, respectively, giving net social losses of $683 million and $893 million for the two years. These social losses are more than twice the size of the estimated losses in the first GAO study and more than twenty times Haley's estimate. The benefits of the program flow entirely to sugar producers and processors because, the authors argue, technology changes in corn wet-milling have allowed HFCS producers to lower prices so that a fall in sugar prices no longer leads to substitution of sugar for HFCS.

The social loss is divided between losses due to economic inefficiencies (deadweight losses) and transfers of quota rents to foreign producers. For 1996, the authors estimate inefficiency losses of $273 million, compared to $410 million in quota rents. The figures for 1998 are $532 million and $361 million, respectively. The loss of quota rents benefits U.S. consumers but lowers producer welfare in exporting countries. In this study, the world price for raw sugar is predicted to increase about 1.2 cents per pound, and U.S. imports are some 1.1 to 1.6 million short tons higher without the program. World trade in raw sugar was about 22 million short tons in 1996 and 1998. If increased U.S. imports are an addition to this total, then 23.1 to 23.6 million tons would be traded at the slightly higher price. This translates into a gain of $554 million to $566 million for sugar exporters against the loss of quota rents of $410 to $532 million. It would appear that the elimination of the U.S. sugar program would raise welfare for foreign producers, although the net effects in exporting countries may well be negative when consumer losses are added.

Another important element that is not included in these studies is the impact of sugar production on the environment, both within the United States and in the exporting countries where sugar production may expand. Sugar beet production probably has little environmental impact other than the common effects of various cultural practices (e.g., fertilization) on such resources as ground and surface water. In contrast, Florida cane production has been implicated as a source of the excess phosphorous that is damaging the Everglades, an immense wetland of great ecological importance. Haley predicts that eliminating the program would cause a reduction of Florida cane production of around 26 percent, and this could reduce the negative externality associated with Florida sugar production. Internationally, expanded production could require additional land and may lead to further clearing of rain forests. This would have a negative effect on biodiversity and global warming. Without a great deal of further study, it is impossible to draw any reasonable conclusions about the overall environmental impact of eliminating the U.S. sugar program.

Sugar is not a particularly healthy food, and increased consumption in the United States could lead to public health problems, while the reduced consumption in the exporting countries could have a positive impact on health. As in the case of the international and environmental impacts, much more information would be needed to estimate the impacts of eliminating the U.S. sugar program in these areas. Finally, there could be a variety of less direct effects that would need to be taken into account in making a full assessment of this program. For example, the predicted decline in the Louisiana sugar industry could lead to social impacts associated with job losses and the abandonment of plants and equipment that may have little salvage value and few alternative uses. Both domestically and internationally, changes in the U.S. sugar program are likely to have distributional implications that may need to be taken into account. Gains to low-income farmers in developing countries might need to be weighted more heavily than the losses to reasonably well-to-do owners of processing plants in the United States. International, environmental, health, indirect, and distributional impacts of the sugar program could be sufficiently large to completely alter the net effects of the program as measured by Haley and the GAO.

CONCLUSION

Although the ERS model described in this case study and the models used for the two GAO studies are relatively complex, they do not include enough detail to provide predictions about some significant effects of the U.S. sugar program. They do, however, provide information that is relevant to the policy debate within the United States on the future of the sugar program. The authors of the GAO studies argued that the sugar program was extremely costly to U.S. consumers and that it should be eliminated.[14] Haley's study suggests that the costs estimated by the GAO may be exaggerated. While information from these studies is of use in the political debate over agricultural policy in the United States, that debate would be better informed if information on the international, environmental, health, and other impacts of the program were included. Information on the international repercussions of the program could be used in foreign policy discussions as well as in negotiations on trade policy reform conducted within the WTO. Still, these studies attempt to answer questions that are central to the U.S. political debate and therefore constitute useful policy analysis.

If one limits attention to the relatively narrow question of the net cost of the U.S. sugar program to U.S. consumers and sugar and HFCS producers, the main result of Haley's study is that these costs are less than the initial GAO estimates. What should policy makers conclude from this about the advisability of continuing, discontinuing, or modifying the program? The fact that both Haley and the two GAO studies find that the program generates consumer costs that are greater than the benefits to sugar and HFCS producers would seem to be a fairly strong argument that this program should be discontinued. From this perspective, the crucial result is that the program gives rise to net costs, and the question of whether these costs are large or small is of limited interest. On the other hand, if one believes that an important social objective is fulfilled by providing support to sugar and HFCS producers, it might be easier to justify the program if the net costs are $34 million rather than $291 million or $700 to $900 million.

As noted in Chapter 10, there has been a long history of providing financial support to farmers in the United States. Much of the original rationale for this support was the recognition that there is inherent instability in agriculture, which can be socially harmful. In addition, farm incomes were often lower than the incomes of workers in other sectors with equivalent levels of education and training. Programs that benefit farmers could be justified on the grounds that they correct the instability and unfairness in the agricultural system. With modern technology and other changes, however, these arguments have lost some of their force. While there is still weather-induced instability in agriculture, it should be possible for farmers to purchase insurance against these risks. In addition, as noted in Chapter 10, farm incomes appear to have caught up with incomes in other sectors. For many, sugar and HFCS producers are simple rent seekers lobbying for pork barrel programs. Archer Daniels Midland (ADM), a large agribusiness firm with substantial corn wet-milling capacity, has actively lobbied for continuation of the sugar program even though ADM produces no sugar. The high sugar prices induced by the sugar program allow ADM to sell more HFCS at higher prices. ADM has contributed heavily to political candidates and, according to Bovard, "is the driving force behind the sugar lobby in [the] battle over the future of the sugar program."[15]

Others believe that farmers, including cane and beet growers, merit support because of the intangible contributions that family farmers make to national well-being. For many, the true value of family farms is far greater than any measure of their economic importance would indicate. Family farmers are thought to be better stewards of the land, and family farms are often seen as sources of traditional moral values without which the United States is likely to become corrupt and venal. Finally, it is often argued that all U.S. citizens benefit from a well-tended countryside complete with red barns and other attributes that make the countryside more attractive.[16]

These observations lead to the idea that an agricultural system based on family farms gives rise to positive externalities that have both monetary and aesthetic value to society. Family farms often fail because market prices are too low for them to cover their costs. The low prices result from stable demand and growing supplies due in large measure to the output of the large, efficient, corporate farms that seem to dominate rural America. Farm price supports could be seen as compensation for the positive externalities generated by family farms. Such compensation would help them survive and prevent the complete industrialization of U.S. agriculture by corporate farms that place profits ahead of community, the environment, and rural people. This argument is prominent in Europe, where it is referred to as multifunctionality (see Chapter 12).

Such arguments often grow out of communitarian ethical principles. From this perspective, family farms are essential to the well-being of rural communities, which have intrinsic value. An empirical question that divides advocates of family farms concerns their technical efficiency relative to corporate farms. For some, family farms are capable of being just as efficient as the industrial operations. The reason that they are unable to compete is simply that the system is biased against them.[17] For others, family farms are unable to realize the economies of scale that make the larger corporate farms more efficient and are likely to disappear, along with the positive externality they provide, if they do not receive government subsidies. All these arguments involve empirical claims that are controversial. More fundamentally, they tend to grow out of the philosophical split between those who are more individualistic and those with communitarian leanings. In other words, much of the debate about agricultural policy and the need to protect family farms or farms in general has to do with conflicting moral visions.

There is an empirical fact that undermines the justification for the sugar program and other government farm policies on the basis of the need to protect family farms. Most of the subsidies associated with these programs accrue to the large corporate farms and agribusiness firms, such as ADM, rather than to firms that the family farm advocates would identify as family farms.[18] If farm programs could be focused so that subsidies would flow only to family farmers (assuming that there is some agreement on just what a family farm is), their cost would be much less than is now the case. This suggests that current programs are highly inefficient as mechanisms for compensating family farmers for any positive externalities they may generate, because they transfer substantial amounts of income to nonfamily farms.

Two other considerations are relevant here. First, family farmers may generate negative externalities, such as pollution of the Everglades, as well as the positive externalities previously noted. Should compensation to family farmers be based on the net effects of their activities, computed as the value of the positive effects minus the negative externalities? Second, whether one wishes to justify transfers to all farmers or just to those designated as family farmers, it is clear that a more efficient way to provide support or compensation is through direct payments rather than price supports. The sugar program costs consumers more than the total benefits flowing to cane, beet, and corn growers combined. Direct payments from the federal budget equal to these benefits would save consumers and taxpayers from $34 million to $900 million, depending on which set of results is used.

Of course, producers correctly realize that direct payments would be more vulnerable to political attack than the relatively well-hidden subsidies generated by trade barriers and price supports. In Chapter 5, we reported Stiglitz's explanation of why it was impossible to replace dairy support prices with direct payments when he was on the Council of

Economic Advisers. The government was unable to make a credible commitment that the payments would be maintained indefinitely, and the same problem arises when policy makers consider reforming the sugar program. These political realities generally mean that only small changes in policy will be possible. Although the analysis of the sugar program in this case study has been based on comparing the economic impacts of continuing the program with eliminating it, neither of these options is likely to be implemented. The program will probably be modified, at least to some extent, to comply with future international trade agreements. It is even possible that the ongoing public debate will eventually force fairly substantial reductions in the public subsidization of this industry. Any such modifications will be guided by analytical results reported in studies such as the one described in this chapter. Such policy analyses can play important roles in helping to define the issues, identify the important relationships, provide information on the costs and benefits of the program, and thereby support and advance democratic deliberation.

DISCUSSION QUESTIONS

1. Why is it necessary to include corn sweeteners in a model of the U.S. sugar industry? Discuss the effects of the sugar program on the prices, supply, and demand of corn.

2. In Haley's model, the elasticity of demand for sugar by the beverage industry is estimated to be -3.43. Explain the implications of this estimate for sugar and HFCS demand.

3. The sugar program imposes a net cost on U.S. society, and its elimination would be a Pareto better change. Why would it be difficult to eliminate this Pareto inferior program?

4. Aside from the direct costs and benefits to sugar and HFCS producers and consumers, what other costs and benefits should be included in the analysis of this program?

5. Explain the use of counterfactuals in policy analysis. What is the counterfactual in Haley's study (and in the GAO studies), and how is it used in the analysis?

SUGGESTIONS FOR FURTHER READING

General Accounting Office. *Supporting Sugar Prices Has Increased Users' Costs While Benefiting Producers*. Washington, D.C.: GAO, June 2000.

Mintz, S. W. *Sweetness and Power*. New York: Penguin Books, 1986.

Polopolus, L. C., and J. Alvarez. *Marketing Sugar and Other Sweeteners*. New York: Elsevier, 1991.

APPENDIX 11.1: TECHNICAL PRESENTATION OF THE ERS MODEL

The ERS model described by Haley has two primary components, production and demand, as well as several equations to define the links between various prices in the system (world prices, raw sugar, wholesale and retail prices).[19] The production component includes nine submodels, one for each of five beet-producing regions (Far West, Northwest, Great Plains, Great Lakes, and Red River Valley) and four cane-producing states (Florida, Louisiana, Texas, and Hawaii). Each of these regional or state models includes two statistically estimated equations and three identities. The structure of Florida cane sugar production, for example, is:

(11.2) Capacity constraint: $CAP_t = \text{Max } (CP_{t-1}, CP_{t-2}, CP_{t-3})$

(11.3) Costs: $CPR = CPP + CPRP$
$$CPA = CPT + YLD$$

(11.4) Capacity equation: $CAP_t = Cap(P_{t-n}, CPR_{t-n}), n = 1, ..., 5$

(11.5) Acreage equation: $ACR_t = Acr(CAP_t, P_{t-n}, CPA_{t-n}), n = 1, ..., 5$

(11.6) Production: $CP_t = ACR_t \times YLD \times R$

where

CAP = capacity constraint (maximum of three previous years of production)
CP = sugarcane production in Florida
CPR = costs per pound of producing and processing cane sugar in Florida
CPP = costs per pound of producing cane
$CPRP$ = costs per pound of processing cane
CPA = costs per acre of cane production
CPT = costs per ton of raw cane
YLD = cane yield (tons per acre)
ACR = acreage harvested
P = price (cents per pound) of raw sugar
R = recovery rate (amount of refined sugar obtained from a ton of cane)
t = year index (t = 1981 to 1995)

Equations (11.4) and (11.5) are estimated statistically. Equation (11.4) shows sugarcane processing capacity as a function of raw sugar prices and the costs of sugar cane production and processing. All monetary values (prices and costs) are corrected for inflation. Equation (11.5) represents the acreage planted to cane in Florida as a function of the state's capacity constraint, raw sugar prices, and the per-acre costs of producing cane. The other equations are definitions. Equation (11.2) defines the capacity constraint as the maximum production during the three previous years. Equations (11.3) define the two cost variables that enter equations (11.4) and (11.5). The first is the cost of producing sugarcane and processing it into raw sugar. The second is the per-acre cost of producing sugarcane. The final equation computes raw sugar production in Florida as the product of predicted acreage, average historical yields, and the recovery rate, defined as the amount of raw sugar that can be obtained from a ton of cane.

The other eight systems are identical to the Florida system just described, although minor adjustments are made to account for differences in the way cane and beets are measured. Each of these submodels gives predictions of sugar production in particular regions or states. The predictions from the nine submodels are added up to obtain an estimate of total U.S. production. The capacity and acreage equations are estimated with a second-degree polynomial distributed lag specification. Such specifications are used frequently to account for the fact that producers may only be able to adjust output over a number of years in response to changing economic conditions. The results of the statistical estimation are not reported here but can be found in the study. In general, the parameters of the equations appeared reliable, and the statistical properties suggest that the equations are fairly good representations of the relationships being modeled.

The demand model includes demand by industries (bakery, beverages, canned/bottled/frozen foods, confectionery, dairy, and other foods) as well as household demand referred to as nonindustrial demand. For each industrial sector, demand for sugar is modeled as a share of total sweetener demand by that sector. A typical sectoral model (e.g., bakery) has the following structure:

(11.7)　　Sweetener price: $RPS_{ti} = ws_{ti}PS_t + wf_{ti}PF_t$

(11.8)　　Sweetener demand: $SWT_{ti} = swti(RPS_t, Q_i)$

(11.9)　　Share of sugar in sweetener demand: $ws_{ti} = wsi(PS_t, PF_t, SWT_{ti})$

(11.10)　Share of HFCS in sweetener demand: $wf_{ti} = 1 - ws_{ti}$

where

RPS = sweetener price (weighted average of sugar and corn syrup prices)

PS = sugar price

PF = HFCS price

ws = share of sugar in sweetener demand defined as the quantity of sugar demanded divided by the total quantity of sweeteners demanded [sugar use/(sugar use + HFCS use)]

wf = HFCS share [HFCS use/(sugar use + HFCS use)]; $ws + wf = 1$

SWT = demand for all caloric sweeteners (sugar and HFCS)

Q = total sectoral production

t = year (1981–1995)

i = sectors: bakery, beverages, etc.

Equation (11.7) defines a price for sweeteners as a weighted average of the prices for sugar and HFCS. The weights are the shares of sugar and HFCS in total sweetener demand. Sweetener demand in a given sector is assumed to be a function of the retail sweetener price and the quantity of output from that sector. In the text of the report, Haley suggests that sectoral output is explained by an estimated equation relating bakery production, for example, to real bakery prices and real food expenditures. In the technical appendix, however, sectoral output appears to be introduced exogenously into the model. In any case, the more goods the bakery sector produces, the more sweetener it will wish to purchase. In contrast, sweetener demand is expected to be inversely related to its price given the fact that demand curves are downward sloping. Equation (11.9) is estimated with the prices of sugar and HFCS as well as total sweetener demand used to explain the share of sugar in sectoral demand for sweeteners. The share of sugar in sweetener demand is inversely related to its real price and positively related to the price of HFCS. If sugar prices increase relative to HFCS, then industrial users are likely to substitute HFCS for sugar. If HFCS prices increase relative to sugar prices, then sectoral users would decrease HFCS use in favor of the relatively cheaper sugar. The final equation simply computes the share of HFCS in sweetener demand.

Equation (11.8) is estimated for all six sectors. Equation (11.9) represents an Almost Ideal Demand System (AIDS) model. It is modified slightly for the beverage equation. In all cases, the dependent variable is the share of sugar in the sector's sweetener demand. Nonindustrial sugar demand is also modeled in two stages. In the first stage, household sweetener demand is explained by a weighted average real price for sweeteners embodied in various food products and sugar, and real food expenditures. In the second stage, real prices of sugar and sugar-containing products are used along with total sweetener demand to explain the share of sugar in total sweetener use by households. The equations for non-industrial use are:

(11.11) Household sweetener demand: $SWTH_t = swth(RPS_t, FX_t)$

(11.12) Sugar share of sweetener demand: $wsh_t = wsh(RP_{ts}, RP_{tb}, RP_{tv}, RP_{tf}, RP_{tc}, RP_{td}, SWTH_t)$

where

$SWTH =$ household sweetener demand (sugar and HFCS in processed foods and direct sugar consumption)

$RPS =$ weighted average real sweetener price

$FX =$ real food expenditures

$wsh =$ sugar's share of household sweetener consumption

$RP =$ real consumer price of sugar (s), bakery goods (b), beverages (v), canned-bottled-frozen (f), confectionery (c), and dairy (d)

$t =$ year (1981–1995)

Total sugar demand is then computed as the sum of the shares of sugar in predicted total sweetener demands by industrial sectors and households. To link the production and demand components, the model employs a set of margin equations. Retail demand is the sum of sectoral and household demand for sugar. Supply at the retail level depends on the amounts provided from the wholesale level and real retail prices. At the wholesale level, demand depends on the demand of retailers for sugar and other inputs and real wholesale prices. Wholesale supply is determined by the raw sugar supplied from the production component of the model and real wholesale prices. Finally, demand for raw sugar is derived from wholesale demand for sugar and other inputs along with real raw sugar prices, while supply is determined by the production component of the model. These relationships are brought together in four log-linear equations:

(11.13) Retail-wholesale linkage: $lnP_{rt} = a_0 + a_w lnW + a_z lnZ + a_q lnQ_{wh}$
$$lnP_{wh} = b_0 + b_w lnW + b_z lnZ + b_q lnQ_{wh}$$

(11.14) Wholesale-raw linkage: $lnP_{wh} = c_0 + c_w lnW + c_z lnZ + c_q lnQ_{rs}$
$$lnP_{rs} = d_0 + d_w lnW + d_z lnZ + d_q lnQ_{rs}$$

where

P = price
W = marketing cost index
Z = sector-specific demand shift variables
Q = quantities demanded as inputs by the retail and wholesale sectors
rt = retail level
wh = wholesale level
rs = raw sugar level

The model is closed by the determination of the raw sugar price, which depends on world market conditions and the U.S. TRQ, as illustrated in Figure 11.1. The TRQ raises raw sugar prices in the United States and depresses world prices. Independent estimates of the price responsiveness of world excess supply and demand are used to predict U.S. and world price after removal of the TRQ. In the baseline scenario, raw sugar prices as determined by the wedge between U.S. and world prices are used. Price linkage equations are then used to predict wholesale and retail prices. Raw sugar prices enter directly into the production component, while retail prices are used to predict demand. The model is structured so that the difference between raw sugar supply and demand will be equal to the quota. In the liberalization scenario, independent elasticities are used to determine the world price of raw sugar after elimination of the TRQ. This price enters the production simulations and is translated by the price-linkage equations into retail prices that drive the demand component. The model is structured so that the difference between supply and demand for raw sugar is equal to the predicted amount of imported sugar. The results of the two simulations provide predictions of all the prices and quantities shown as dependent variables in Equations (11.2) through (11.14) and these predictions can then be used to compute changes in producer and consumer surplus in the sugar sector. The information needed to compute changes in producer surplus for the HFCS sector is derived from independent assumptions.

NOTES

1. Stephen L. Haley, "Modeling the U.S. Sweetener Sector: An Application to the Analysis of Policy Reform," working paper no. 98-5, International Agricultural Trade Research Consortium, University of Minnesota, St. Paul, Minn., August 1998.

2. See S. W. Mintz, *Sweetness and Power* (New York: Penguin Books, 1986).

3. L. C. Polopolus and J. Alvarez, *Marketing Sugar and Other Sweeteners* (New York: Elsevier, 1991), 7.

4. Ibid, 17.

5. Ibid.

6. *Sugar and Swedtener*, SSS-226, Washington, D.C.: ERS/USDA, September 1999, Table 22, page 40.

7. Polopolus and Alvarez, *Marketing Sugar*.

8. Ibid., 225.

9. Author's calculations, using data from International Monetary Fund, *International Financial Statistics Yearbook* (Washington, D.C.: IMF, 1998).

10. W. W. Koo, "Major Issues for the Sugar Industry," in *2000 WTO Negotiations,* ed. L. M. Young, J. B. Johnson, and V. H. Smith (Bozeman: Montana State University Trade Research Center, 1999).

11. Stephen L. Haley, "Sugar Program: Changing Domestic and International Conditions Require Program Changes," CAO/RCED-93-84, April 16, 1993.

12. General Accounting Office, *Sugar Program: Changing Domestic and International Conditions Require Program Changes*, GAO/RCED-93-84 (Washington, D.C.: GAO, 1993).

13. General Accounting Office, *Supporting Sugar Prices Has Increased Users' Costs While Benefiting Producers* (Washington, D.C.: GAO, June 2000). Also available at the Internet Web site www.gao.gov/new.items/rc00126.pdf.

14. Other analysts have estimated even higher net costs. A Cato Institute study cites one estimate that the costs to consumers are $3 billion, more than twice the GAO estimate. James Bovard, "Archer Daniels Midland: A Case Study in Corporate Welfare," Cato Policy Analysis No. 241 (Washington D.C.: Cato Institute, 26 September 1995).

15. Ibid. ADM has also played a major role in promoting continued subsidization for ethanol, another output from the corn wet-milling industry. The company and its chief executive officer settled charges of fixing HFCS prices and other illegalities out of court.

16. See Gary Comstock, ed., *Is There a Moral Obligation to Save the Family Farm?* (Ames: Iowa State University Press, 1987).

17. For arguments of this nature, see K. M. Thu and P. Durrenberger, eds., *Pigs, Profits and Rural Communities* (Albany, N.Y.: SUNY Press, 1998). Many of the writers in this book suggest, for example, that agricultural research is biased against family farms, developing technological innovations that work only on large farms and neglecting technical options that would benefit smaller, family farms.

18. Deciding what is or is not a family farm is not obvious. Such large agribusiness firms as ADM or Cargill are involved in farming as well as processing activities and are run by families that continue to control the firms. Other large farms are also family enterprises. D. W. Allen and D. Leuck note that the 1997 U.S. Census of Agriculture reports that more than 86 percent of U.S. farms are classified as family enterprises ("Family Farm Inc.," *Choices* 1 [2000]: 13–17). Family farm advocates such as Wendall Barry (in *Is There a Moral Obligation to Save the Family Farm?* ed. Gary Comstock [Ames: Iowa State University Press, 1987]) are irritated by questions of definition, claiming that everyone knows what family farms are without trying to specify the precise characteristics of such entities.

19. This description of the ERS sugar model is based on a working paper written by Dr. Stephen Haley ("Modeling U.S. Sweetener Sector"). The working paper is fairly brief, and some of the intricacies of the actual model are somewhat unclear. The description in this appendix is an attempt to figure out and describe as accurately and clearly as possible the model's structure and parameters. It is quite possible that I have misunderstood some of the details in the model, given the rather summary documentation at my disposal. I believe that the general description is fairly accurate and that any errors of interpretation that I have made will not seriously misrepresent the work done by Dr. Haley and his colleagues at ERS.

12

Conflicts between the United States and the European Union over Food: Hormones and Genetically Modified Organisms

INTRODUCTION

The case study presented in this chapter concerns ongoing conflicts between the European Union (EU) and the United States over the way in which food is produced.[1] In 1986, the EU banned the use of growth-promoting hormones in livestock production. In the United States, Canada, and several other countries, hormones are administered to beef cattle to increase the efficiency with which feed is converted into meat and to produce a leaner final product. The use of growth-promoting hormones lowers the cost of producing high-quality meat. The EU ban meant that EU producers would not be able to use this cost-reducing technology, placing them at a disadvantage relative to producers in other countries where the use of hormones is allowed. To protect EU producers from lower-cost imported products as well as to fully respond to the concerns that had led to the ban in the first place, the European Commission extended it to imported livestock products.

When this measure took effect in 1988, U.S. exports to the EU of high-quality beef and edible offal (liver, tongue, kidneys, etc.), valued at about $100 million per year, were eliminated altogether. From the U.S. perspective, the European policy was a simple barrier to trade aimed at protecting European livestock producers from foreign competition. EU policy makers, on the other hand, argued that the ban had been introduced to protect consumers from what was considered a potentially unsafe product. From the EU perspective, the extension of the ban to imported products was necessary to fully protect EU consumers, and the protection afforded to EU livestock producers was incidental.

The hormone conflict has raged ever since. In addition, new sources of disagreement have arisen with the widespread adoption of biotechnology innovations such as genetically modified food products and bovine somatotropin, another hormone used in dairy production. Since the earliest domestication of wild plants and animals, humans have selected particular strains or varieties with desirable characteristics. Early farmers retained seeds from the most productive plants for planting in the next season. Over time, this type of selection caused the plant varieties used by farmers to become quite different from their wild ancestors. Eventually, people discovered that plants and animals could be crossbred to produce better offspring. The greater understanding of genetics ushered in by Mendel's research in the nineteenth century led to further advances in plant and animal breeding. In the twentieth century, scientific plant breeding, including hybridization, led to extraordinary increases in crop yields, and animal breeders have been just as effective at developing new animal lines with desirable characteristics.

Until the biotechnology revolution of the 1980s, however, breeders were limited to combining genetic material from plants or animals of the same species. The development of techniques for separating, cutting, and splicing DNA sequences opened the door for the mixing of characteristics from different species. In the 1990s, many farmers in the United States began to plant corn varieties that had been engineered to include DNA from a type of bacterium (*Bacillus thurengiensis*, or Bt) that produces certain toxic chemicals. The genetically engineered corn differs from other corn only in its ability to resist damage from the European corn borer, an insect that is susceptible to the Bt toxins.

Or so it is claimed by proponents in the United States. For many Europeans, Bt corn and other genetically modified organisms (GMOs) are seen as dangerous to humans and other living things. Foods containing GMOs are often referred to as "Frankenfoods," after Dr. Frankenstein's monster in the horror story by Mary Shelley. Europeans and Americans seem to have very different perceptions of the risks and benefits of these innovations. In many ways, these differing perceptions are driven by different sets of values, but they also stem from ambiguities and confusion in the scientific and factual grounding for these technologies. Devising sensible national and international policies in this setting is extremely difficult. The case study presented in this chapter focuses on the analysis of these complex policy issues.

THE INSTITUTIONAL SETTING FOR INTERNATIONAL TRADE

The U.S.-EU policy conflicts over hormones and GMOs arise because the goods involved are traded. If there were no trade in beef or products that contain GMOs, each side in this debate could regulate food production technologies as it wished (see Box 12.1). As illustrated by the hormone ban, conflicting national food regulations may act as barriers to trade, leading to disagreements on appropriate regulation of the production and exchange of food products. To fully understand these intricate relationships, it is important to develop some background information on the institutional setting for international trade in general and agricultural trade in particular.

Box 12.1. The problem of food safety

In traditional food systems, most of the production, processing, and preparation of food is carried out within the household. Thus, a traditional peasant household processes and stores the crops it harvests and the livestock products it obtains from its animals. The stored and processed food is then used as needed over the course of the year. In such a setting, the problem of food safety falls primarily within the domain of the household itself. There may be a perception that the public sector has an educational role to play, particularly in terms of providing information that will assist households to avoid such deadly diseases as botulism.

Box 12.1. The problem of food safety (continued)

As the food sector is modernized, many of the functions that previously were carried out within the household begin to be handled by private firms through market transactions. Such food systems benefit from the efficiencies associated with specialization and the division of labor but may generate greater risks as food products that can harbor toxic organisms are transported and processed through ever more complex and convoluted arrangements. Market failures can arise in such systems and there is a general perception that assuring a safe food supply is an appropriate function for the government.

The primary market failure associated with food safety is the problem of information. Consumers may be unable to detect pathogens in the food they purchase from the supermarket. Retail food outlets and restaurants clearly have an interest in providing safe and healthy food to their customers and can be relied upon to take measures to avoid selling harmful products. Yet they too may suffer from the problem of detecting dangerous pathogens, particularly as food moves greater distances from its origin to its final consumers.

In many industrialized countries, the main food safety problem concerns bacteria that may grow in fresh and processed meat products. In the United States, salmonella and *E. coli* bacteria have been found in ground beef and chicken. In Europe, there have been outbreaks of illness associated with another bacterium, listeria, often found in processed meat products. These bacteria cause severe illness and even, on occasion, death. As they are invisible to the human eye, consumers cannot know of their presence, and because the risks are so great, some public regulation is warranted. The best control for them is usually some form of inspection (see Chapter 5 for a description of meat inspection in the United States).

With international trade, the same kinds of information problems may arise, and the public sector can be thought to have a responsibility to inspect and control imported products as well. Of course, such inspections add to the transaction costs of conducting trade and can thus act as barriers to trade. The case study presented in this chapter illustrates the problem of distinguishing regulations aimed at food safety from those that are put in place mainly to prevent imports and protect a domestic industry.

At the end of World War II, leaders from the allied nations of North America and Western Europe met in Bretton Woods, Vermont, to discuss the establishment of international organizations and institutions to oversee the world economy in the postwar period. The delegates to this meeting were intensely aware of the destructive national policies implemented in the 1930s. After the stock market crash in the United States, the U.S. government adopted the infamous Smoot-Hawley tariff in a misguided effort to increase domestic employment. Trade barriers generally do not reduce unemployment, because they lead to the destruction of more jobs than are created by the protected industries. When other nations retaliated by implementing trade barriers of their own, the world economy slumped badly. There was a dramatic decline in world trade, which is one of the mechanisms through which recovery from economic recessions can be brought about. The tariffs thus prolonged and deepened the Great Depression.

Conscious of this history and aware that stable financial and trade regimes would be essential for the reconstruction of the war-ravaged world, the Bretton Woods delegates set about constructing three international organizations to administer new regulations for international trade and finance. The first was the International Monetary Fund (IMF) established to oversee a new world monetary system. This system was based on an exchange-rate regime under which countries fixed the exchange rates of their currencies in relation to the U.S. dollar, which in turn was tied to gold by a U.S. guarantee to redeem dollars at a rate of one dollar equal to one thirty-fifth of an ounce of gold.[2] The financial system established at Bretton Woods remained in place until 1972, when various financial crises led to its abandonment. It was replaced by a regime in which major currencies are allowed to float, albeit with occasional interventions by national central banks. An exchange rate is the price of one currency in terms of some other currency. When exchange rates are allowed to float, this price is determined by supply and demand in currency markets. In a regime of fixed exchange rates, the price is set by governments rather than the market. Fixed exchange rates can be adjusted in light of economic events, but they do not vary on a daily basis as do floating exchange rates. As the international monetary system evolved, the role of the IMF has also evolved. Today, its main function is to provide financial support for countries with persistent imbalances in their foreign accounts (the balance of payments) and to assist these countries in the reform of their macroeconomic policies.

Many of the countries in need of financial assistance today are developing countries in Asia, Africa, and Latin America, and the new nations that have emerged from the collapse of the Soviet Union. The second organization set up at Bretton Woods, the International Bank for Reconstruction and Development (the World Bank), also deals with these countries, although its original purpose was to finance European reconstruction in the aftermath of World War II. Today, the main function of the World Bank is to provide loans and other support for development projects. In addition to the two financial organizations, the delegates to the Bretton Woods conference also wished to create a world trade organization. The United States opposed it, however, and the delegates settled for the General Agreement on Tariffs and Trade (GATT). The GATT is similar to an international treaty and has been incorporated into the national legal systems of the participating countries. The GATT was subsumed under the World Trade Organization (WTO) created in 1995. In the rest of this chapter, I will generally refer to the WTO, even though it is more accurate to speak of the GATT for events prior to 1995.

The primary objective of the WTO is to provide a means for countries to agree on the reduction and elimination of barriers to trade. Trade liberalization is pursued in periodic meetings of the member states in which the rules and regulations governing international commerce are debated and adopted. These meetings are known as "rounds" in which multilateral trade negotiations (MTN) are conducted. Rules are adopted under a unanimity voting rule, although there are ways in which countries that disagree can be exempted from a particular discipline. We noted in Chapter 5 that unanimity gives each party to an agreement a veto over any proposed action. This voting rule helps insure that the provisions adopted will not harm the national interests of a particular country, at least as the government of that country perceives those interests. Of course, the compromises required to achieve consensus often dilute the impact of the negotiations, and reaching agreement at all is a complex diplomatic affair.

At this writing, there have been eight negotiating rounds, the most recent of which, the Uruguay Round, took place between 1986 and 1994.[3] The early MTN were very effective in getting the member states to agree to reduce industrial tariffs. One reason for this is that membership in the early years was relatively small, making it easier to reach consensus. Only twenty-three countries participated in the first round, the Geneva Round of 1947, and, for the most part, these countries all had similar economic and political systems. The fifth round, the Dillon Round of 1960–61, involved thirty-nine countries. These early rounds generally only lasted about a year and accomplished substantial tariff reductions. Many economists believe that the trade liberalization accomplished in these early rounds set the stage for the impressive economic growth recorded in the second half of the twentieth century.

One reason so few countries were involved in the early MTN was that most parts of the world were colonies of the European powers at that time. Beginning with the independence of the Indian subcontinent in 1947, a wave of decolonization swept the world and led to the creation of dozens of new nations. The first thing most of these nations wished to do was to join the United Nations and the Bretton Woods organizations, all of which are part of the UN. Delegates from 74 countries attended the Kennedy Round (1963–67), while representatives of 99 countries participated in the Tokyo Round (1973–79). By the beginning of the Uruguay Round (UR) in 1986, 103 nations had joined the GATT, and this number had risen to 128 by 1995, when the UR was finally ratified. The later rounds all lasted several years, in contrast to the much shorter earlier rounds.[4]

In addition to periodic negotiations over the rules and regulations for international trade, the WTO also operates a permanent dispute-resolution system. Suppose, for example, that it has been agreed that the maximum tariff that can be applied to imported steel is 6 percent, and this tariff restriction has become part of the WTO rules on international trade. The agreement to adopt this maximum tariff means that all members have taken on a commitment to limit tariffs on steel to 6 percent or less. If a member country elects to charge a 10 percent tariff to assist its domestic industry in the face of intense competition from steel imports produced in countries with modern steel plants, it will be in violation of its commitment. Under these circumstances, the country facing the discriminatory tariff could file a complaint with the WTO, which would determine whether the complaint had any merit and, if so, would set up a dispute-resolution panel. Dispute-resolution panels hear arguments from both sides and then render a judgment on the charge that a country has violated its commitments under the WTO agreements. If the panel rules against a country, the government of that country has three options: (1) it can change its policy so that it is in compliance with the agreement, (2) it can retain its policy and negotiate acceptable compensation with the other party, or (3) it can refuse to change its policy or to compensate the other country. If it chooses the third option, the country that filed the complaint can implement a retaliatory tariff without being charged with violating its WTO commitments.

One of the principles at the heart of the WTO is a principle of nondiscrimination. Originally known as the most-favored-nation (MFN) principle, this concept is now referred to as normal trade relations (NTR). Members of the WTO are generally bound to grant NTR to all other members. What this means in practice is that if a country has agreed to set its tariffs on imported steel at 4 percent in compliance with the general WTO rule that steel tariffs be limited to 6 percent or less, then it must apply this tariff to all steel-exporting members of the WTO. In other words, the country would be in violation of its

WTO commitments if it attempted to apply a higher tariff to imports from one or more countries while maintaining its lower rate for others, even if the rates applied were within the WTO limit. NTR is designed to prevent countries from pursuing trade policies that discriminate in favor of or against particular countries. NTR is partially suspended for regional trade agreements such as the North American Free Trade Agreement (NAFTA) and the EU, which by definition are based on discrimination between members and nonmembers of the agreements even when both sets of countries belong to the WTO. There are also exemptions for trade between low- and high-income countries under the generalized system of preferences (GSP).

An example of the importance of NTR is offered by the case of China, whose application for membership in the WTO is pending. From the late 1970s, when trade relations between China and the United States were resumed, the U.S. Congress held an annual debate and vote on whether to grant China MFN/NTR for another year. While MFN/NTR status is automatic for members of the WTO, these countries can discriminate against non-members such as China with impunity. In fact, the U.S. Congress always approved MFN/NTR for China in its annual vote, although there was occasional opposition from various groups. In 2000, Congress ratified an agreement to grant permanent NTR in anticipation of China's eventual entry into the WTO and in return for Chinese tariff reductions on its imports from the United States.

There was strong political opposition from labor unions and various environmental and human rights groups. Unions feared increased imports of Chinese goods, even though granting permanent NTR actually opened the Chinese market to U.S. exports while granting China nothing that it had not been getting each year in the annual votes. Other groups critical of China's human rights and environmental records joined the unions in an unsuccessful effort to prevent permanent NTR. The Chinese government was willing to grant trade concessions to the United States in return for permanent NTR to avoid the annual debate over its human rights record and as a measure of its acceptance into the world community. This is another example of the kind of trade controversies that often arise in a world of conflicting political and moral visions.

Prior to the Uruguay Round, agriculture had largely been left out of the GATT system. The United States had insisted on exempting agriculture from the early liberalization efforts, because removal of trade barriers in agriculture would have interfered with the operation of its domestic agricultural programs (see Chapter 10). The development of the EU's Common Agricultural Policy (see Appendix 10.3) transformed the EU from a region that imported large amounts of food to a major agricultural exporter. By the 1980s, the United States and the EU were engaged in an all-out subsidy war, and there was widespread recognition that rules for international agricultural trade were needed. The UR negotiations on agricultural trade were extremely controversial and almost derailed the entire MTN. The initial U.S. position called for the complete elimination of all subsidies, including those that might be thought of as purely domestic programs (e.g., government-funded agricultural research). The EU called for managed agricultural markets, which essentially meant freezing world market shares at existing levels. A group of exporting countries known as the Cairns Group (Australia, Canada, New Zealand, Thailand, Brazil, and others) had a separate proposal but eventually threw its support to a revised U.S. position. Japan sided with the EU, setting up an intense debate between those who favored more protectionist policies led by the EU and those who favored more liberal policies led by the United States.

The Agreement on Agriculture that was eventually crafted was relatively modest, calling for reductions in export subsidies, the transformation of nontariff import barriers into equivalent tariffs that are to be reduced, and modification of trade-distorting domestic policies. The tariff and subsidy reductions ranged from 20 to 36 percent, so the Uruguay Round agreement on agriculture left a substantial amount of protectionism in place. Perhaps the most important aspect of the agricultural agreement was that it brought agriculture under WTO discipline for the first time. The agreement included a provision for new negotiations on agricultural trade to begin in 2000. These negotiations, which are aimed at further liberalization of agricultural markets, began on time but little progress had been made by the middle of 2001. The negotiations have been somewhat unproductive because the U.S. executive branch lacks fast-track negotiating authority (see Box 14.1 for an explanation of fast-track authority), the issues are controversial, and some countries have little incentive to make concessions on agriculture in the absence of potential concessions in other sectors. Such concessions could not be tabled because WTO members failed to launch a new MTN during the Seattle meeting in 1999 so the only trade negotiations taking place in 2001 were those on agriculture.

In addition to the measures noted earlier, the agricultural agreement also included new rules on the use of sanitary and phytosanitary (SPS) measures. The SPS agreement is of particular relevance for the case study that is the focus of this chapter. Its most important provision is that SPS measures have to be based on scientific evidence. In the case of the hormone ban, the new rules mean that the EU cannot block imports of livestock products from animals that have been treated with hormones unless it can be shown scientifically that such products are harmful to human or animal health.

THE EU HORMONE BAN

The EU ban on the use of hormones in livestock production grew out of a 1981 directive adopted by the Commission of the European Community (EC) banning the use of certain kinds of hormonal substances.[5] This initial regulation stemmed from the discovery of the hormone diethylstilbestrol (DES), known to be carcinogenic, in Italian baby food. In 1985, a second directive was adopted banning other kinds of hormones, notably those used in the United States and other countries to lower production costs in the livestock sector. The 1988 ban was applied to imported products preventing the sale of U.S. meat and meat by-products (edible offal) in the EC. This was not the first dispute with the EC over livestock products. In 1980, a dispute between the United States and the United Kingdom arose concerning the processing of chickens. The EC-approved method for cooling poultry after slaughter required water to flow in the opposite direction as the movement of the poultry line. The practice in the United States was for water and poultry to flow in the same direction. The U.K. banned U.S. chicken imports on the grounds that U.S. procedures were not in compliance with EC regulations.

The chicken-chilling controversy and the hormone ban both have to do with technical standards applied to processing and production methods (PPM). In classic trade disputes, trade barriers are applied to final products with no concern for the way in which they are produced. The GATT was ill prepared to deal with cases where the trade barriers apply to the way in which goods are produced. The GATT did include articles aimed at preventing technical standards from being used to protect domestic firms from import competition, and there were specific articles that allowed for the use of food-safety standards to protect public health. In 1980, the Agreement on Technical Barriers to Trade was adopted to clarify the GATT principles in this area. The agreement requires that countries choose the least trade-restrictive measures for accomplishing the objective, that imports be treated the same as domestic production, that clear and transparent procedures be used to notify exporters of the rules, and that international standards be used where possible.

The GATT provisions on technical standards mainly concerned final product characteristics. Thus, for example, a ban on the import of meat from countries in which hoof-and-mouth disease had not been eradicated was acceptable under the codes because the final product could be deemed harmful. This disease, which harms livestock but not human beings, is highly contagious and can be transmitted by processed and fresh products as well as by humans and through other means. In 2001, a strain of foot-and-mouth disease that had originated in India was inadvertently introduced into the United Kingdom from which it spread to the rest of Europe causing extensive damage to livestock.[6] The United States banned the importation of meat from Europe to protect domestic livestock industries.[7] This disease is caused by an airborne virus not by production and processing methods. In the case of the hormone ban, the EC claimed that the GATT rules simply did not apply because the rules were directed only at final products, not PPM. This is, of course, somewhat disingenuous because the main grounds for banning hormones, as claimed by the EC, were that hormone-treated meat might be harmful to human health due to the characteristics of the final product.

The EC argument that the use of hormones is harmful to human health is not supported by scientific evidence despite extensive testing by both European and American scientists. In the United States, hormones are administered through time-release implants placed in the animal's ear. At slaughter, the implant no longer contains hormones, and the ear is discarded in any case. Both natural and synthetic hormones are used, but until recently, it was virtually impossible to distinguish meat from animals with the hormone implants from meat from untreated animals. At about the time of the initial hormone ban, the EU began to apply another barrier to meat imports, the Third Country Meat Directive, in a more rigorous manner. This directive requires that packing plants in other countries be certified in compliance with EU norms before being allowed to sell their products in the EU. Because the procedures for obtaining this certification are cumbersome and often arbitrary, the Third Country Meat Directive amounts to a significant trade barrier. These factors lend some weight to the U.S. contention that the hormone ban is really protectionism disguised as a food safety standard.

It is worth noting that different beliefs about the way in which goods should be produced have become increasingly prominent in controversies over international trade. The United

States banned the import of canned tuna from Mexico and several other countries on the grounds that the way in which tuna were hunted by fleets from these countries led to excessive dolphin deaths. (See Box 14.2 in Chapter 14 for details of the dispute.) In essence, the United States was attempting to apply U.S. laws embodied in the Marine Mammal Protection Act—designed to protect whales, dolphins, and other marine mammals—to foreign countries. In part because it applied to PPM and in part because it was drafted in a way that made it impossible for Mexican fishing fleets to comply, the U.S. tuna ban was ruled to be in violation of U.S. GATT commitments. Mexico subsequently withdrew its complaint and joined with other countries to negotiate an international agreement on tuna fishing in the South Pacific. The number of dolphins killed in the course of capturing tuna has fallen to levels that are insignificant.

In addition to the tuna-dolphin case, there are many actual and potential disputes over the use of environmental and labor standards, for example, as barriers to trade. It is not because Nike tennis shoes are harmful that many labor groups wish to see trade barriers on their import. Rather, these groups argue, it is because they are produced in sweatshops. The problem for the WTO is to decide whether trade barriers based on beliefs about the correct way to produce goods should be allowed.[8] This type of question was one of the reasons WTO delegates were unable to agree on an agenda for the next round of trade negotiations.

The SPS agreement adopted by the UR delegates expands the provisions of the technical standards code by giving the WTO authority to rule on trade barriers that relate to PPM. It requires transparency in national regulations and encourages the use of international standards where possible. The Food and Agriculture Organization (FAO) of the UN administers a set of international food standards, the Codex Alimentarius, and there are international standards for plant and animal protection. Standards that are stricter than the international rules can be applied, but only if they are based on scientific evidence.

Under the new SPS agreement, the United States and Canada in 1996 filed a complaint with the WTO on the EU hormone ban. The dispute-resolution panel ruled in favor of the U.S. and Canadian position, suggesting that a labeling program be initiated so that consumers could identify meat from treated animals and make their own choices. The EU rejected this conclusion but lost its appeal of the decision. When the EU failed to comply with the panel's decision, the United States and Canada were allowed to retaliate by placing tariffs on goods imported from the EU.[9] The effect of these tariffs, of course, is to raise the price that U.S. and Canadian consumers have to pay for imported EU food items, such as French Roquefort cheese or Italian wine. Thus, ironically, in order to penalize the EU for its persistence in banning imports of their meat products, the United States and Canada ended up adopting a policy that actually lowers the welfare of their own consumers.

A MODEL OF THE EC MARKET FOR EDIBLE OFFAL

The EC hormone ban was analyzed in our study with a simple partial-equilibrium model of the EC market for edible offal.[10] The bulk of U.S. beef exports to the EC was made up of edible offal (also known as "variety meats," such as tongues, livers, kidneys, hearts, and so on). Such products are highly prized in Europe and many other countries, although not in the United States. Because U.S. consumers prefer other kinds of meat, much of the edible offal that comes as the inevitable by-product of beef production is used for pet food. The ability to sell such products at higher prices to Europeans and other foreign consumers adds to the profit margins in the U.S. beef industry. Still, the value of this trade was only about $100 million in the mid-1980s, out of total agricultural exports of about $30 billion and total exports of more than $300 billion at that time. Despite the relatively small value of this trade, the U.S. government and livestock industries reacted very strongly to its disappearance.

The first step in conducting the analysis was to develop a graphical representation (Fig. 12.1) of the relationships in this market. Because edible offal is a by-product of the beef-packing industry, its supply is expected to be perfectly inelastic with respect to its own price. The number of animals slaughtered depends on prices in the market for meat. The amount of edible offal contained in each animal is a more or less fixed proportion of its weight, so once the number of animals to be slaughtered is determined in the meat market, the supply of edible offal is also determined. This supply varies with conditions in the meat market, not with the price of edible offal. In other words, supplies are perfectly inelastic, as shown in Figure 12.1.

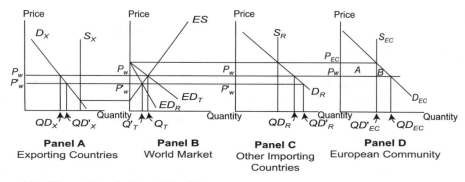

Figure 12.1. The world market for edible offal.

In the early 1980s, total world trade in edible offal was on the order of 700,000 metric tons per year. The EC imported a total of around 200,000 metric tons, about half of which came from the United States. Total U.S. exports were around 230,000 metric tons, so somewhat less than half of these exports were destined for the EC. The EC was the leading importer of U.S. offal, with Japan, Egypt, and Mexico accounting for most of the rest. Canada, Australia, and Latin America were other sources of exports. The structure of world trade suggested that it would be important to distinguish non-EC imports from those sold to the EC, as shown in Figure 12.1 in Panels C and D. The market for edible offal in the EC is illustrated in Panel D. EC demand intersects its perfectly inelastic supply at P_{EC}.

To keep the graphs simple, it is assumed that the autarky equilibrium price in the EC (P_{EC}) is the same as the autarkic equilibrium in other importing countries, as shown in Panel C. Excess demand is defined as the horizontal difference between supply and demand (see

Appendix 10.1), and the excess demands below P_{EC} from both importing regions are added horizontally to obtain the excess demand curve ED_T in Panel B. All exporting countries are treated as a group, with supply and demand shown in Panel A. Excess supply (ES) in Panel B is derived as the horizontal difference between D_x and S_x. The intersection of ES and ED_T determines the world price (P_w) at which the quantity Q_T is traded. P_w is the free trade price in the EC, other importing countries, and the exporting countries. EC imports are equal to the difference between QD_{EC} and domestic production of QD'_{EC}.

For this analysis, we assumed that all exporting countries use hormones, so the EC ban effectively eliminates all imports from that market. The counterfactual, the free trade situation prior to the EC ban, is compared with the circumstances predicted to prevail after the EC ceases to import edible offal. Because of the ban, the EC price increases to P_{EC}, where domestic supply intersects demand. The withdrawal of the EC from the world offal market means that excess demand in that market is defined exclusively by non-EC importers in Panel C and shown as ED_R in Panel B. The intersection of ES and ED_R defines a new, lower world price, P'_w. Demand in the other importing countries has increased (from QD_R to QD'_R) as a result of the lower world price. The demand from these other countries absorbs some of the shock of the EC withdrawal from the market. Consumers in the EC lose consumer surplus equal to the areas labeled A and B. Area A is transferred to EC producers, and area B is an efficiency loss. Consumers in the other importing countries and the exporting countries benefit from the lower price, although producers of edible offal in both regions are worse off.

EC demand and a model of the world market were estimated statistically using data on the quantities produced, consumed, and imported in the nine countries belonging to the EC at the time. The estimated equations are presented in Appendix 12.1. Data on the supply of edible offal to the world market and an estimate of the world price obtained by dividing the dollar value of world trade by the quantities traded were also used. The EC price was obtained by applying real exchange rates for the EC to the world price, with an adjustment to reflect the impact of a small tariff applied to imported offal. Before estimating the EC demand equation, we estimated a supply equation for the EC and determined that although supply did not appear to be perfectly inelastic, the supply elasticity was so close to zero that the assumption of a vertical supply schedule would be appropriate.

The EC demand equation was estimated with price as the dependent variable and per capita offal consumption and real per capita income as the explanatory variables. (Statistical results for all the estimated equations are presented in Appendix 12.1). Such a specification provides an estimate of a price flexibility, showing the percentage change in price given a 1 percent change in the quantity demanded. Flexibilities are similar to elasticities except that they measure the impact of quantity changes on prices rather than the reverse. The data used for this analysis were available for only a short time period and were probably quite inaccurate. Both of these factors reduce the reliability of the statistical results obtained. In addition, econometric developments since this study was done have shown that time-series data should be tested to determine whether they have certain properties. These tests often show that the kind of specification used in this analysis is inappropriate. Although the statistical results in this case were less robust than would have been desirable, they were not unreasonable, and the model seems to have provided fairly accurate predictions. To mitigate the statistical defects, we used econometric methods designed to correct for statistical problems, along with sensitivity analysis based on several alternative specifications. From these results, we concluded that the true price flexibility was probably between -2.80 and -3.60. The actual estimate from the model was -3.21.

This information was used to predict the impact in the EC of eliminating imported offal. Imported offal made up about 12 percent of total offal consumption in the EC prior to the ban. Based on the flexibilities noted earlier, a 12 percent fall in the quantity consumed would lead to a price increase of between 34 and 43 percent. This result was used to determine P_{EC}, the EC price following elimination of imported offal. Because the average quantities produced and imported before the ban are known, it is a matter of simple geometry to compute the gain in producer surplus, the loss in consumer surplus, and the net effect of the ban. Results based on the three flexibility estimates are shown in Table 12.1. On the basis of this analysis, producer surplus increases by between 720 and 925 million real ECU (European Currency Units, which were approximately equal to one dollar in the 1980s) as a result of the ban. However, consumer surplus is reduced by greater amounts, and the net cost to the EC is estimated to be between 50 and 64 million ECU.

Table 12.1. Welfare effects of EC hormone ban in EC (millions of 1980 European currency units)

Measure	Flexibility estimate of -2.80	Flexibility estimate of -3.21	Flexibility estimate of -3.60
Change in Producer Surplus (PS)	719.7	825.0	925.3
Change in Consumer Surplus (CS)	-769.6	-882.3	-989.6
Net Welfare Change in EC (PS + CS)	-49.9	-57.3	-64.3

Source: "Quality Restrictions as Barriers to Trade: The Case of European Community Regulations on the Use of Hormones," by E. Wesley F. Peterson, M. Paggi and G. Henry, *Western Journal of Agricultural Economics*, 13-1 (1988): 82-91.

These results may be sensitive to some of the assumptions made. If producers in some of the exporting countries do not use hormones or if producers in the United States and Canada could be certified as being in compliance with the EC ban, then EC imports need not fall to zero, and this would lessen the impact of the policy. On the other hand, experience with the Third Country Meat Directive suggests that obtaining certification from the EC is often difficult and expensive. At least one other study predicted similar effects to those reported here. In any case, it is not unreasonable to think that the reduction of supplies by 12 percent would have a significant impact on prices, particularly as one would expect the offal demand schedule to be fairly steep (inelastic), as is the case with most food products. An important question to be addressed subsequently concerns whether conventional welfare measures have any meaning in this context. After all, if consumers have lobbied for the exclusion of hormone-treated meat, does it make any sense to conclude that their welfare has been lowered by the success of their lobbying efforts?

Of course, the hormone ban had wider repercussions than the effects within the EC analyzed earlier. As shown in Figure 12.1, the ban leads to a fall in world prices, with impacts on welfare in exporting and other importing countries. One way to measure these effects would be to build a model of these markets with price equations to link the three sets of countries. Given the limited amount of data available and its low quality, that was not a practical option. As an alternative, we attempted to estimate a model of Panel B in Figure 12.1 (see Appendix 12.1). This model was specified as two simultaneous equations, one for excess supply and one for excess demand. These equations were estimated in both linear and logarithmic forms with statistical methods appropriate for the estimation of simultaneous equations. Excess supply was modeled as a function of the real-world price, world meat production, and a time trend. Excess demand was assumed to depend on the real-world price, real-world gross domestic product (GDP) and consumption of a substitute good (poultry). The statistical results were not very good. The best model was in logarithms, with meat production dropped from the excess supply equation and prices dropped from the excess demand equation. The results from this model indicated that the excess supply elasticity was 0.2. The impact of the hormone ban can be estimated by subtracting average EC imports from the average total excess supply to the world market and using this number to represent world trade volume after the ban. The excess supply equation is then solved for the real-world price, holding the other variable constant. Based on this procedure, world price was predicted to fall by 78 percent.

This estimate appeared somewhat extreme. As an alternative, we decided to directly estimate a world-price equation based on the variables from the excess supply and excess demand model that had appeared to have a statistically significant impact on the world market. The equation was specified with real-world price as the dependent variable and the volume of traded offal, world GDP, a time trend, and per capita poultry consumption as explanatory variables. The flexibility from this equation was 1.3. Based on this estimate and the reduction of world trade in edible offal following the EC hormone ban, it was predicted that world prices would fall about 35 percent. A price fall of this magnitude would reduce the value of annual world edible offal trade from over a billion dollars to around $500 million. A few years after the ban went into effect, a U.S. government official suggested in a private conversation that the world price for edible offal had indeed fallen by about 35 percent. Other analysts predicted a world price decline of 14 percent, with a contraction in world edible offal trade of 39 percent.[11]

FURTHER OBSERVATIONS ON THE EU–U.S. FOOD CONFLICT

The results presented in the preceding section highlight some of the costs to the EU of its policy on hormones. Because edible offal is only a part of the EU's total meat imports, it is likely that the full welfare loss is much greater. In addition, the ban means that a significant cost-reducing technology cannot be used in the EU, with the result that consumers have to pay higher prices for all livestock products than would otherwise be the case. For those who question the economic significance of this technology, it is worth pointing to a perverse side effect of the EU policy. The banned hormones as well as other substances that have similar effects on feed conversion are widely traded on a black market that has developed within the EU. The economic incentive to purchase these substances illegally and administer them clandestinely must be very great given the severe penalties imposed on those caught violating the EU regulations. These substances are often administered by injections rather than through the use of ear implants, and such procedures can in fact lead to harmful concentrations in meat that is actually consumed. It could be argued that EU policies implemented in the name of food safety have resulted in a less healthy food supply. The safety of EU meat supplies was brought into further question by the unfortunate experience with bovine encephalopathy, or "mad cow" disease. This disease is caused by peculiar elements known as prions that distort the way in which protein molecules are folded. The disease is thought to have been spread through the use of sheep remains in cattle feed, a practice that was approved in the EU prior to the outbreak of bovine encephalopathy. "Mad cow" disease, which is unrelated to the use of hormones, has been linked to a devastating human variant known as Creutzfeldt-Jakob disease.

If we ignore these perverse effects for the time being, the central question in the hormone conflict seems to concern the significance of the economic impacts analyzed earlier. Is the loss in consumer surplus an appropriate indicator of the welfare effects of the ban, given that consumers appear to have wanted the ban to be put in place? One way of looking at this question would be to suggest that the social costs of the ban are an indication of the amounts Europeans are willing to pay to avoid the risk of hormones in their food supply. In other words, consumers could be thought to value the elimination of this risk as being at least equal to 770 to 980 million ECU. The EU might justify its policy by suggesting that governments have a responsibility for protecting citizens from potentially harmful substances in the food supply, even if the true risks are unknown and particularly in cases where the public places a high value on this protection. This argument can be grounded in standard welfare economics (see Box 12.1). Markets fail to appropriately signal the presence of harmful substances that are not visible to the naked eye. Consumers are thus unable to make choices that reflect their true preferences without government protection.

The difficulty with this argument, however, is that it seems to require some agreement that meat from hormone-treated animals is indeed a risky product. As noted previously, the scientific evidence shows that hormones can be used in livestock production with no serious threat to human health. EU consumers appear to be skeptical about the validity of such evidence, given the frequent changes in what passes for scientific truth and the uncertainty of our understanding of complex medical and biochemical processes. Of course, this kind of uncertainty is present in everything we do, including the act of consuming meat from animals that have not been treated with hormones—as illustrated in the EU, for example, by the outbreak of "mad cow" and Creutzfeldt-Jakob diseases. The EU might still argue that national sovereignty requires that nations be allowed to respond to the demands of their citizens, even if these citizens are exceedingly risk-averse or misinformed. Scientific evidence might be seen as irrelevant as long as consumers want their governments to take a certain action. Because consumers have expressed clear preferences for the ban, the EU has the right and the obligation to implement it.

Even if one accepts this argument, it does not necessarily justify a government ban. The market fails because of lack of information. Consumers are unable to distinguish hormone-treated meat from meat produced in traditional ways. Many would argue that if information is the problem, the appropriate government response is to insure the provision of the missing information. This could be accomplished through a system of labeling allowing producers to certify that their product is "hormone-free" or using testing procedures to identify meat from animals treated with hormones and labeling those products as such. All things equal and in the absence of peculiar agricultural policies, meat from the hormone-treated animal will be less expensive. An important advantage of providing information rather than banning the product is that it leaves consumers with a greater range of choices. As noted before, however, the EU has rejected the use of labels in place of its ban on hormone-treated meat. Because there is no clear danger associated with the hormone levels found in meat from treated animals, the ban amounts to preventing those who wish to purchase the lower-cost meat from doing so.

Suppose, on the other hand, that EU policy makers wish to adopt a prudential stance according to which it is better to err on the side of caution, even if the potential harm of a product is uncertain or unlikely. Is a government ban justified in these cases? If European consumers elect to purchase meat from hormone-treated animals, does the government have an obligation to protect them from themselves? Is this a case of a paternalistic state acting on nosy preferences (assuming that some Europeans would like to buy meat from hormone-treated animals)? In what precise circumstances should governments intervene to prevent individuals from expressing preferences that appear, from the point of view of the government, not to be in the true interest of the individuals themselves? Are these circumstances met in the case of the hormone ban? In the United States, motorcycle helmets are mandatory in some states, and many states also have mandatory seat belt laws. These policies are also paternalistic in the sense that they aim to constrain individual choices deemed to be misguided by the general public. They do differ in one sense, however: there is abundant evidence that helmets and seat belts save lives. So far, at least, that is not true for hormone bans.

A second line of argument has developed around the indirect effects of the use of hormones. According to this argument, the EU would be justified in banning hormones even if no harmful residues could be found in meat from hormone-treated animals, because their use exacerbates other problems, such as disposal of surplus beef, survival of family farms, or protection of animal rights. Within the EU, environmental and animal rights groups have been influential in establishing regulations on agricultural production practices (e.g., minimum cage size for laying hens in Denmark). As in the case of the recombinant bovine growth hormone (rBGH), some may argue that growth promotants place undue stress on the animals. Another argument about the indirect effects of the technology concerns survival of family farms. The ban on growth-promoting hormones will lead to less total output, higher prices, and improved chances of survival for family farms. Finally, many Europeans feel that modern, industrialized agriculture produces foods that are of poor quality, even if they are not unsafe. They often fear that modern agribusiness practices will overpower traditional producers, leaving consumers with nothing but hard, tasteless tomatoes and McDonald's hamburgers.

EU policy makers argue that such social criteria should be specifically recognized along with scientific considerations in judging whether a technical standard can legally be applied to traded goods. In recent years, they have also promoted a related concept known as "multifunctionality." According to this idea, European farms have multiple functions, with food production being but one. Such farms also protect fragile ecosystems, maintain an attractive countryside, and provide other rural amenities. Yet they receive no compensation for these positive externalities. High support prices allow them to continue producing food and the positive externalities that accompany their farming activities. Some economists have noted that this story can be turned on its head. If the true purpose of these farms is to provide rural amenities, then the food they produce could be seen as an undesirable by-product (it adds to costly food surpluses) of providing rural amenities. The appropriate policy for this negative externality would be to tax farmers the marginal cost of disposing of the surplus food or to subsidize their amenity-producing activities under a requirement that they not produce food beyond the amounts they can consume themselves. Direct payments to farmers for their contribution to the rural setting would clearly cost less than supporting milk and cereal prices at levels that allow marginal farmers to remain in business while generating a windfall for the more efficient EU farmers.

The main EU argument for the inclusion of social criteria in evaluating the hormone ban stems from the existence of costly surplus beef production. The argument is similar to the one made by the Wisconsin dairy farmers who opposed the introduction of rBGH on the grounds that milk in the United States was already in surplus (see Chapter 10). For EU policy makers, large beef surpluses are a good reason to ban the use of the productivity-enhancing hormone technology. The problem with this argument is the existence of agricultural policies without which there would be no surpluses in the first place. The beef surplus is caused primarily by price supports and export subsidies, not the use of hormones. It is true that productivity increases will add to surplus production and the cost of disposing of these surpluses as long as the support prices and subsidies remain in place. But the appropriate response to the surplus problem is to change the policies, not to repress effective technologies.

This observation leads to the heart of the U.S. argument against the EU hormone ban. The United States maintains that the true motivation for the ban is maintenance of the highly protectionist system for agriculture in the EU. Widespread consumer support for the ban is thought to be an artifact of a misinformation campaign. One of the things that makes this case so intriguing is the difficulty of determining whether the beneficial effect of the ban on EU producers is simply a by-product of a consumer-protection initiative or is really the primary target of the policy. It is quite possible that consumers were only mildly concerned in 1986 when the directive was adopted but that the subsequent public debate has led them to become increasingly adamant in their opposition to industrialized agriculture. One of the reasons U.S. policy makers have reacted so strongly to the hormone ban is the fear that these hardening European attitudes will spread to other countries and even, perhaps, to the United States. This fear is also part of the newer conflict over GMOs.

CONTROVERSIES SURROUNDING GENETICALLY MODIFIED ORGANISMS

The use of genetically engineered corn and soybean varieties had become extremely widespread by the late 1990s in the United States. Because products derived from these crops are widely used in the food manufacturing industry, few processed food products are completely free of GMOs.[12] Bt corn is less subject to damage from the European corn borer, which actually causes only a modest yield reduction in untreated non-Bt corn. Glyphosate-resistant soybeans can be sprayed with a popular herbicide that is highly effective in controlling weeds. It turns out that one or two sprayings with the glyphosate herbicide control weeds at lower cost and with less environmental harm than do conventional methods that require repeated sprayings with alternative herbicides. Glyphosate-resistant varieties are popular because they reduce the time required to care for the crop and lower pesticide costs.

Most of the benefits of these crops are realized at the producer level, where yields are increased or costs reduced. Concerns about their safety, as well as the safety of other GMO crops, have to do primarily with their impacts on consumers and the environment. Bt corn, for example, has been criticized because its use may lead to increased insect resistance to Bt, which is used extensively by organic gardeners to control insect pests. Because the bacterium is a naturally occurring organism, farmers can use it without losing their organic certification. In addition, it was discovered that the pollen from Bt corn can be toxic to monarch butterflies and their larvae if it lands on the milkweeds upon which the larvae feed and which often grow near cornfields. The experimental results that led to this discovery have been questioned, and it is not clear just how serious a threat to monarch butterflies the extensive planting of Bt corn would be. Many suggest that appropriate management of Bt cornfields would eliminate any risk to butterflies as well as significantly reducing any problems associated with resistance. While some consumers fear that these corn and soybean varieties may contain elements that would be harmful to human health, that does not appear likely.

A greater potential health threat is the possibility that allergy-causing genetic material may be inadvertently introduced in the process of engineering new varieties. For example, Nelson and associates note that a seed company was not allowed to release a soybean variety genetically engineered to include DNA from Brazil nuts because a gene from the Brazil nut causes allergic reactions in some people.[13] The possibility that harmful allergens may wind up in foods that are unrelated to the usual vectors may be a real problem with GMOs. People who are allergic to peanuts, for example, avoid products that contain peanuts or peanut oil. If the allergy-causing DNA from a peanut plant has been spliced into a tomato, for example, the risk of serious harm to unsuspecting consumers is increased.

Of course, most allergies are well understood, and it is not impossible to establish regulatory systems to prevent this kind of problem, as shown by the case of the soybean variety that included allergy-causing Brazil nut genes. In the United States, the Food and Drug Administration (FDA) is the primary agency charged with assuring food safety, including the safety of foods containing GMOs. The Environmental Protection Agency (EPA) and the U.S. Department of Agriculture (USDA) are also involved in regulating GMOs because of their responsibilities in the areas of pesticides and animal and plant health. The FDA treats foods containing GMOs in the same way it treats other food products. New food products that are substantially equivalent to existing products do not have to be tested prior to their release onto the market. Novel foods must be shown to be safe before they are allowed to be commercialized. The FDA also has the authority to remove foods from circulation if it is found that they are unsafe. Labeling of products that contain GMOs is not required unless they differ substantially from similar, non-GMO products. Labeling is required for products that may contain genetically engineered proteins that cause allergies, and some products have not been approved because of the presence of potential allergens. According to Nelson and associates, the FDA believes that genetically engineered foods cannot be distinguished from foods derived from conventional plant breeding methods, so there is no reason for special labeling requirements for these food products.[14]

Unlike the United States, the EU does not have an agency to oversee food safety. Instead, food safety regulations, including those pertaining to GMOs, are set out in directives adopted by the EU bureaucracy but implemented through the national legal systems of the member states. The EU has adopted directives on research on GMOs and the release of such products. GMOs cannot be released without a risk assessment and approval from national authorities. Another directive requires that all foods that are genetically modified or that contain at least 1 percent genetically modified products be so labeled. Some European policy makers have called for segregating imported corn and soybeans into GMO and non-GMO lots. In contrast to the United States, EU policy makers view GMOs as different from ordinary foods and in need of special regulations. Because of the controversies over GMO crops, some private firms in both the EU and the United States have announced that they will only buy non-GMO crops or that they will pay a premium for crops certified to be free of GMOs. Some farmers have decided not to plant GMO crops because of the uncertainty surrounding the market for these products.

In many respects, the issues that arise with respect to GMOs are the same as those encountered in the hormone conflict. In both cases, technological change is being resisted because of worries about its long-term effects on food safety, food quality, and a variety of related concerns that seem to have their roots in a distaste for modernity, industrialization, and the free play of capitalist markets. In both cases, U.S. policy makers, farmers, and food manufacturers fear that these attitudes will spread from Europe to Japan and other countries, eventually infecting U.S. consumers, who seemed largely indifferent to the entire question, at least up until the late 1990s. Those who favor the development and use of GMOs and other agricultural technologies see these innovations as crucial for insuring that world food output is able to keep up with population growth without destroying the environment. Glyphosate-resistant soybeans are less harmful to the environment than conventional varieties. Genetically engineered crops have the potential for offsetting recent trends that have seen a slowing of the rate of growth in yields. They also offer possibilities for crops that contain medicinal agents or particular vitamins and minerals that may be lacking in the diets of people in certain regions. It is interesting to note that U.S. supplies of insulin for the treatment of diabetes are produced by genetically engineered bacteria at lower cost and with greater purity than could be achieved with other methods.

For supporters of the new technologies, scientific evidence is all that matters in assessing their costs and benefits. Because the potential benefits for world food security are so great, proponents believe that GMOs and other modern agricultural technologies should be embraced with minimal public regulation. For those who oppose the new technologies or wish to see their introduction slowed and subjected to debate and evaluation before allowing them to be introduced, science is not always right, and there are other considerations. For example, opponents may feel that preserving multifunctional family farms is more important than any beneficial consequences of the new technology and that suppressing the technology, by maintaining high food prices, will help insure that such farms continue to thrive.

CONCLUSION

Political interests that figure prominently in the design and implementation of national policies (see Chapter 5) also play important roles in the creation of agreements on international institutions, such as those that regulate food and agricultural trade. Most national governments place higher priorities on protecting narrow national interests than achieving economically efficient international regimes for trade, development, finance, environmental protection, or public health. Rules and regulations in all these areas are international public goods and are unlikely to be supplied in optimal amounts in the absence of some governmental authority. The problem, of course, is that there is no international government to provide them. Organizations such as the WTO are the only instruments available for addressing the problem of international public goods.[15] As in national settings, decisions on the nature and amounts of the public goods to supply are subject to intense political pressure, and the potential for stalemate and inaction is great. The problem of constructing regimes to provide international public goods, such as a smoothly functioning world trade system, is compounded by the great variety of ideological and moral visions that animate the different peoples of the world. At its root, the controversy over hormones and GMOs reflects different beliefs about the right way for people to live. Even if all the parties to this conflict could reach general agreement on the facts of the matter, their conflicting positions on the use of hormones in livestock production or the introduction of GMOs into the food supply would probably change very little. Economic analyses, as described in this chapter, can contribute to clearer understandings of the factual bases for such controversies. In addition, it is often possible to find reasonable ways to resolve the issue without forcing anyone to give up cherished beliefs. In both examples in this case study, labels indicating whether food was produced with technologies such as hormones or genetic engineering may offer a way to maintain the liberal trade regime that has been of such benefit to the world without requiring the acceptance and consumption of products some people dislike.

It is interesting that the EU resists labeling in the case of hormones while requiring it in the case of GMOs. The United States is more consistent, opposing labeling in both cases. It is true, of course, that there are many practical problems with segregating GMO and non-GMO crops as well as labeling foods to indicate the presence or absence of GMOs, hormones, or whatever else may come up in the future as technological innovation marches on. Adding labeling or segregation requirements increases the transaction costs of international trade and may in some cases make continued trade uneconomic. Nevertheless, such procedures may be the only practical solution to the conflict, and they have the ethical advantage of respecting such individual rights as the right to decide what one wishes to consume. While economic analysis can help clarify the issues, options, and consequences of policies in such contested areas as food safety and international trade, the final decisions about what to do are more likely to be the products of ethical, legal, and political argument than of technical or economic analysis.

DISCUSSION QUESTIONS

1. The EU hormone ban reduces consumer welfare but is strongly favored by EU consumers. Are EU consumers irrational? Discuss.

2. Does the state have a responsibility to implement policies responding to the desires of the public when the public is misinformed? Should the government pursue paternalistic policies that protect individuals from their own bad judgment? Explain.

3. Why might edible offal supplies be perfectly inelastic?

4. Discuss the moral issues involved in the use of genetic engineering techniques allowing genetic transfers across species.

5. What might account for the different attitudes of Europeans and Americans regarding food, trade, and scientific evidence?

6. Can partial equilibrium modeling and welfare analysis be of use in designing and evaluating domestic regulations and trade policies for goods produced with technologies such as hormones or GMOs? Can quantitative analysis help foster an understanding of these issues, or are they simply matters of opinion and conflicting values? Explain.

SUGGESTIONS FOR FURTHER READING

Hoekman, B., and M. Kostecki. *The Political Economy of the World Trading System.* Oxford: Oxford University Press, 1996.

Nelson, Gerald C., Timothy Josling, David Bullock, Laurian Unnevehr, Mark Rosegrant, and Lowell Hill. *The Economics and Politics of Genetically Modified Organisms in Agriculture: Implications for WTO 2000.* Research Bulletin 809. Champaign-Urbana: Illinois Office of Research, University of Illinois, November 1999.

Thompson, P. B. *Food Biotechnology in Ethical Perspective.* Wilmington, Del.: Aspen Publishers, 1997.

Vogel, David. *Trading Up: Consumer and Environmental Regulations in a Global Economy.* Cambridge: Harvard University Press, 1997.

APPENDIX 12.1: ESTIMATED EQUATIONS FOR THE EDIBLE OFFAL MODEL

The data used in this study were available only for the period 1972 to 1984.[16]

Equation 1: price-dependent EC demand equation for edible offal

$$ECP = 5,595.46 - 814.72PCOC + 0.067RECY$$
$$(1,805.2)\quad (348.1)\qquad\qquad (0.39)$$

where *ECP* is the EC price for edible offal in real European Currency Units (ECU) per metric ton, *PCOC* is per capita offal consumption in kilograms, and *RECY* is per capita income in real ECU. The figures in parentheses underneath the equation are standard errors, which can be used to test the estimates for statistical significance. The estimated parameters are judged to be significantly different from zero at the 10 percent level of confidence. Equation 1 showed evidence of serial correlation. The estimates have been corrected for this problem using the Cochrane-Orcutt procedure. The estimated value for rho is 0.54, and R^2 is 0.55.

Equations 2 and 3: Excess supply and demand for edible offal

$$WSO = 10.38 + 0.221RWP + 0.46T \qquad R^2 = 0.97 \qquad rho = 0.463$$
$$(0.734)\ (0.077)\qquad (0.054)$$

$$WDO = -1.63 + 1.583WY + 0.557PCP \qquad R^2 = 0.95 \qquad rho = 0.246$$
$$(4.51)\ (0.579)\qquad (0.472)$$

where *WSO* is world supply of offal in metric tons, *RWP* is the real-world price of offal in dollars per ton, *T* is a time trend, *WDO* is world demand for offal in metric tons, *WY* is real-world income in dollars, and *PCP* is world per capita poultry consumption in kilograms. All the variables are expressed in natural logarithms, so the coefficients are elasticities. The two equations were estimated with two-stage least squares corrected for first-order serial correlation. Standard errors are shown in parentheses.

Equation 4: reduced form price equation for world edible offal

$$RWP = -45.53 + 1.289VT + 4.061WY - 2.227T + 1.870PCP$$
$$(19.36)\ (0.755)\qquad (2.525)\qquad (0.616)\qquad (1.177)$$

$$R^2 = 0.81 \qquad\qquad\qquad rho = 0.588$$

where *VT* is the volume traded (equal to *WSO* = *WDO*) and the other variables are as defined before. As with the simultaneous system, all the variables are in natural logarithms, so the coefficient for the volume traded is the price flexibility. This equation has also been corrected for first-order serial correlation.

NOTES

1. This case study is based on research done in the 1980s. That work is updated and expanded to include consideration of genetically modified organisms (GMOs). Discussion is built around the following work: E. Wesley F. Peterson, M. Paggi, and G. Henry, "Quality Restrictions as Barriers to Trade: The Case of European Community Regulations on the Use of Hormones," *Western Journal of Agricultural Economics* 13, no. 1(1988): 82–91.

2. A full account of the history and significance of different exchange-rate regimes can be found in textbooks on international economics, such as P. Krugman and M. Obstfeld, *International Economics: Theory and Policies*, 5th ed. (Longmont, Calif.: Addison-Wesley, 2000).

3. In 1999, the members of the WTO met in Seattle, Washington, to decide on an agenda for a new round of MTN. The delegates deadlocked over the issue of labor and environmental standards. The United States and some European countries called for universal environmental and labor standards on the grounds that developing countries, where such standards may be lacking, have an unfair trade advantage. The developing countries correctly saw these calls as a new form of trade barrier aimed at closing their access to the markets in the developed countries. This disagreement led to the collapse of the talks, and the new round of negotiations was left up in the air.

4. See B. Hoekman and M. Kostecki, *The Political Economy of the World Trading System* (Oxford: Oxford University Press, 1996). There were 136 members of the WTO and 188 members of the United Nations as of April 2000. Earlier rounds were simpler to complete not only because fewer countries were involved but also because the United States provided effective support for trade liberalization. In addition, the early GATT had less effective dispute-resolution mechanisms, so countries may have felt better able to circumvent rules that might turn out to cause domestic political difficulties.

5. Prior to 1992, the European Union (EU) was known as the European Community (EC). The two expressions are used interchangeably in this chapter.

6. See "Plague Island," *The Economist*, March 3–9, 2001, p. 51–52.

7. See Jill Carroll and Sarah Lueck, "Disease Spurs Ban on European Imports," *The Wall Street Journal*, March 14, 2001, Section A, p. 3.

8. In the context of the tuna-dolphin case, Vogel notes that it makes as much sense for the United States to impose its values concerning dolphins on other countries as it does for India to boycott the beef-eating West because of its beliefs about sacred cows. David Vogel, *Trading Up: Consumer and Environmental Regulations in a Global Economy* (Cambridge: Harvard University Press, 1997).

9. See M. R. Reed, *International Trade in Agricultural Products* (Upper Saddle River, N.J.: Prentice-Hall, 2001), 73–75.

10. At the time of the study, the EC included France, Germany, Italy, Belgium, the Netherlands, Luxembourg, the United Kingdom, Ireland, and Denmark. Although Greece had joined in 1980 and Spain and Portugal joined in 1986, these countries are not included in the analysis due to lack of consistent data. As of 2000, the EU was made up of these twelve countries plus Finland, Sweden, and Austria.

11. J. Ginzel and B. Krissoff, "An Assessment of the Economic Effects of a Ban on Beef Trade," (Economic Research Service, USDA, Washington, D.C., 1987).

12. Much of this discussion is based on G. C. Nelson et al., *The Economics and Politics of Genetically Modified Organisms in Agriculture: Implications for WTO 2000*, Research Bulletin 809 (Champaign-Urbana: Illinois Office of Research, University of Illinois, November 1999). This bulletin includes a summary of the regulatory environment for GMOs as well as much additional information.

13. Ibid, 51.

14. Ibid., 51.

15. This theme is developed more fully in E. Wesley F. Peterson, "The Design of Supranational Organizations for the Provision of International Public Goods: Global Environmental Protection," *Review of Agricultural Economics 22*, no. 2 (2000): 352–66.

16. Details of the study appear in Peterson, Paggi, and Henry, "Quality Restrictions."

13

Multisector, Systems, and Economy-Wide Models for Policy Analysis

INTRODUCTION

In the preceding chapters, benefit-cost analysis and partial equilibrium models for policy experimentation have been described and illustrated with a series of case studies. The purpose of this chapter is to provide an inventory of some of the more complex modeling approaches used in policy analysis. The goal is not to present recipes for building such models but rather to offer general descriptions of a few basic approaches and their use in the analysis of public policies. Entire books have been written to describe particular kinds of policy models, and it would be impossible to reproduce the contents of this extensive literature in a book of this nature. It is possible, however, to provide an introduction to these more complex modeling approaches with a view toward developing a broad understanding of different kinds of policy models, their limitations, and the trade-offs associated with their use. The chapter will also serve to set up the final case study on the North American Free Trade Agreement (NAFTA), agriculture, and the environment.

MULTISECTOR AND SYSTEMS MODELS

Any classification scheme for the different kinds of analytical frameworks used in policy analysis will be somewhat artificial. The distinction between partial and general equilibrium models, for example, may simply be a matter of degree. A general equilibrium model includes more relationships between different parts of the economy than does the relatively simple model used to analyze the U.S. sugar program in Chapter 11. For our purposes, however, it may be useful to distinguish models that focus on some subset of the general economy from those that are designed to represent the economy as a whole. Partial equilibrium, multisector, and systems models may be thought of as falling into the first category, while such approaches as general equilibrium or input-output analysis can be seen as economy-wide approaches. One way to formulate this distinction is to introduce the concepts of "exogeneity" (factors produced or occurring within a system) and "endogeneity" (factors introduced from outside the system). Models are made up of equations and identities that specify the relationship between such variables as prices, income, supply, demand, and so forth. The predicted values for *endogenous variables* are computed within the model itself, using technical relationships, and the values of *exogenous variables* introduced from outside the model. The purpose of the model is to predict the values of the endogenous variables in light of alternative policy scenarios.

Consider the sugar model described in Chapter 11. The important relationship between sugar and high-fructose corn syrup (HFCS) was included in the model through equations that predicted the share of each type of sweetener in the total demand for caloric sweeteners. The shares are endogenous in the sense that they are calculated within the model. On the other hand, the price of HFCS, used to calculate the sweetener price and the sugar and HFCS shares, is exogenous. It is not determined within the sugar model but rather brought in independently to calculate the other variables. It would have been possible to develop a full model of the HFCS industry with supply, demand, and endogenous prices. Adding a component of this nature to the sugar model would have allowed more precise predictions about the impact of liberalizing sugar policies on the production, consumption, and price of HFCS. The analysis described in Chapter 11, however, focused on the sugar sector, so a relatively simple link to HFCS was all that was needed.

For broader questions, models may need to include more endogenous variables to take full account of the interaction between different sectors and the feedback from changes in one variable to other variables considered to be part of the system. In many cases, it is possible to simply add other sectoral models along the lines of the sugar model, with provisions for any important relationships between these sectors. Thus, a model of the agricultural sector might include the major field crops (grains, oilseeds, sugar, cotton), livestock enterprises, and appropriate links to reflect the fact, for example, that the output from the feed grain sector is an input into livestock and poultry production. *Multisector models* are extensions of the partial equilibrium models described in the last three chapters. *Economy-wide models*, in contrast, aim to represent the entire economy. This means that they must include the links between labor and capital markets and goods or output markets, as well as such macroeconomic variables as money and interest rates. In a sense, each of these extensions endogenizes markets and variables that are treated as exogenous in the simpler models.

Consider a model of U.S. demand for gasoline. From economic theory, we would expect that gasoline demand would depend on its price, the prices of substitutes and complements, and income. Most motor vehicles in the United States run on one type of fuel—gasoline or diesel, for example—and there is little scope for substituting alternative sources of fuel once an individual has purchased a particular kind of vehicle. It is also difficult to imagine what kind of complementary product (engine oil?) would influence consumer decisions on gasoline. Of course, petroleum products such as gasoline (of various octanes), diesel fuel, engine oil, and so on are all derived from the same raw product, so it is likely that their prices would move together. For these reasons, it seems unlikely that the prices of substitute fuels or complementary inputs such as engine oil would possess much explanatory power in the gasoline demand equation.

The most likely short-term effect of changes in gasoline prices would be adjustments in the amount of driving done. In addition, however, in any given year, many drivers will replace their vehicles. Their choices for replacement transportation will be influenced by the price of fuel, the prices of different kinds of vehicles, and their incomes. A simple model might treat gasoline demand as an endogenous variable dependent on gasoline prices, relative prices of different kinds of vehicles (perhaps in a previous year to take account of the fact that adjustment to the new prices will be fairly slow), and income, with all three of these variables treated as exogenous. This simple model could be used to examine the impact of a tax on gasoline by predicting demand at the higher price given any expected increases in income or changes in relative vehicle prices.

As shown in Chapter 10, this simple model may lead to inaccurate predictions, because the tax will affect supply as well as demand. A partial equilibrium model including the supply side of the market might be structured so that both the quantity of gasoline exchanged on the market and its price are endogenous with income, vehicle prices, and other variables introduced exogenously. A more complex model might include different types of fuel and different types of consumers (truckers, business travelers, households) as well as a more precise model of fuel production that would include all the possible responses of fuel producers to price changes. Such responses may involve different fuel compositions, different mixes of final products (e.g., more heating fuel, less diesel), alternative marketing strategies, and so on. Such a model could lead to better predictions of the overall effects of the tax as well as additional detail on how various industries might react to the change. Because gasoline is derived from petroleum, which is a source of energy for a wide variety of uses (electricity generation, steel production) as well as a raw material for such industrial inputs as plastics and fertilizers, a gasoline tax could have widespread repercussions as economic agents in all these different industries respond to the changing economic conditions induced by the tax. Moreover, petroleum-based energy sources compete with other fossil fuels (coal and natural gas) as well as nuclear or wind power and other sources of energy. The gasoline tax could affect these sectors as well.

Consider the following sketch of economic adjustment to a tax on gasoline. Households consume less gasoline in response to the tax, and the price received by refiners falls, as shown in Figure 10.1. Refiners, finding that gasoline is less profitable, shift their output to produce more diesel fuel, which leads to a fall in the price of diesel and increased demand by truckers and farmers. Over time, households find that they wish to purchase more fuel-efficient vehicles, and these demand shifts induce the automobile makers to adjust their vehicle mix with respect to smaller fuel-efficient cars and larger trucks and SUVs. Such impacts will be compounded by any adjustments in industries producing alternative sources of energy as well as industries that use energy as an input. The adjustments in the petroleum refining industry, the automobile industry, and other related industries will affect demand for labor and capital, and changes in these factor demands will affect incomes. Of course, changes in income also affect household demand for gasoline and new vehicles. Moreover, all these changes can be expected to alter the amounts of greenhouse gases produced, and these changes in turn could have an impact on global warming, which could affect agriculture and other industries, and so on ad infinitum. A complete model of all these relationships would allow an analyst to make more precise predictions about the impact of the gasoline tax on gasoline consumption, greenhouse gas emissions, climate change, refinery output, coal supply and demand, agricultural and industrial outputs, incomes, prices, employment, and so on.

The final model sketched in the preceding paragraph would be very complex. It would have to include submodels of climate change as well as models of the entire economy, including all input and output markets and other income-generating activities. While such complexity is more realistic, it comes at a high price. Large, complex models often include so many equations that they become unmanageable, difficult to parameterize, costly to maintain, and confusing to interpret. This last problem can be particularly significant in policy models, because such models are only useful if policy makers can follow the logic of the analysis, understand the results, and draw useful conclusions for the problem with which they are confronted. It is not uncommon for such models to take on the appearance of a *black box*. A question is fed into the left-hand opening of the box, an answer pops out of the opening on the right-hand side, and it is next to impossible to know how the answer was generated by the model inside the black box or whether it is an answer that makes sense.

Because multisector and economy-wide models usually include many economic activities, it is usually necessary to specify the model variables at a fairly high level of aggregation. For example, it is not uncommon in large agricultural models for all feed grains (corn, barley, oats, and sorghum) to be treated as a single variable rather than including submodels of each type of feed grain. The more comprehensive the model, the less industry-specific detail can be included. The loss of detail may be compensated for by the increase in scope, but there is clearly a trade-off between the greater range of the predictions from a large model and the various disadvantages noted earlier. Effective policy analysis is usually based on the simplest analytical framework that is consistent with the kinds of experiments required to address the policy issue under consideration. If the analyst is interested in the impact of NAFTA on the U.S. market for corn, a relatively simple partial equilibrium model along the lines of the one used to examine the sugar program in Chapter 11 may be perfectly adequate. If the question being addressed concerns the impact of NAFTA on the environment, however, a much more elaborate model will be required.

Multisector Models

The most obvious way to expand the range of issues that can be analyzed is to develop a series of partial equilibrium models of different sectors and link them with appropriate equations and equilibrium conditions. Such multisector models have been widely used to analyze agricultural policies in the United States. For example, the Economic Research Service (ERS) of the U.S. Department of Agriculture (USDA) maintains and operates such a model.[1] ERS has helped fund numerous studies of particular agricultural markets, and these studies provide a rich database of elasticity estimates and other technical relationships of significance to the U.S. agricultural sector. The general framework for the model allows ERS analysts to use results from these studies along with their own estimates as sources of parameters. Another large agricultural model is operated by two universities (Iowa State and the University of Missouri) that established a policy research group known as the Food and Agricultural Policy Research Institute (FAPRI).[2] The FAPRI model includes numerous econometric equations to represent various agricultural sectors in the major regions of the world. Both of these models can be used to answer a broad range of policy questions, such as the impact of the 1995 Federal Agriculture Improvement and Reform Act (FAIR) on prices and quantities in the agricultural sector, the environment, agricultural trade, and so on.

Systems Models

Most multisector models focus on economic relationships, treating such technical aspects as the link between agricultural production and soil erosion, for example, as exogenous. The FAPRI or Swopsim models might be used to predict the amounts of land planted to different crops under FAIR as contrasted to some alternative policy, and these results could then be used in conjunction with exogenously determined technical parameters to predict changes in soil erosion or groundwater contamination. If the analytical focus is on environmental impacts or some other problem area that is not fully described by the models of market behavior, then it may be necessary to employ an alternative model that incorporates physical, chemical, or biological relationships. Such models are sometimes referred to as *systems models*. Economists are not the only scientists engaged in model building. Climatologists, wildlife biologists, agronomists, engineers, and many others use mathematical models to analyze issues such as climate change, the management of wildlife populations, soil-water relationships, or any of a wide variety of technical and scientific relationships. These models can be linked to economic models to broaden the range of predicted impacts that may follow from a change in policy.

Bouzaher and Shogren describe a systems model used to study the impact of alternative policies on nonpoint source pollution from agricultural production.[3] "Nonpoint source pollution" is pollution that does not come from a specific place, such as an industrial plant. For example, the nitrate contamination of groundwater described in Chapter 9 is not generated by a particular farm but arises rather from the agricultural practices of large numbers of farmers. With nonpoint source pollution, it is difficult to trace particular pollutants back to their original source. The model used in the study by Bouzaher and Shogren is named the Comprehensive Environmental Economic Policy Evaluation System (CEEPES). It is designed to analyze the environmental impacts of alternative weed control strategies. The authors note that there are two agencies with regulatory authority for these issues: the Environmental Protection Agency (EPA) and the USDA. The EPA regulates pesticides that may be nonpoint sources of pollution, while the USDA sets the agricultural policies that influence decisions on production, input use, and so on. Lack of coordination between these separate agencies may lead to conflicting policies. Bouzaher and Shogren note, "The EPA can unilaterally restrict or ban selected agri-chemicals, while the USDA can unilaterally set price subsidies to output such that intensive agri-chemical use is promoted."[4]

The CEEPES model described by Bouzaher and Shogren links six submodels together. The first of these is a model that simulates the effects of weed competition on crop yields, given a set of technical parameters describing the relationship of plant growth to sunlight, nutrients, and water and the impact of weed populations on the availability of these resources. The output of the weed competition model is an input into the second model, which simulates the effects of various weed control strategies that farmers might follow. The output from the model of weed control strategies is then used in an economic model of producer decision making to predict the particular herbicides that will be chosen for various crops and in various regions of the country, given specific agronomic characteristics, crop biology, prices, and costs. The policy experiments are introduced at the stage where producers are deciding on their weed treatments. The authors of this study test the impact of a policy to ban the herbicide Atrazine and an alternative policy of banning a different herbicide, Triazine, against a baseline scenario of no policy change. The predictions of herbicide use from the producer decision model are used in conjunction with various economic variables to compute the net returns for the farms.

The fourth and fifth components of the model are technical models that convert the chemical usages predicted by the producer decision model into specific contamination levels for groundwater and surface water. The outputs of these models are then used in a model of human chemical exposure to determine the risk of negative effects on human health. The final step is to do a risk-benefit analysis, comparing the benefits of agricultural output with the associated health risks in light of the alternative herbicide policies. The model can be calibrated to reflect soil, water, and other agronomic conditions in different states, with the final results reported for states, regions, or the nation as a whole. Each of the six models linked together for this analysis is written as a computer program that draws on data from the study area and the relevant technical parameters. The entire system is solved using a complex computer algorithm. The authors find that the ban on Atrazine would be ineffective because producers would be able to substitute other chemicals. As a result, there would be no impact on human health, and the only effect of the ban would be to lower producer income. The Triazine ban resulted in chemical exposures consistent with health standards, although it also lowered producer returns.

Models of this nature allow much more precise estimates of the overall impacts of alternative policies. While such models would be very costly to build if the analyst had to start from scratch, the existence of large numbers of specialized models often means that the problem is reduced to one of finding ways to link the submodels together and to operationalize the analysis. Because many of the potential submodels have been developed by specialists from a variety of disciplinary fields, this type of policy analysis is inherently interdisciplinary. Clearly, such models would be useless without high-speed computers to perform the extraordinarily large number of calculations required in their application. Even with modern computing capabilities, a complex model can prove difficult to set up and use. In fact, Bouzaher and Shogren modified some of the submodels to reduce the need for certain technical parameters that were difficult to locate. Such models also suffer from the black-box problem, because they are so intricate that it is difficult for anyone who has not actually worked with them to fully grasp how they are structured and applied. Nevertheless, they can be extremely useful for analyzing problems that involve highly complex causal relationships, such as the link between production decisions on the farm and pesticide-induced cancers in distant cities.

ECONOMY-WIDE MODELS

The multisector and systems models described in the previous sections tend to focus on only part of the general economic system. For some analytical purposes, it may be necessary to take account of the interrelationships between the various parts of the economy. Consider the classic representation of the *circular flow* found in most textbooks on macroeconomics and illustrated in Figure 13.1. The core of the circular flow is the link between households and firms. In the general equilibrium model developed in Chapter 3, Sam and Mary could be taken to represent households, with the production of food and shelter being carried out by firms. In that theoretical model, the only link between these households and firms was the flow of the two goods from firms to households and the flow of unowned labor and capital to the firms. In Figure 13.1, the flow of goods is matched by an expenditure flow in the opposite direction. Moreover, the diagram endogenizes the labor and capital used by firms by vesting their ownership with the household sector. In return for supplying these labor and capital services, households receive payments in the form of wages and profits. These payments constitute the income used by households to purchase goods and services from the firms. Note that in this representation, real flows (final goods and factor services) are matched by monetary flows (income and expenditures).

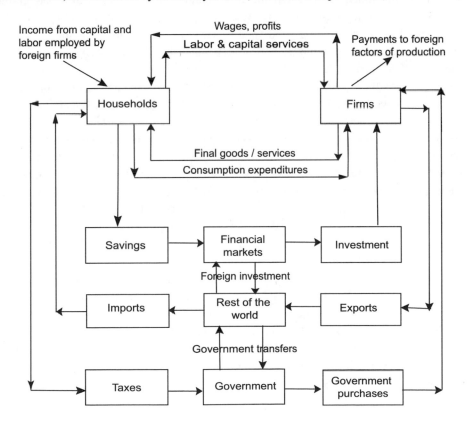

Figure 13.1: The circular flow.

The flows between households and firms are modified by various leakages and injections. Households do not consume all their income when they receive it. Some is set aside as savings to be used for future consumption. Through the mediation of financial markets, these savings are translated into investments, which allow firms to expand their productive capacity so that the economy can grow over time. Population growth may also cause an increase in the number of households or consumers. Investments in physical capital as well as human and social capital (education, informal institutions) can lead to productivity increases that allow the growing population to consume more or better goods. In addition to savings, the taxes that households pay to the government for the supply of public goods and other government services are also deducted from household income. Government expenditures are represented as purchases from the private sector (firms). Finally, some of the production generated by firms is sold to foreigners, and some of the goods consumed by households are purchased from foreign firms. These links are shown in Figure 13.1 as a flow running in the opposite direction to the financial (savings and investment) and government flows.

It should be noted that the leakages and injections shown in the lower part of Figure 13.1 all generate complementary flows in the opposite direction that are not shown in the diagram. Thus, investments generate dividends and interest payments that flow back through the financial sector to households. Likewise, government purchases from the private sector are used to provide public goods and services to households, and some of the tax receipts are redistributed in the form of money rather than goods purchased from the private sector. The links to the rest of the world are even more complicated. Imports are like other consumer goods and services in that they generate an expenditure flow similar to the one in the upper part of the diagram. Likewise, exports generate payments to domestic firms. In addition, both the financial and government sectors are linked to the rest of the world. Governments provide official transfers in the form of foreign aid, government borrowing, and purchases of foreign goods. Household savings can be invested in foreign firms, and foreign households may invest in domestic firms. Investments generate dividends, profits, and interest payments to and from foreigners, as shown by the two arrows at the top of the diagram.[5] The arrows also reflect payments to foreign labor employed by domestic firms and income paid to domestic workers employed by foreign firms.

The economy-wide model illustrated in Figure 13.1 highlights the links between the main aggregates commonly found in national economies. The box labeled "Firms" contains all the industries in an economy, including those that specialize in the production of goods used as inputs by other industries that produce goods destined for consumer markets. The circular-flow model can be translated into a series of equations showing the relationships among these aggregate variables. Such *macroeconomic models* will include supply and demand equations for all labor, all industrial goods, all financial or governmental services, and so on. Macroeconomic models are useful in analyzing the impact of changes in monetary or fiscal policy on employment levels, inflation, interest rates, exchange rates, and the balance of payments. Monetary policies are usually run by treasuries and central banks and involve regulating the rate of growth of the money supply to maintain interest rates at levels that keep inflation in check. Fiscal policy is set by the executive and legislative branches, which may attempt to use taxation and spending policies to stimulate or slow the economy in an effort to avoid excessive unemployment and inflationary spending. Appropriate macroeconomic policies are the subject of much controversy, and there is much less agreement among economists in this area than is commonly found in microeconomic analysis. Some macroeconomists have suggested that large models based on the linkages illustrated in Figure 13.1 do not allow for behavioral responses that may well offset the intended effects of monetary and fiscal policies, and they argue that attempts to use such policies to fine-tune the economy are misguided.[6] In the following sections, we will concentrate on some specific economy-wide models that include various macroeconomic components but are not structured at the level of aggregation usually found in large macroeconomic models.

Input-Output Analysis

Many of the basic elements of *input-output analysis* can be found in the *Tableau économique* developed by the French economist François Quesnay in 1758. The *Tableau économique* showed how an economic activity, such as the payment of rent, generated further economic transactions. The recipient of a rent payment spends most of the money on goods, services, or additional land purchases, and these transactions lead to further expenditures by the recipients of the payments, which would constitute income for another set of recipients, and so on. Modern input-output analysis further develops the basic ideas set out in Quesnay's table. The originator of this method is Wassily Leontief, who won the Nobel Prize in economics for this work in 1973. In their simplest form, input-output models are designed to take account of the fact that the outputs of some industries are used as inputs by others.

For example, corn is an input for industries manufacturing livestock feed, ethanol, corn syrup, corn oil, breakfast cereals, corn meal, and other food products. The final demand for corn is the sum of the intermediate demands by all these industries and any final demand for the raw product that may exist. In many cases, an industry uses some of its own output as an input into its production activities. Such would be the case, for example, with seed corn used in the next production cycle or steel used by the steel industry itself in the construction of its plants and equipment. Input-output analysis has been widely used as a planning tool at both the regional and national levels. Thus, for example, if an agency in charge of regional development in southern France wishes to increase tourism through an advertising campaign, an input-output model of the region could be used to determine the changes in demand for final products and industrial inputs that would be brought about by the increased tourism.

Consider a simple economy that includes three industries—corn, fuel, and transportation—and one factor of production—labor. Each industry produces only one output using inputs in fixed proportions. Thus, each unit of corn output might require fixed amounts of labor, corn, fuel, and transportation. Fuel output uses labor inputs and inputs from the corn (for ethanol) and transportation industries as well as from itself, since fuel production requires some energy. Suppose that the outputs of each industry are measured in dollars and that a dollar of corn output requires 10 cents in seed, 25 cents in fuel, 35 cents in transportation, and 30 cents in labor. Similar relationships exist for the other industries. In addition to the industrial output used as an input to other production processes, households may wish to purchase these goods as final consumer items. Suppose that final demand is projected to be $100 million for corn, $80 million for fuel, and $50 million for transportation. For corn, total production has to be large enough to satisfy the final demand as well as the needs of the other industries. We know that the corn industry needs to retain 10 percent of its total output as seed. There is also a demand from the fuel industry for corn used in ethanol production. Thus, total corn production will have to be greater than $100 million to satisfy these needs.

Suppose that the relationships between the different sectors are as shown in Table 13.1. The columns reflect the needs of each sector for labor and inputs from itself and the other industries. The rows show the demands for the output of each sector by the other industries. Thus, 10 cents per dollar of corn production will be used by the corn industry itself, 15 cents per dollar of corn production will be used by the fuel industry, and no corn will be used to produce transportation. The outputs of the fuel and transportation industries are used by themselves as well as the other two sectors as shown in Table 13.1. With these assumptions, the economy can be represented as a set of three simultaneous, linear equations. Let C, F, and T represent corn, fuel, and transportation output, respectively. Then,

(13.1) $C = 0.10C + 0.15F + 0T + 100$

(13.2) $F = 0.25C + 0.30F + 0.40T + 80$

(13.3) $T = 0.35C + 0.25F + 0.10T + 50$

This is a system of three linear equations in three unknowns, which can be solved using matrix algebra (see Appendix 13.1) or through successive substitutions of the three equations.[7] The results show that it would be necessary to produce $158 million worth of corn to satisfy total demands. Of this $158 million, about $16 million would be used by the corn industry itself, $42 million would go to the fuel industry, and the remaining $100 million is household demand. Fuel production would have to be about $283 million: $80 million in final demand, $40 million for the corn industry, $85 million used by the fuel sector itself, and $78 million for transportation. Transportation production of $196 million is divided among the four uses according to equation (13.3) as follows: $55 million for corn, $71 million for the fuel industry, $20 million for itself and $50 million in final demand.

These results can be used to determine whether there are enough resources to produce the required amounts. In this example, the only productive resource is labor, and the bottom row in Table 13.1 can be used to figure out how much labor will be required in total and in each industry. For example, each dollar of corn production requires 30 cents of labor, so production of $158 million worth of corn means that labor worth about $47 million will be needed in this industry. For fuel and transportation, labor requirements are $85 million and $98 million, respectively, so total labor requirements are $230 million. This amount turns out to be exactly equal to the total value of the final output ($100 + 80 + 50 = 230$), so the wages paid to workers will just cover their expenditures for the final goods.

Table 13.1. Inputs and outputs in a three-industry
economy (in dollars)

Input	Output Users		
Producers	Corn	Fuel	Transportation
Corn	0.10	0.15	0.00
Fuel	0.25	0.30	0.40
Transportation	0.35	0.25	0.10
Labor	0.30	0.30	0.50
Total	1.00	1.00	1.00

A more formal presentation of input-output analysis using matrix algebra can be found in Appendix 13.1. Input-output models for real-world policy analysis are much more complex than this simple three-industry model, which is formally solved in Appendix 13.1. In fact, input-output tables for the United States and other countries generally include several hundred industries and many factors of production. Nevertheless, the simple model presented in this chapter can be used to illustrate how such models are used in policy analysis. Suppose that the government of the region wishes to implement a policy that would increase final demand for corn by 20 percent. For this exercise, the government is treated as exogenous because there is no government sector in the model. Thus, it is assumed that the government can magically implement the policy without dipping into the labor and other resources already being used. The policy to increase final corn demand can be analyzed by re-solving the system using $120 million for final corn demand instead of $100 million as in the previous example. Using the computational procedures described in Appendix 13.1, it turns out that it would be necessary to produce $183 million of corn, $304 million of fuel, and $212 million in transportation to meet this higher demand. These production levels require an increase in labor of about 10 percent, from $230 million in the first example to $252 million following the new policy. If the additional $22 million in labor is not available to this economy, the consumption target cannot be met and more modest goals may have to be set.

This type of analysis is often based on matrices of *multipliers* derived from the basic input-output relationships described to this point. The derivation of input-output multipliers is briefly discussed in Appendix 13.1. The basic procedure is analogous to the methods used to derive income multipliers as described in most macroeconomics textbooks. The idea is that an exogenous shock to an economic system induces a series of changes that feed back on each other so that the final result is greater than the initial impact. An exogenous increase in investment will lead to an increase in GNP. An increase in GNP, in turn, stimulates an increase in consumption, and this consumption increase has a further impact on GNP, which then causes additional changes in consumption, and so on. The end result of all this is an increase in GNP that is greater than simply adding in the exogenous change in investment. The matrix of multipliers developed from basic input-output relationships can be used to predict the final impact of changes in complex combinations of the exogenous variables.

Real input-output tables allow for many interindustry relationships, multiple factors of production, international trade, and several categories of final demand. Without attempting to reproduce a complete input-output table, the basic structure of such models is illustrated in Table 13.2.[8] In this model, there are four industries (A, B, C, and D) producing inputs for the other industries as well as outputs for final demand. The *interindustry transactions matrix* is simply a four-industry version of the industrial structure described in the preceding example. In that example, however, there was only one input (primary sector), which we named labor. The labor input shows up as labor services provided by households in Table 13.2, which also includes other inputs such as government and capital services as well as imports. In addition, final demand has been divided among households, government, exports, and investments.

Table 13.2. Basic structure of an input-output table

	Production				Final Demand	Gross Output
	A	B	C	D	HH Gov. Exports Invest	
Production						
A						
B		Inter-Industry			Household, Government,	Total outputs
C		Transactions			Foreign, and Investment	(equal total
D						inputs)
Primary Sectors						
Households (labor)		Resource				
Goverment services		Inputs =				
Imports		Value Added				
Capital inputs						
Total inputs (equal total outputs)						

HH = households, Gov. = government services.

The model in Table 13.2 includes many of the elements of the circular flow. The financial, government, and foreign sectors are all specifically included. Notice, however, that the lower right-hand corner (the part that would represent the link between the primary sectors and the final demands) of the input-output model has been left blank. This means that the input-output model has not fully endogenized the link between input payments and output expenditures. Thus, there is no indication of how household income generated by payments for labor services is distributed among purchases of domestic or foreign goods and services, taxes, and savings. In the next section of the chapter, we will introduce the concept of social accounting matrices (SAMs), which are structured to remedy this deficiency.

Input-output models can be developed for regional, national, or international administrative units. In many U.S. states, input-output models of the state economy are maintained by local government agencies to plan development activities and to analyze the implications of changes caused by shifts in policy or new technologies. An important development in recent years has been the creation of computer software that provides the basic structure and technical coefficients for input-output analysis in a spreadsheet format. These software programs make it very easy for policy makers and analysts to customize the basic model in the software package to fit local or regional conditions. Input-output models can be used to measure such impacts as the effect on the local economy of a state-funded university or of a trade agreement that leads to increased demand for local goods. A group of English and Welsh researchers has made extensive use of the input-output approach to analyze agricultural policies and rural development in the United Kingdom.[9] National input-output models have been used fairly extensively in developing countries as planning tools, and national and regional models have been linked together to create multinational and world input-output models.[10]

Despite their usefulness and relative simplicity, input-output models suffer from a number of limitations. First, it is difficult to include industries that produce joint products. More significantly, they are based on the assumption that industries use inputs in fixed proportions, eliminating the possibility of substituting one input for another. Variants of the basic input-output model that make the productive activities somewhat more realistic have been developed, but such refinements increase the computational complexity and may not fully overcome the basic limitations. As noted earlier, they do not include specific links between payments for inputs and final demand for outputs. Input-output analysis focuses primarily on interindustry transactions in the production of goods and services. Analysis using such models is generally based on exogenous changes in demand. Thus, for example, in the simple model used to illustrate this method, we simply introduced an exogenous change in the final demand for corn to predict the impact on the three industries and their needs for labor. In the next section, the basic input-output model illustrated in Table 13.2 is extended to incorporate specific links between the primary sectors and final demand.

The Social Accounting Matrix (SAM)

Social accounting matrices (SAMs) were developed as a way to systematically represent the structure of an economy at a point in time.[11] They are a natural extension of input-output models and found their first applications in policy analysis and planning in developing countries. SAMs are based on double-entry bookkeeping but are set up in the form of a table for greater efficiency of exposition. Thus, a single entry in the SAM can be thought of as both an asset and a liability. The terminology used in discussing SAMs reflects their origin in the double-entry bookkeeping of the national accounts. They are divided into accounts, with the flows into and out of the different accounts required to balance. That is, payments to households for labor services are a flow into the household account that is balanced by a flow from the production account. As with input-output tables, row entries are receipts and column entries are expenditures. This is sometimes confusing, because a single cell in the table is part of both a row and a column. If one is reading the table across the row, the entry should be understood as a receipt by the account shown for that row. If one is reading down a column, the same entry represents an expenditure from the account shown at the top of the column.

There are five major accounts that are generally included in a SAM: production, factors of production, institutions, combined capital, and rest of the world. Each of these accounts can be divided into subaccounts, depending on the level of detail that is both feasible and desirable given the availability of reliable statistical data. The production account is the basic interindustry transactions matrix of the input-output table. Such an account could include all the hundreds of industries in a country, although in practice, it is generally necessary to structure the production activities at a fairly high level of aggregation if the SAM is to be kept small enough to be useful. The factors account is generally subdivided into labor (sometimes further disaggregated into skilled, semiskilled, and unskilled labor categories) and capital, although alternative subdivisions are possible, depending on the application. For example, Pyatt and Round develop a SAM that divides labor into urban, rural, and labor from large plantations and estates, with capital divided into public and private capital.[12] The institutions account includes firms, households, and the government. As with the other accounts, these categories may be further subdivided if needed. For example, an analyst might wish to disaggregate households into those with low, middle, and high incomes.

In describing the structure and use of SAMs, it is convenient to base the discussion on a specific example. King presents a SAM for Sri Lanka in 1970 in his introduction to this method, and I will draw on that discussion in what follows. The basic structure of the Sri Lanka SAM is shown in Table 13.3. This SAM is the same as the one shown in the article by King,[13] although I have combined two of the accounts and rearranged some of the rows and columns to make it more consistent with the input-output tables already presented.

Table 13.3. Basic structure of a simplified SAM (receipts and expenditures in millions of Sri Lankan rupees)

Receipts	Expenditures					Total
	1	2	3	4	5	6
1. Production	4,660		9,350	1,962	2,113	18,085
2. Factors of Production	11,473					11,473
3. Institutions	885	11,360	3,496	313	97	16,151
4. Combined Capital			2,214		425	2,639
5. Rest of the World	1,067	113	1,091	364		2,635
6. Total	18,085	11,473	16,151	2,639	2,635	

Source: Benjamin B. King, "What Is a Sam?" in Social Accounting Matrices: A Basis for Planning, ed. G. Pyatt (Washington, D.C.: World Bank, 1985), 26, with minor modifications. Copyright 1985 by the World Bank. Reproduced with permission of the World Bank, via Copyright Clearance Center.

In the Institutions row in Table 13.3, households and the government receive 11,360 million rupees (MR) in return for supplying labor services and capital. This income is a receipt from the account for Factors of Production. This latter account registers an expenditure as the income is transferred to the Institutions account. The Factors of Production account receives 11,473 MR from the Production account. This payment to the Factors of Production account is an expenditure from the Production account. The payment from the Factors of Production account to Institutions is slightly less than the amount received from the Production account. The difference is a payment to the Rest of the World.

Institutions also receive income from themselves (e.g., taxes paid by households to the government), from the Combined Capital account, and from the Rest of the World. The total receipts by the Institutions account are 16,151 MR, which is equal to the expenditures from that account.

The totals for receipts for the various accounts are all equal to the totals of the expenditures from the corresponding accounts. The accounts for Production, Factors of Production, and Institutions make up the core of the SAM accounting for most of the receipts and expenditures. The institutional account has already been described, along with a few elements from the other accounts. The Production account receives income from itself (this is the interindustry transactions matrix of the input-output table), Institutions (final demand for consumption goods), Combined Capital (investment), and the Rest of the World (export receipts). Reading down the Production column, labeled "1," we find expenditures for inputs from other industries (the interindustry transactions again), payments to Factors of Production, payments to Institutions (taxes), and payments to the Rest of the World (expenditures on imports). The account for Factors of Production is simpler, showing receipts from the Production sector (payments for labor and capital) that are equal to the transfers to Institutions (households and government) plus the Rest of the World (foreigners).

The other two accounts, Combined Capital and the Rest of the World, are needed to handle savings and investments and transactions with the rest of the world. The Combined Capital account receives 2,214 MR in savings from the Institutions account and 425 MR in borrowed capital from the Rest of the World. Investments from the Combined Capital account are distributed among the accounts for Production, Institutions (government taxes), and the Rest of the World. The Rest of the World receives payments from the Factors of Production account (income to factors that is transferred abroad) and payments for imports by the Institutions, Production, and Combined Capital accounts. Expenditures from the Rest of the World account consist of transfers from the Combined Capital and Production accounts to domestic Institutions and exports.

Table 13.3 sets out the basic structure of a typical SAM. For practical applications, this model would be disaggregated to show how various components of the five major accounts are related. The disaggregated SAM presented by King includes five industrial sectors in the production account (agriculture, industry, construction, trade and transport, private services, and government services), two factors of production (labor and capital), and four institutions (households, corporations, government, and a separate category for indirect taxes). Along with the combined capital and rest of the world accounts, the final SAM is made up of fourteen rows and columns. The entries in the SAM correspond to figures found in the national accounts, which include the gross national product accounts and the balance of payments account. The SAM is a numerical representation of the circular flow shown in Figure 13.1.

The construction of a SAM is not a trivial exercise. The obvious question that must be asked is what these models can be used for and whether the results that can be obtained from their application to a problem are worth the effort required to put such models together. King notes that SAMs can be particularly useful in developing consistent statistical data on a country's economy and presenting these data in a manner that can be used by planners. In addition, SAMs can be used in much the same way as an input-output table to derive a matrix of multipliers (see Appendix 13.1). These multipliers can then be used to predict the impact of exogenous shocks to the economic system. For example, Pyatt and Round develop a set of multipliers for the Sri Lanka economy based on a more elaborate version of the SAM presented in this section and use the model to explore the relationship between the structure of production in Sri Lanka and the distribution of income. Bell and Devarajan use a variant of the SAM approach to do a social benefit-cost analysis of an irrigation project in Malaysia. The authors identify a set of exogenous variables that will be affected by the project and then predict the impact of changes in these variables, using the multipliers derived from the SAM. As in all good policy analysis, they compare the predicted outcomes with the project to those predicted to prevail if the project is not undertaken.[14]

Computable General Equilibrium (CGE) Models

SAMs can be structured to include all the relations shown in the circular flow diagram and thus take account of links that are not included in the input-output model in Table 13.2. At the same time, SAMs share certain characteristics with input-output models that limit their usefulness in certain applications. Both approaches assume fixed coefficients for the relationships between the various components of the economy. For example, a policy to increase tourism in a developing country would lead to increased demand for the services of hotels, restaurants, local transportation, and all the other industries and service providers that are the source of inputs to these activities, including labor and capital services. The use of input-output or SAM multipliers to predict the magnitude of these impacts means the analyst is assuming that the changes in wages and prices induced by the shifts in demand do not lead to the substitution of cheaper inputs for those that have become more expensive.

Suppose that the tourism policy leads to an increase in wages for unskilled workers. Industries employing these workers, including the ones most directly affected by tourism such as hotels and restaurants, may find that they will earn greater profits if they substitute capital inputs for the more costly unskilled labor they employed in the past. Thus, for example, they might elect to purchase mechanical cleaning tools operated by one skilled worker rather than continuing to employ numerous unskilled workers using hand tools. Input-output and SAM multipliers would show that if the output of a given industry doubles, it will simply use more of the same inputs in the same proportions as it did before the demand shock. That may not be the case if firms in the industry can change the combination of inputs used to produce their products.

Because firms and households are likely to adjust their behavior in light of changes in prices and other economic variables, neglecting this response could lead to predictions that are considerably off the mark. *Computable general equilibrium (CGE) models* remedy this defect by replacing the fixed multipliers of the SAMs with equations aimed at capturing economic responses to exogenous shocks to the system. A CGE model is a set of equations that is structured to replicate the five general accounts described by the SAM. According to Dervis, de Melo, and Robinson, CGE models include the basic feedback mechanisms described by the SAM, along with endogenous prices.[15] Price endogeneity is achieved by solving the model equations to find the set of wages and prices that are consistent with equality between the amounts and combinations of goods and services produced by firms and the amounts that households and other demanders wish to purchase.

To see how this works, consider the general CGE model described by Dervis, de Melo, and Robinson. The model includes equations that calculate import prices on the basis of exogenous world prices and policy interventions, as well as other equations to calculate export prices on the basis of domestic prices and domestic policies. These and other prices are used to explain production and consumption decisions. On the production side, the model includes production functions for the various goods and services that can be used to derive expressions for input demand. Input demand along with exogenous capital supplies and endogenous supplies of different categories of labor describe factor markets which can be solved, given a set of wages and prices, to predict factor use. The amount of factors employed then determines, via the production functions, how much of each good or service will be supplied. The amount of different kinds of labor and capital that are employed at different wage and interest rates not only determines output but also the incomes to the households that supply the factor inputs. There are also equations to explain exogenous export demand by the rest of the world, as well as endogenous import demand and the balance of payments (exports minus imports plus exogenous net foreign capital inflows). Finally, domestic consumption by households depends on prices and incomes, and final demand is made up of household consumption, demand for inputs (interindustry transactions as in input-output analysis but now based on production functions that allow substitution among the inputs), investment demand, government demand, and export demand. All of these equations combine to generate domestic supplies and total demands for the various goods and services in the economy.

Equilibrium requires that domestic supply be equal to total demand. Once the model has been set up and the parameters for all the equations have been estimated, it is solved for the set of prices that will cause domestic supply to equal total demand. The procedure is to specify some initial set of prices and wages and insert them into the model equations to predict supply and demand. These will not normally be equal on the first try, so the prices and wages are adjusted according to guidelines determined by the solution algorithm, and the model is solved again. This process continues until the difference between domestic supply and total final demand is zero (or very close to zero). The term for the procedure of repeatedly solving the model after adjusting the prices and wages is *iteration*. Thus, the CGE model is solved through iteration until the equilibrium set of prices and wages that cause supply to be equal to demand is discovered.

The final set of prices and wages is endogenous because it is determined within the model. The basic CGE described by Dervis, de Melo, and Robinson treats foreign prices, policies, foreign capital inflows, total capital, savings rates, and other policy variables as exogenous. This specification means that the analyst can use the model to discover equilibrium prices and quantities given a set of exogenous variables designed to represent the current situation and then compare these results with new equilibrium prices and quantities that result from altering some of these exogenous variables.

Thus, an analyst could predict the impact of eliminating tariffs on imported manufactured goods by comparing the model predictions with the tariff in place, with predictions generated by the model when the tariff is set at zero in the relevant equations. Note that the results of this analysis will allow measurement of the effects of removing the manufacturing tariffs not only on the manufacturing industries themselves but on government revenues, the balance of payments, supply and demand of agricultural goods, and so on. One of the great advantages of general equilibrium analysis is that it takes into account the fact that everything in an economy is related to everything else. Because of these intricate relationships within an economy, a relatively minor policy change—as raising the support price on corn, for example—may have a measurable impact on seemingly unrelated economic variables, such as demand for plastic containers or the supply of computer monitors. The nature and magnitude of such impacts is not intuitively obvious, as they depend on complex sequences of causes and effects.

There are, of course, many ways to specify CGE models as well as to determine the values of the parameters in the model. In general, they are used in conjunction with a SAM that serves to specify the initial conditions in the economy. The analyst can then use the equations to predict the impact of a policy change in terms of how these initial conditions will be modified. The CGE model can be structured to include equations corresponding to all the relationships in the SAM, or it can be specified to focus on certain sectors of the economy, relying on the SAM multipliers to predict the impacts of a change in sectors that are not specifically modeled in the CGE. Thus, for example, it would be possible to develop a full model of the agricultural sector and measure the impacts of changes in agricultural policy on nonagricultural industries, incomes, and investments, for example, through the linkages shown in the country's SAM. CGE models have been widely used to analyze major trade initiatives, such as NAFTA or the Uruguay Round agreement that created the World Trade Organization (WTO). Some of the studies of NAFTA will be reviewed in the next chapter. CGE models have also proved particularly useful for analyzing development strategies in low-income countries. Such models usually include a large number of equations arranged in a complicated structure to insure that the fundamental concept of all economy-wide models—that is, the requirement that flows into one account be balanced by flows out of other accounts—is preserved.

An example may help further an understanding of the nature of these complex models. Liu, Yao, and Greener use such a model to analyze the advantages and disadvantages of alternative agricultural policy reforms in the Philippines.[16] From the 1960s to the early 1980s, many developing countries pursued development strategies based on extensive government intervention in the economy. This was the approach taken by the Philippine government, which used taxes on food exports to lower food prices, an overvalued peso to keep import prices down, and high import tariffs in certain sectors to implement an industrialization strategy based on *import substitution*. Import substitution aims to replace certain imported goods with similar goods produced locally. Such substitutions can arise only if domestic firms are protected from foreign competition by trade barriers.[17] Import tariffs raise the internal prices high enough to allow the domestic industries to survive despite their relative inefficiency. The overvalued exchange rate acted as a subsidy for imported goods, while the tariffs used to promote import substitution raised import prices. These policies penalized the agricultural sector (farmers received low prices for their output and had to pay higher prices for manufactured inputs produced by the protected domestic industries). They also caused unemployment, debt and burdensome debt-service payments, and balance-of-payments deficits.

The Philippine economic problems were the same as those of a great many developing countries in the 1980s. Under pressure from the World Bank and the International Monetary Fund (IMF) and faced with the imminent exhaustion of their foreign exchange reserves and capacities to service their debts, many developing countries initiated policy reforms known as *structural adjustment programs* (SAPs). In the Philippines, government policy shifted to an emphasis on economic growth, external balance (reducing the balance-of-payments deficit), and increased incomes for farmers and other low-income households. The authors of this study note that this policy mix is problematic. Progress in one area could make achievement of other policy goals more difficult. Balance-of-payments equilibrium, for example, may require devaluation of the currency, raising prices for imported food and other consumer items that are important for low-income urban households. Efforts to stimulate agriculture could slow industrial development, and the objective of greater food self-sufficiency is in conflict with efforts to promote agricultural exports as a means to reduce trade deficits. Because of these inherent conflicts, policy reform was potentially quite tricky. In this context, analysis of the likely effects of alternative reform scenarios could be of great use in determining how to manage the reform process. CGE models are designed to provide such analytical insights.

Liu, Yao, and Greener focus on agricultural policy reforms in their analysis. The Philippine National Food Authority (NFA) played an important role in the marketing of rice, the main food crop. Although private firms are the main actors in rice markets, the NFA serves as an alternative buyer of rice and controls rice imports. The NFA sold rice at market prices as long as those prices were below a government-set price ceiling. Philippine policies also included tariffs on imported agricultural inputs, export taxes, and an overvalued exchange rate. The CGE model constructed to analyze policy alternatives includes a production component to represent production of rice, corn, sugar, and other agricultural products in various regions of the country. Output depends on factor use, which is predicted by factor market models. The marketing component is designed to model the behavior of private traders, the NFA, and private importers and exporters. Finally, the equations for consumption are structured to capture differences between rural and urban consumers, consumers in the northern islands as opposed to those in the south, and the various income levels found in the country. The model includes 250 equations linked to a SAM. The equations are used to simulate the effects of alternative food policy reforms on participants in the food system, while the SAM allows predictions of the more general impacts of these alternatives on the broader economy.

The authors use this model to conduct several experiments related to potential policy reforms in the Philippines. They identify seven policy initiatives targeting the supply side, the demand side, and trade. Each policy initiative is translated into an exogenous shock to the system so that the effects of that initiative can be compared to baseline predictions designed to represent the counterfactual of no policy reform. In addition, the authors simulate four alternatives that are combinations of the seven policy changes. The types of reforms analyzed in this study include such measures as raising the price floor for rice, eliminating a rural income tax, and increasing income taxes for high-income urban consumers, eliminating tariffs on imported agricultural inputs, eliminating agricultural export taxes, and devaluation. The simulations generate predictions of rice, corn, and sugar production, incomes of various social groups (urban poor, landowners, tenants, etc.), food consumption, government finances, economic growth, and the balance of payments.

The results suggest that the supply-side policies analyzed would lead to increased rice output compared to the baseline but lower the incomes and consumption of all consumer groups and worsen the budget deficit and the balance of payments deficit. The demand-side reforms, which include redistributional policies, are predicted to increase the incomes and consumption levels of the poor and to improve the government budget and balance-of-payments deficits but slow overall economic growth (per capita income) and lower food production. The trade policies are likely to have positive impacts on production, incomes, consumption, and growth but worsen the budget and balance-of-payments deficits. Devaluation by itself is the most effective reform in improving the budget and payments deficits. Overall, the authors conclude that a "combination of tax reform, trade liberalization and devaluation of the domestic currency produces a set of balanced results."[18] They suggest that such a combination will help raise food production and incomes without causing severe negative effects on the government budget and external balance.

The analysis of Philippine policy reform illustrates the usefulness of CGE models in providing predictions that can be of use to policy makers. They allow a wide range of experiments to be conducted, and the alternatives analyzed can be specified at a level of detail that is highly realistic. Of course, such models are costly to build and complicated to use. They are subject to the black-box problem, and policy makers are likely to be highly skeptical if the model predictions differ substantially from their intuitions or biases. These models are also subject to maintenance costs, as is the case with the large multisectoral and systems models described earlier. It is not the case that a CGE model for a particular country can be carefully constructed and then used for the indefinite future without reestimating the parameters, adding equations, and adjusting the macroeconomic accounts. In other words, the shelf life of these models may be relatively short, given the amount of resources required to put them together. In addition, a CGE model built to analyze a specific set of issues may be ill suited for use in analyzing some other set of problems. This would seem to indicate that it may be impossible to construct a kind of universal CGE model that could be used to address any policy question that might arise.

Of course, these limitations may become somewhat less severe as computing capacities continue to expand. It is possible that CGE templates similar to the input-output software described earlier will be developed. Such innovations could eventually become sufficiently user-friendly to allow policy makers to carry out sophisticated policy analysis using highly complicated models, the precise nature of which remains hidden in the black box. Such tools would have to demonstrate a high degree of reliability to convince practitioners of their usefulness. Until such innovative analytical tools are invented, realistic economy-wide models are likely to remain expensive undertakings. Whether the cost of applying such frameworks to policy problems is justified by the benefits of more complete and precise predictions depends on the problem, its importance, the resources that can be mustered for its analysis, and other factors that may vary from case to case.

CONCLUSION

The three core chapters in Part II of this book (Chapters 7, 10, and 13) provide a broad inventory of some of the methods used to predict the effects of alternative public policies. Benefit-cost analysis and comparative static analysis using partial equilibrium, multisector, systems, and economy-wide models are among the most common approaches, but they do not exhaust all the possibilities. Some policy models incorporate dynamic elements related to time, risk, and uncertainty. Others focus on strategic interactions among economic agents using the methods of game theory. Advanced time-series econometric methods can also be applied to policy issues. Clearly, a wide array of analytical techniques and models can be called into service to refine predictions of the effects of alternative public policies. The purpose of the three core chapters has been to provide students with a general overview of these methods as an introduction to their use and as an aid in understanding some of the analytical results that may be put forward in policy discussions.

Two themes have been emphasized throughout the descriptions of these methods and in the illustrative case studies. The first has been that the economic analysis of public policies always requires the comparison of predicted effects of a policy with the likely evolution of the economy in the absence of the policy. Economic analysis of public policies is analogous to a controlled experiment in which the model provides a method for predicting the state of the world with and without the particular policy initiative under consideration. Thus, for example, in the Philippine study described in this chapter, the authors use their CGE model to predict baseline outcomes that are expected to prevail if no policy reform is undertaken and compare these outcomes with those predicted by the model in light of the exogenous policy changes. It is this comparison that is enlightening, because it helps policy makers to decide whether implementing any of the available policy alternatives is better than doing nothing at all.

The other theme that has been emphasized throughout is the idea that analytical results describing the costs and benefits of alternative policies should always be seen as just one input into the decision-making process. Evidence that the economic impacts of a given policy option are broadly beneficial is clearly an important piece of information in trying to decide whether that option should be exercised. It is unrealistic, however, to think that political considerations will ever be absent from policy making and intellectually dishonest to leave the ethical presuppositions that underlie any policy analysis unarticulated. In all the case studies included in this book, ethical judgments about the best course of action are present, although in some cases, they are not well developed or even acknowledged. Likewise, the political realities in which policies are made and implemented influence the final decisions taken as well as the nature of the economic analysis that is appropriate. In the next chapter, a case study of the use of CGE models in the analysis of the expected impact of NAFTA on agriculture and the environment will provide a final illustration of these two themes.

SUMMARY

1. Partial equilibrium models include a limited number of endogenous variables. Model complexity is increased by including more endogenous variables. In all cases, policy analysis is done by varying the exogenous variables and measuring the impact of these changes relative to an appropriately designed baseline.

2. More complex models may provide more complete and more precise predictions, but they are costly to build and maintain and may be difficult to interpret (the black-box problem).

3. Multisector models typically link together a relatively large number of partial equilibrium models to provide insights on the broad implications of alternative policies.

4. Systems models often include noneconomic components, such as biological or engineering models, combined with economic models.

5. Economy-wide models replicate some or all of the relations in the circular flow at various levels of aggregation. They include macroeconomic models, input-output analysis, social accounting matrices, and computable general equilibrium models.

6. Input-output analysis grows out of Quesnay's *Tableau économique* and is based on the idea that total output of a given industry has to be large enough to satisfy its own demands and the demands of other industries as well as final demand.

7. Social accounting matrices (SAMs) extend input-output tables to include links between primary sectors and final demands. As with input-output, SAMs can be used to derive matrices of multipliers.

8. Computable general equilibrium (CGE) models use sets of equations to predict important economic relationships coupled with SAMs to capture the more important aggregate impacts and to assure balance between expenditures and receipts. CGEs have proved extremely useful for analyzing broad development policies, international trade agreements, and many other policy initiatives.

KEY CONCEPTS

endogenous variables	*Tableau économique*
exogenous variables	multipliers
multisector models	interindustry transactions matrix
economy-wide models	social accounting matrix (SAM)
black box	computable general equilibrium (CGE)
systems models	models
circular flow	iteration
macroeconomic models	import substitution
input-output analysis	structural adjustment programs (SAPs)

DISCUSSION QUESTIONS

1. Discuss the advantages and disadvantages of using more complex models for policy analysis.

2. Explain the basic procedures for using models to analyze alternative policies. Describe how a model might be used to analyze the impact of a policy change such as the imposition of a cigarette tax or the elimination of agricultural import tariffs.

3. How do SAMs differ from input-output tables?

4. Explain the basic nature of structural adjustment programs (SAPs), and discuss why CGE models are particularly appropriate for analyzing the welfare effects of these policy reforms.

5. Describe the general procedure for determining equilibrium prices in CGE models.

6. Discuss the role of models in policy analysis.

SUGGESTIONS FOR FURTHER READING

Dervis, K., J. de Melo, and S. Robinson. *General Equilibrium Models for Development Policy*. New York: Cambridge University Press, 1982.

Miller, Ronald E., and Peter D. Blair. *Input-Output Analysis: Foundations and Extensions*. Englewood Cliffs, N.J.: Prentice-Hall, 1985.

Pyatt, G., and J. I. Round, eds. *Social Accounting Matrices: A Basis for Planning*. Washington, D.C: World Bank, 1985.

APPENDIX 13.1: A BRIEF INTRODUCTION TO MATRIX ALGEBRA AND FORMAL INPUT-OUTPUT ANALYSIS

Matrix algebra provides a way to represent and solve systems of simultaneous equations. Suppose, for example, that one wished to write down a system of demand equations for 50 categories of food (fluid milk, ice cream, beef, pork, potatoes, carrots, tomatoes, apples, wheat flour, sugar, etc.). Economic theory suggests that demand for each of these items is dependent on the price of each item, all other prices, and income. Such a system could be written out as follows, with D representing demand, P price, Y income, and α_{ij} the coefficients or parameters of the system ($i,j = 1, ..., 50$):

$$(13.4) \quad D_1 = \alpha_{11}P_1 + \alpha_{12}P_2 + \alpha_{13}P_3 + \ldots + \alpha_{150}P_{50} + \alpha_{1y}Y$$
$$D_2 = \alpha_{21}P_1 + \alpha_{22}P_2 + \alpha_{23}P_3 + \ldots + \alpha_{250}P_{50} + \alpha_{2y}Y$$
$$D_3 = \alpha_{31}P_1 + \alpha_{32}P_2 + \alpha_{33}P_3 + \ldots + \alpha_{350}P_{50} + \alpha_{3y}Y$$
$$\vdots \qquad\qquad\qquad\qquad\qquad \vdots$$
$$D_{50} = \alpha_{501}P_1 + \alpha_{502}P_2 + \alpha_{503}P_3 + \ldots + \alpha_{5050}P_{50} + \alpha_{50y}Y$$

The dots between variables and equations in (13.4) indicate that some of these items have been omitted to avoid having to write out 50 equations with 51 variables on the right-hand side. Clearly, working with this type of notation would be extremely cumbersome. With one small modification, this system can be written out much more simply. Assume that the 50th price is a price index for all consumer goods and that this price index can be used to express all the food prices as relative prices (or as real prices, that is, prices corrected for inflation in the general economy). In other words, let P_i be the price of item i divided by the 50th price. This reduces the number of variables on the right-hand side to 50 (49 relative prices and income). We can retain the equation for demand for the 50th good, which in this case represents all nonfood items. With this modification, the system is said to be conformable (there are 50 equations and 50 unknowns) and can be rewritten using matrix notation.

A vector is either a column or a row of figures or variables. The vector of food demands is shown as a column vector, while the price and income vector is shown as a row vector:

$$\begin{bmatrix} D_1 \\ D_2 \\ \cdot \\ \cdot \\ \cdot \\ D_{50} \end{bmatrix} \qquad [P_1 \, P_2 \ldots P_{49} \, Y]$$

A matrix is an array of ordered figures or numbers:

$$\begin{bmatrix} \alpha_{11} & \alpha_{12} & \cdot \cdot \cdot & \alpha_{149} & \alpha_{1y} \\ \alpha_{21} & \alpha_{22} & \cdot \cdot \cdot & \alpha_{249} & \alpha_{2y} \\ \cdot & & & & \cdot \\ \cdot & & & & \cdot \\ \cdot & & & & \cdot \\ \alpha_{501} & \alpha_{502} & \cdot \cdot \cdot & \alpha_{5049} & \alpha_{50y} \end{bmatrix}$$

It is customary to denote matrices and vectors with capital letters, often written in bold type. Let D be a vector of food item demands, P a vector of relative prices and income, and A the matrix of coefficients shown above. Both D and P are column vectors. The demand system can be written:

(13.5) $D = AP$

A typical equation from this system is: $D_i = \sum \alpha_{ij} P_j + \alpha_{iy} Y$. The advantage of this notation is clear. Rather than writing out large numbers of equations, the entire system can be reduced to the simple expression in (13.5).

Notational simplicity is not the only reason for using matrix algebra when faced with large systems of equations, however. It turns out that the basic algebraic manipulations used to solve single equations can be applied, under certain circumstances, to matrix equations. Many of the operations required to apply matrix algebra to a system of equations are complicated and require numerous calculations. Such problems can be overcome easily with the use of computers. I will not attempt to provide a course on matrix algebra in this appendix. Rather, I will offer a few general ideas on the use of these techniques, with reference in particular to input-output analysis. There are numerous textbooks and other sources of information that can provide the details of matrix algebra for those who may be interested.

Matrices and vectors can be subjected to the basic arithmetical operations of addition, subtraction, multiplication, and division and satisfy the standard commutative ($a + b = b + a$; and $ab = ba$), associative [$a + (b + c) = (a + b) + c$; and $a(bc) = (ab)c$] and distributive [$a(b + c) = ab + ac$] laws. For these laws to hold and to carry out normal arithmetical operations, it is generally necessary, however, for the matrices and vectors to be conformable. For example, conformability requires that the two vectors in (13.5), D and P, have the same number of rows as A and that A be a square matrix (same number of rows and columns). The conformability requirement was the reason for expressing the first 49 prices in the demand system in relation to the 50th price. Adding or subtracting matrices simply means adding or subtracting the corresponding elements in the two matrices. Thus,

$$
\begin{bmatrix} 3\,3\,7 \\ 2\,6\,4 \\ 1\,8\,1 \end{bmatrix} + \begin{bmatrix} 5\,3\,1 \\ 7\,2\,2 \\ 3\,1\,2 \end{bmatrix} = \begin{bmatrix} 8\,6\,8 \\ 9\,8\,6 \\ 4\,9\,3 \end{bmatrix}
$$

Subtracting the second matrix from the first results in a 3×3 matrix with the first element in the upper left-hand corner equal to $-2 (= 3 - 5)$ and the other elements calculated in the same way. Multiplication and division of matrices are quite a bit more complicated. Consider the following two matrices:

$$
A = \begin{bmatrix} 5\,4\,3 \\ 3\,2\,1 \\ 6\,1\,5 \end{bmatrix} \qquad B = \begin{bmatrix} 2\,1\,4 \\ 6\,2\,7 \\ 3\,4\,1 \end{bmatrix}
$$

They are conformable, and multiplication will lead to a new 3×3 matrix. The elements of this matrix are computed as follows:

Multiply the first row of A by the first column of B:
$(5 \times 2) + (4 \times 6) + (3 \times 3) = 33$
Multiply the first row of A by the second column of B:
$(5 \times 1) + (4 \times 2) + (3 \times 4) = 25$
Multiply the first row of A by the third column of B:
$(5 \times 4) + (4 \times 7) + (3 \times 1) = 51$
Multiply the second row of A by the first column of B:
$(3 \times 2) + (2 \times 6) + (1 \times 3) = 21$
Multiply the second row of A by the second column of B:
$(3 \times 1) + (2 \times 2) + (1 \times 4) = 11$
Multiply the second row of A by the third column of B:
$(3 \times 4) + (2 \times 7) + (1 \times 1) = 27$

Multiply the third row of A by the first column of B:
$(6 \times 2) + (1 \times 6) + (5 \times 3) = 33$
Multiply the third row of A by the second column of B:
$(6 \times 1) + (1 \times 2) + (5 \times 4) = 28$
Multiply the third row of A by the third column of B:
$(6 \times 4) + (1 \times 7) + (5 \times 1) = 36$

The new matrix will look like this:

$$\begin{bmatrix} 33 & 25 & 51 \\ 21 & 11 & 27 \\ 33 & 28 & 36 \end{bmatrix}$$

To illustrate the equivalent of division in matrix algebra, it is useful to start with an example based on a single equation and standard algebraic division. Consider a simple macroeconomic identity, with income divided between consumption and investment (there is no government or foreign sector):

(13.6) $Y = C + I$

where Y is GNP (income), C is consumption, and I is investment. Consumption is generally thought to depend on income, and we will assume a simple linear relationship:

(13.7) $C = a + bY$

Substitute (13.7) into (13.6) to obtain:

(13.8) $Y = (a + bY) + I = a + bY + I$

Subtract bY from both sides of the equation and factor out Y:

(13.9) $Y - bY = a + I$

$$(1 - b)Y = a + I$$

Now, divide both sides by $(1 - b)$:

(13.10) $(1 - b)/(1 - b)Y = a/(1 - b) + 1/(1 - b)I$
          or $Y = a/(1 - b) + 1/(1 - b)I$

Note that the final expression in (13.10) is the familiar expression for the income multiplier. An exogenous increase in investment will lead directly to an increase in GNP, and the increase in GNP will feed back through consumption to a further increase in GNP. The total effect is given by the multiplier. Suppose that b (the marginal propensity to consume) is equal to 0.6 and that the exogenous increase in investment is $100 million. Equation

(13.9) suggests that GNP will increase by $250 million [1/(1 − 0.6) = 2.5].

Now consider the system of equations presented earlier in the text as equations 13.1 through 13.3 and reproduced here as:

(13.11) $C = 0.10C + 0.15F + 0T + 100$
$F = 0.25C + 0.30F + 0.40T + 80$
$T = 0.35C + 0.25F + 0.10T + 50$

Recall that C, F, and T stand for corn, fuel, and transportation, respectively. Let D be used for demand in general, Y for total output, L for labor requirements, and X for total inputs. These same variables can be used with subscripts to indicate, for example, demand for fuel or total output of corn. Based on the results of the earlier analysis, the following table of resource flows can be constructed:

	C	F	T	D	Y
C	16	42	0	100	158
F	40	85	78	80	283
T	55	71	20	50	196
L	47	85	98		
X	158	283	196		

Input-output analysis begins by defining a matrix of technical coefficients showing the proportion of the total inputs (X_j for $j = C$, F, and T) accounted for by each industry. Let X_{ij} represent the figures in the top three rows and first three columns of the resource-flow table and denote the technical coefficients as α_{ij}. The technical coefficients are defined as:

(13.12) $\alpha_{ij} = X_{ij}/X_j$

For example, the technical coefficients in the row for corn would be computed as 16/158, 42/158, and 0/158, or 0.101, 0.148, and 0.0. The full matrix of technical coefficients, A, is:

(13.13) $A = \begin{bmatrix} 0.101 & 0.148 & 0.000 \\ 0.253 & 0.300 & 0.408 \\ 0.348 & 0.251 & 0.102 \end{bmatrix}$

With these definitions, the original system of equations can be represented in matrix notation as:

(13.14) $X = AX + D$

Equation (13.14) can be solved algebraically in a manner analogous to the methods used for single equations. First, subtract AX from both sides of the equation. Then, factor out X to obtain an expression similar to $(1 - b)$ in equation (13.9). For matrix algebra, the equivalent to the number 1 is known as the identity matrix (I). The identity matrix has the number 1 on the diagonal and zeros everywhere else. For example, a 3×3 identity matrix is defined as:

$$I = \begin{bmatrix} 1 & 0 & 0 \\ 0 & 1 & 0 \\ 0 & 0 & 1 \end{bmatrix}$$

Using the identity matrix, equation (13.14) can be written as:

(13.15) $(I - A)X = D$

In the case of a single equation, the next step in this process is to divide both sides by the expression in parentheses. In the case of equation (13.9), $(1 - b)/(1 - b) = 1$, and we are left with the variable of interest on the left-hand side. Dividing $(I - A)$ by $(I - A)$ will do the same thing for the left-hand side of equation (13.15). But how do we divide D by $(I - A)$? Recall that the inverse of a number multiplied by the number itself always gives a product of one: $3 \times 1/3 = 1$. The same holds true in matrix algebra, where the inverse of a matrix is written M^{-1}. In other words, $[M] \times [M^{-1}] = [I]$ by definition. The trick is to find a matrix $[M^{-1}]$ that when multiplied by the original matrix $[M]$ will generate an identity matrix.

Locating such a matrix is known as inverting a matrix. The procedures are fairly complicated and for larger systems of equations can only be done practically by high-speed computers. Not only is a computer necessary for the large number of calculations that need to be done, but computers carry more decimal places than most human beings are willing to carry through a long series of calculations done by hand. Even with relatively small problems, significant inaccuracies are introduced as a result of rounding errors when the matrix inversion is done by hand. Details of matrix inversion can be found in any textbook on matrix algebra or mathematical economics. For our purposes, it is sufficient to note that if $(I - A)$ has certain properties, it can be inverted to give a solution to the original problem:

(13.16) $X = (I - A)^{-1}D$

$(I - A)$ is often referred to as the *Leontief matrix*. The technical coefficients matrix in (13.13) can be subtracted from a 3×3 identity matrix to obtain the Leontief matrix for the three-industry input-output system described earlier in the text. Note that the diagonal elements are all calculated as 1 minus the coefficient shown in (13.13), while all but one of the off-diagonal elements are negative because they are calculated by subtracting the positive technical coefficients from zero.

(13.17) $(I - A) = \begin{bmatrix} 0.899 & -0.148 & 0.000 \\ -0.253 & 0.700 & -0.408 \\ -0.348 & -0.251 & 0.898 \end{bmatrix}$

The inverse of (13.17) is:

$$(13.18) \qquad (\boldsymbol{I} - \boldsymbol{A})^{-1} = \begin{bmatrix} 1.2578 & 0.3178 & 0.1444 \\ 0.8831 & 1.9304 & 0.8776 \\ 0.7343 & 0.6626 & 1.4154 \end{bmatrix}$$

Multiplying $\boldsymbol{D}$ by $(\boldsymbol{I} - \boldsymbol{A})^{-1}$ gives the final output required in the three sectors.

$$C = 1.2578(100) + 0.3178(80) + 0.1444(50) \approx 158$$
$$F = 0.8831(100) + 1.9304(80) + 0.8776(50) \approx 286$$
$$T = 0.7343(100) + 0.6626(80) + 1.4154(50) \approx 197$$

Note that these numbers differ slightly from the original numbers in the table of resource flows. The reason for these differences is the rounding errors due to the fact that the results were calculated by hand. Note that the inverted Leontief matrix is a matrix of multipliers similar to the multiplier derived in equation (13.10). This matrix of multipliers could be used to predict total output and input use that would arise from changes in demand. If, for example, the final demand vector were given by [120 105 76], these numbers could be used in the previous equations to determine total output. These calculations are left to the student as an exercise.

Notes

1. See the Economic Research Service Internet Web site at http://usda.mannlib.cornell.edu/data-sets/trade/92012 for descriptions of the Swopsim model. Further background can be found in V. O. Roningen, *Documentation of the Dynamic World Policy Simulation (DWOPSIM) Model Building Framework,* Staff Report 9226 (Washington, D.C.: USDA/ERS, October 1992).

2. The FAPRI model was used for the most recent assessment of the U.S. sugar program, as described in Chapter 11. Development of the FAPRI model is described by the Center for Agricultural and Rural Development (CARD) in CARD Report 11, no. 2 (summer 1998), and further information is available at the CARD Internet Web site: www.card.iastate.edu.

3. Aziz Bouzaher and Jason Shogren, "Modeling Nonpoint Source Pollution in an Integrated System," in *Modeling Environmental Policy*, ed. W. E. Martin and L. A. McDonald (Boston: Kluwer Academic Publishers, 1997).

4. Ibid., 9.

5. The difference between gross national product (GNP) and gross domestic product (GDP) is that GNP includes the net payments to foreign factors of production and net income from capital and labor employed by foreign firms; GDP does not.

6. See Arjo Klaamer, ed., *Conversations with Economists: New Classical Economists and Opponents Speak Out on the Current Controversy in Macroeconomics* (Totowa, N.J.: Rowman and Allanheld, 1984).

7. There are many ways to solve this system. One way is to begin by deriving an expression for C from equation (13.1). Subtract $0.1C$ from both sides and solve to obtain $C = 0.167F + 111.1$. Next, substitute this expression for C into equation (13.3), collect the terms, subtracting those for T on the right-hand side from both sides, and solve to obtain $T = 0.34272F + 98.77$. Now substitute these

expressions for C and T into the fuel equation (13.2), collect terms, and solve for a numerical value for F of about 283. Substitute this result into the two expressions for C and T and solve them to obtain $C = 158$ and $T = 196$ (all results are rounded).

8. For a description of tables similar to the one in Table 13.2, see E. M. Hoover and F. Giarratani, *An Introduction to Regional Economics* (New York: Alfred A. Knopf, 1984).

9. See P. Midmore, ed., *Input-Output Models in the Agricultural Sector* (Aldershot, U.K.: Avebury Press, 1991); and P. Midmore and L. Harrison-Mayfield, *Rural Economic Modeling: An Input-Output Approach* (Wallingford, U.K.: Cab International, 1996).

10. F. Duchin, "The World Model: An Interregional Input-Output Model of the World Economy," and D. E. Nyhus and C. Almon, "Linked Input-Output Models for France, the Federal Republic of Germany, and Belgium," both in *Global International Economic Models*, ed. B. G. Hickman (New York: North-Holland, 1983).

11. The description of social accounting matrices in this section draws heavily on Benjamin B. King, "What Is a SAM?" in *Social Accounting Matrices: A Basis for Planning*, ed. G. Pyatt and J. I. Round (Washington, D.C.: World Bank, 1985). Another useful source is D. Roberts, "A Comparison of Input-Output and the Social Accounting Methods for Analysis in Agricultural Economics," in *Input-Output Methods in the Agricultural Sector*, ed. P. Midmore (Aldershot, U.K.: Avebury Press, 1991).

12. G. Pyatt and J. I. Round, "Accounting and Fixed-Price Multipliers in a Social Accounting Matrix Framework," in *Social Accounting Matrices: A Basis for Planning*, ed. Pyatt and Round (Washington, D.C.: World Bank, 1985).

13. King, "What Is a SAM?" 26.

14. Pyatt and Round, "Accounting and Fixed-Price Multipliers;" C. Bell and S. Devarajan, "Social Cost-Benefit Analysis in a Semi-Input-Output Framework: An Application to the Muda Irrigation Project," in *Social Accounting Matrices: A Basis for Planning*, ed. G. Pyatt and J. I. Round (Washington, D.C.: World Bank, 1985).

15. K. Dervis, J. de Melo, and S. Robinson, *General Equilibrium Models for Development Policy* (New York: Cambridge University Press, 1982), 136.

16. A. L. Liu, S. Yao, and R. Greener, "A CGE Model of Agricultural Policy Reform in the Philippines," *Journal of Agricultural Economics 47*, no. 1 (1996): 18–27.

17. This has to be the case, because if the domestic firms could compete without protectionist trade barriers, they would already be producing the imported goods and there would be no need for any policy intervention. The fact that such goods are not being produced domestically means that local firms lack the technology or resources (skilled labor, capital, etc.) to be able to operate as efficiently as the foreign firms supplying the imported goods.

18. Liu, Yao, and Greener, "CGE Model," 27.

14

Case Study: The North American Free Trade Agreement (NAFTA)

INTRODUCTION

The North American Free Trade Agreement (NAFTA) entered into force in January 1994 after intense public debate in the three countries involved and a difficult ratification process in the United States. A free trade agreement between Canada and the United States had already been implemented in 1989.[1] In 1990, the governments of Mexico and the United States announced their intentions to begin trade negotiations, and Canada subsequently joined these talks. A tentative agreement was reached in 1992, and after further refinements, the treaties establishing NAFTA were ratified by the three governments in late 1993. NAFTA added bilateral agreements between Mexico and Canada and between Mexico and the United States to the existing Canada-U.S. agreement.

NAFTA is a trade agreement. Its primary purpose is to eliminate trade barriers among the three countries. According to the theory of customs unions, the static welfare effects of this type of trade agreement depend on the extent of trade creation and trade diversion. Economic integration is also likely to generate dynamic processes with potential repercussions for foreign investment, sectoral (e.g., agricultural) adjustment, jobs, rural economies, migration, the environment, and so on. The central empirical issues are whether these expected impacts are positive or negative and whether they are large or small. At the time NAFTA was being negotiated, the U.S. economy was over twenty times the size of Mexico's and almost seven times the size of the economies of Mexico and Canada combined. Given these differences in economic size, one would expect the impact of the trade policy changes, whether positive or negative, to be small in the United States and perhaps of greater importance in the other countries.

This expectation was largely absent from the highly charged political debate that arose in the United States. Opposition to NAFTA focused almost exclusively on economic integration with Mexico, which was seen as a threat to U.S. prosperity. Independent 1992 presidential candidate Ross Perot claimed that NAFTA would create a "giant sucking sound" as jobs in the United States were moved to Mexico. At the same time that the political rhetoric on NAFTA heated up, policy analysts began to make quantitative predictions of the likely impacts on jobs, economic growth, and so on. Lustig points out that the preferred analytical method was based on CGE models, as described in the last chapter, because these models allow measurement of a broad range of economic impacts while taking account of the complex interactions within an economy.[2]

The results of most of these CGE analyses indicated that NAFTA would have limited effects in all three countries, although they did show that certain sectors could be subjected to severe adjustment pressures. Those who opposed NAFTA attempted to counter such results with predictions of massive negative impacts based on dubious quantitative methods. For example, Choate predicted that six million jobs would be lost by assuming that all U.S. firms in which labor costs make up more than 20 percent of the value of the firm's output would move to Mexico.[3] It should be recalled that multilateral trade negotiations under the auspices of the General Agreement on Tariffs and Trade (GATT)—now the World Trade Organization (WTO)—had begun in 1986 and were being carried out simultaneously with the NAFTA talks. (See Box 14.1). These discussions on free trade provoked a strong countermovement involving organized labor, environmental groups, and others. The changes in trade policy associated with NAFTA are of particular interest because of the intense political debate they set off, the widespread use of economy-wide models to evaluate their impacts, and their potential effects on the general economy, agriculture, and the environment as well as on domestic politics and diplomatic relations among the three countries.

Box 14.1: Fast-track authority and international trade negotiations

International trade agreements are similar to international treaties. The U.S. Constitution grants the U.S. Congress the power to enter into treaties and other international agreements. Congress, however, does not have a permanent trade bureaucracy that could negotiate such agreements, and it is clear that direct participation in trade negotiations by the more than 400 members of Congress would be unworkable.

The solution to this problem has been for Congress to grant the executive branch "fast-track negotiating authority." When it appears that trade negotiations involving the United States are about to begin, the president requests fast-track authority from Congress. If Congress votes to grant this authority, the president's trade representative is free to participate in the negotiations and sign an agreement. Any agreement still has to be ratified by the Congress, but fast-track authority means that the senators and representatives can only vote for or against the agreement; they cannot offer amendments once agreement has been reached.

Fast-track authority is often thought to be necessary for the U.S. government to have any credibility in international trade negotiations. Without it, most countries would be reluctant to enter negotiations because of their expectation that the U.S. Congress would attempt to amend the final agreement. In the 1980s and 1990s, Congress granted fast-track authority to Presidents Reagan, Bush, and Clinton to conduct both the Uruguay Round (GATT) and NAFTA negotiations. The Uruguay Round talks began in 1986 under President Ronald Reagan while the NAFTA discussions began in 1990 under President George H.W. Bush. Both agreements were completed and ratified during the presidency of Bill Clinton.

ISSUES

General Impacts on Trade and Economic Growth

Four sets of issues were at the center of the NAFTA policy debates. The first set of issues concerned the general economic impacts of free trade in North America, particularly in terms of economic growth and trade. Proponents of the agreement argued that removal of trade barriers would stimulate trade and that increased trade volumes would lead to greater economic growth in all three countries. Because of the advantages for a low-income country such as Mexico of being able to sell freely into the vast U.S. market, it was often suggested that free trade would be of particular benefit to Mexico. Those who opposed

NAFTA suggested that the agreement was being written by large corporations, which were likely to be its main beneficiaries. For these individuals, any benefits of the agreement would be inequitably shared, with big business getting the lion's share at the expense of workers and Mexican peasants.

Labor

While some of the NAFTA critics in the United States denied that the agreement would lead to economic growth, most of their opposition had to do with the impact of free trade on unskilled U.S. labor. The U.S. debate over NAFTA developed into an argument about the impact of free trade on jobs. U.S. workers are paid higher wages than are workers in Mexico, and it was argued that this wage disparity would place U.S. firms at a competitive disadvantage relative to firms in Mexico. Cheap imports would put U.S. firms out of business or force them to move their operations overseas to countries whose wage rates were comparable to Mexico's. Either way, the United States would lose jobs. NAFTA supporters pointed out that U.S.-Mexico trade barriers were already fairly low, so further liberalization would not result in enormous job losses. In addition, increased trade flows lead to the creation of jobs, and these new jobs would offset any job losses brought about by lower-cost imports. More important, employment is influenced by many factors, including technological change and macroeconomic conditions such as the business cycle and interest rates. In general, trade liberalization has little effect on the total number of jobs.

One explanation of comparative advantage (see Chapter 2), the Heckscher-Ohlin model, leads to the conclusion that factor prices (e.g., wages) will be equalized under free trade. The factor-price equalization theorem has become a major argument for groups, such as labor unions, that oppose free trade. Unskilled labor is relatively scarce in the United States, while it is relatively abundant in Mexico, and this leads to high wages for unskilled U.S. workers compared with Mexican wage rates. Under free trade, the economies of the two countries are expected to adjust according to their comparative advantages. U.S. industries dependent on unskilled labor will contract, while those relying on skilled labor and capital will expand, leading to higher wages for skilled workers and lower wages for unskilled workers. For unskilled workers, the fact that trade also creates jobs is irrelevant because they are not qualified for the new jobs.

The validity of these arguments depends on empirical evidence, and it is precisely such evidence that the analytical models aim to discover. In the absence of clear evidence, factor-price equalization provides a strong conceptual argument against free trade. Note, however, that the theoretical negative effects of factor price equalization in the United States are mirrored by positive effects in Mexico, where workers with limited skills will be able to find employment at higher wages. It can seem selfish to insist on preserving low-skill jobs in the United States by depriving low-income families in Mexico of the jobs that might result from free trade. Labor unions have recognized this defect in their arguments and added a new twist that allows them to form alliances with groups working for international social justice. These latter groups think it unfair for corporations to pay lower wages to workers in developing countries. They believe that free trade allows firms to exploit poor workers and argue for the maintenance of trade barriers to prevent this exploitation. Thus, protectionist arguments are made to sound like support for the rights and welfare of workers in the low-income countries. As we shall see later in this chapter, environmental activists also joined the anti-NAFTA crusade. From its origins in the NAFTA and GATT debates, this alliance of labor organizations, environmental activists, and social-justice advocates has grown into the antiglobalization movement that is attacking McDonald's restaurants and disrupting the work of the International Monetary Fund (IMF), the WTO, and the World Bank today.

Sectoral Adjustment and Migration

Many opponents and some supporters of NAFTA argued that free trade in such staple foods as corn and dry beans would drive many peasants off their land, leading to increased migration in Mexico from the countryside to the cities and to the United States. Of course, this migration is not necessarily bad if the migrants find jobs in industries that are growing as a result of free trade. Free trade may cause some sectors to expand as others contract. How these sectoral adjustments affect internal and international migration as well as the well-being of others connected to a given sector depends on how industries respond to the changed economic environment. Many representatives of particular industries were concerned that they would be less well-off under NAFTA than with continued protection.

Some of the more contentious debates had to do with adjustments of various agricultural activities, although there were also concerns about such industries as textiles and automotive products. If the pre-NAFTA level of protection was high, liberalization could lead to disruptive shifts in trade flows. In the United States, producers of such horticultural products as tomatoes, asparagus, and avocados were worried that they would be unable to compete with Mexican producers. In Mexico, the concern was that cheap U.S. corn, wheat, and dry beans would flood into Mexico following removal of the high trade barriers, leading to large increases in migration.

Environment

As the negotiations progressed, a new issue—environmental protection—was injected into the debate. That there might be a connection between trade liberalization and the environment was recognized in the early 1970s. Baumol argued that the use of trade barriers by a country suffering the effects of negative environmental externalities originating in another country could lead to a second-best optimum if it proved impossible to correct the problem directly.[4] Economic analysts continued to work on this question, but it had not figured prominently in public policy debates until the NAFTA and GATT negotiations began. The catalyst for rising environmental concerns was a decision by a GATT dispute-resolution panel concerning U.S. restrictions on imported tuna that were put in place in an effort to force Mexico and other countries to alter fishing methods that resulted in the deaths of large numbers of dolphins (see Box 14.2). The panel ruled that the U.S. trade barriers violated the GATT agreement. Many environmentalists concluded that free trade and the international organizations overseeing the trade-liberalization process were inimical to environmental protection.

Box 14.2: The tuna-dolphin dispute

Much of the general opposition to trade liberalization among environmentalists grew out of a case that was brought before a GATT dispute-resolution panel in 1991. In 1972, the United States adopted the Marine Mammal Protection Act (MMPA) in an effort to reduce human-generated mortality among dolphins, whales, seals, and other marine mammals. In the eastern tropical Pacific, dolphins often swim above schools of tuna, and fishing vessels use their presence to locate the tuna they wish to capture. The purse-seine nets used in this type of fishing result in the deaths of large numbers of dolphins caught in the nets along with the tuna. The MMPA initially targeted the U.S. fishing fleet, and efforts by tuna-packing companies to ensure that canned tuna was produced in a dolphin-safe manner resulted in many fewer dolphin deaths as a result of U.S. tuna fishing. Other countries, notably Mexico, were not bound by the MMPA (it is a U.S. law, not an international law), and dolphins continued to be killed by tuna fishers.

The United States was not the only country to regulate tuna fishing, and by the mid-1980s, the number of dolphins killed in conjunction with tuna fishing had declined significantly. The Mexican fleet had also reduced its dolphin kill but had not reached the levels of the U.S. fleet. One reason the U.S. kill rate was so low was that most U.S. vessels had simply moved their operations from the eastern tropical Pacific to areas where the tuna-dolphin association is not a problem. In 1990, the United States banned imports of tuna from Mexico and some other countries following a suit brought by an environmental group calling for the application of the MMPA to these foreign producers. Mexico requested the formation of a GATT dispute-resolution panel to consider this case. From Mexico's point of view, the application of a U.S. law to its fleet and, more important, to the particular way in which the extraterritorial application of the MMPA was implemented, violated several articles of the GATT. The GATT panel agreed with the Mexican position on the grounds that it is a violation of GATT rules to restrict imports because of the way in which the products are produced. In addition, the procedures for determining how many dolphins the Mexican fleet could legally kill under the MMPA made it technically impossible for the Mexican fleet to comply.

The political reaction to this case was dramatic. For many environmentalists, it seemed that important environmental laws such as the MMPA were suddenly under attack by the GATT bureaucracy which was sensitive only to business and trade concerns. The panel ruling meant that the United States would have to amend the MMPA, pay compensation to Mexican and other tuna fishers, or accept that Mexico could legally place retaliatory trade barriers on U.S. goods. In fact, both sides backed off from the confrontation and undertook bilateral negotiations to resolve the issue. In subsequent discussions, several international agreements were reached and the number of dolphins killed in conjunction with tuna fishing declined further. Although the United States was unable to impose its national law on other countries, international rules to protect dolphins were strengthened in the aftermath of the tuna-dolphin dispute through direct international negotiations on the problem.

Source: Most of the information presented in this box comes from David Vogel, *Trading Up: Consumer and Environmental Regulation in a Global Economy* (Cambridge: Harvard University Press, 1997).

In the case of NAFTA, the specific concerns of environmental activists were that free trade with Mexico would allow polluting firms to escape U.S. environmental regulations by moving to Mexico, where environmental standards are less stringent. There was also a fear that U.S. firms would successfully pressure the government to lower environmental standards to improve their competitive position in relation to Mexican firms operating under the less intrusive standards. These environmental concerns are complementary to the labor worries in that both are driven by the expected exodus of U.S. firms seeking the less restrictive regulatory environment thought to exist in Mexico. Proponents of NAFTA pointed out that countries that have attained high standards of living are more likely to implement strict environmental regulations, so the best way to increase environmental protection in Mexico is to liberalize trade, raising incomes and the demand for a cleaner environment.

MAIN ELEMENTS OF THE TRADE AGREEMENT

As noted in Chapter 5, free trade agreements (FTAs) and customs unions both involve the elimination of trade barriers among the countries joining the FTA or customs union. The idea is to create a zone of free trade in a particular region while maintaining whatever barriers to trade with nonmember countries are allowed under the more general GATT/WTO agreements. Regional economic agreements are discriminatory, because countries apply lower levels of protection to partners in the agreement than they do to the rest of the world. Although the ideal is the total elimination of trade barriers within the FTA or customs union, the reality is more often the gradual reduction of most trade barriers and the full elimination of some. The NAFTA treaties specified which tariffs and nontariff trade barriers were to be reduced or eliminated and established a schedule for these changes. For example, Mexico agreed to eliminate its import quota for corn, a nontariff barrier, over a fifteen-year period.

FTAs differ from customs unions in that there is no requirement that the member countries harmonize their external trade policies. Such arrangements give rise to the possibility of trade deflection as imported goods are transshipped through the lower-tariff countries to markets protected by higher trade barriers. To prevent trade deflection, most free trade agreements include elaborate rules of origin to determine which products are eligible to be traded freely within the free trade area. For example, a shirt that is produced in Taiwan and then shipped to Mexico, where the sleeves are sewed on, may not be considered a Mexican shirt when imported by a U.S. firm. The NAFTA treaties include extensive rules of origin that specify the amount of added value required for a good to be considered as originating in one of the three member countries. NAFTA also differs from more conventional FTAs in that it includes rules on investments, financial services, telecommunications, and intellectual property.

In response to the concerns of environmentalists and labor organizations, presidential candidate Bill Clinton raised the idea of side agreements on labor and the environment, and such agreements were added to the negotiating agenda in an effort to win support for NAFTA from labor and environmental groups. The North American Agreement on Labor Cooperation (NAALC) and the North American Agreement on Environmental Cooperation (NAAEC) were adopted in late 1993 as part of the NAFTA package. Both of these supplementary agreements focus on enforcement of existing domestic and international standards. The NAALC calls for the protection of general workers' rights related to strikes, child labor, and so on. The NAAEC is designed to promote enforcement of domestic environmental rules. Both include provisions for cooperation and consultation, with a view toward strengthening labor and environmental standards as well as dispute-resolution mechanisms for the adjudication of conflicts over the application of national standards.

EX ANTE ANALYSES OF NAFTA

Francois and Shiells provide an interesting account of the use of CGE models for the ex ante analysis of the impacts of regional economic integration in North America.[5] They note that one of the lead negotiators on the Mexican side, Jaime Serra-Puche, had built a CGE model for his Ph.D. dissertation at Yale University, and several other Mexican negotiators, including President Carlos Salinas de Gortari, had also completed Ph.D. degrees in economics at prestigious U.S. universities. It was natural for these individuals to use their modeling skills to assess the advantages and disadvantages of the evolving agreement. In addition, the political controversy in the United States created a need for sound analytical results to counter some of the wilder claims being made on both sides of the debate. Francois and Shiells suggest that CGE models were used in Mexico to develop detailed negotiating points, while in the United States such models were used to inform the public debate.

At least ten major studies based on CGE models were done during the negotiations to predict the likely effects of the agreement.[6] These studies were completed before the details of the agreement were known, so the analysts were forced to make assumptions about the timing and extent of the reductions in trade barriers. For example, it was not clear at the outset whether the agreement would cover both tariff and nontariff trade barriers. All of these models followed the general format described in the last chapter but differed in the way various components were modeled, in which of the three countries were included, and in the assumptions made concerning the nature and degree of trade liberalization. For example, a study might be based on a CGE model of Mexico, allowing for imperfect competition and removal of both tariff and nontariff trade barriers. Another study might link together two or three national CGE models, assuming perfect competition and less than total trade liberalization. In some cases, the models emphasized particular sectors, such as agriculture or automobiles, while in others, more general economy-wide effects were targeted. In all cases, the goal was to predict the welfare effects of NAFTA as compared to baseline scenarios designed to predict economic changes in the absence of any trade agreements. Rather than presenting detailed descriptions of each of the models and the results obtained, it may be more informative to summarize the general conclusions on the four issues described earlier.[7]

General Impacts on Trade and Economic Growth

Most of the studies found that the creation of NAFTA would have a positive but limited impact on economic growth. Brown, Deardorff, and Stern report predictions for real Mexican gross national product (GNP) of 0.3 percent to 11.0 percent higher with NAFTA than in the baselines, with most estimates about 4 to 6 percent higher. The estimated GNP increases in the United States are much smaller, ranging from 0.02 percent to 2.6 percent compared with the baselines. For Canada, the estimates range from 0.7 percent to 6.8 percent. Trade among the three countries increases only slightly if only tariffs are reduced. The effects are more substantial if nontariff barriers are also removed. Most of the studies found very little trade diversion from the rest of the world. This result is consistent with the observation that most trade barriers were already low prior to the NAFTA negotiations and the three countries already traded extensively with each other.[8] In general, the results indicated fairly modest positive effects on trade and economic growth.

Labor

The prediction that U.S. wages would fall with the formation of NAFTA was not supported by these studies. In general, the results showed increased employment and wages in all
countries. In three studies, the wages of rural workers in Mexico did decline, and Burfisher,
Robinson, and Thierfelder predicted small wage declines in the United States for unskilled
rural and urban workers as a result of increased international migration.[9] The other studies
generally found wage increases on the order of 5 to 10 percent above the baseline for both
skilled and unskilled workers in all three countries. Note that these results show the
changes in wages and employment that might be attributed to changes in trade policies
under NAFTA. As noted earlier, the total number of jobs in an economy depends more on
macroeconomic conditions than on trade policies. Some of the studies indicate that wages
and employment in particular sectors may fall. For example, Trela and Whalley find that
employment in the U.S. steel industry would be slightly lower with NAFTA than without.[10]

Sectoral Adjustment and Migration

Several of the models focused on particular sectors within the general equilibrium framework. Because large numbers of Mexican peasants are employed in subsistence production
of corn and dry beans, there was intense concern that an influx of cheap U.S. corn would
force many people off the land, leading to increased internal and international migration.
One study predicted that implementation of NAFTA would cause 700,000 people to leave
rural Mexico for the cities, while another estimated the number of migrants at 800,000, of
which 600,000 would end up in the United States as unskilled workers.[11] These dramatic
impacts were mitigated substantially if the trade policy changes were introduced gradually
over a long period of time. In fact, the final agreement specified a fifteen-year transition
period for corn and other sensitive products. The studies predicted that U.S. farmers would
benefit from increased grain and soybean exports to Mexico, while some fruit and vegetable
producers would experience increased competition from expanding imports from Mexico.

Environment

The numerous ex ante studies of NAFTA were largely done before the environmental issue
had come to the fore, so none of them incorporated the NAAEC into their analyses. Several
writers pointed out that rising incomes in Mexico would lead to increased demand for
environmental protection, but there was little formal modeling of the environmental
impacts prior to the ratification of the agreement. A recent analysis of the connection
between trade liberalization and environmental protection shows that free trade without
complementary environmental policies can lead to specialization in polluting industries,
although economic growth is higher than with protectionism. Using a CGE model for
Costa Rica, Dessus and Bussolo show that a combination of free trade and taxation to correct environmental externalities is better than either free trade alone or protectionism coupled with the environmental taxes.[12] If NAAEC is effective in establishing appropriate

environmental policies, the combination of these policies with free trade under NAFTA could lead to similar results for Mexico.

EX POST ASSESSMENTS

As noted earlier, NAFTA took effect in January 1994. The agreement provided for long transition periods for the full implementation of the negotiated policy changes. Because of the long transition periods, the immediate impact of the agreement was quite limited. Mexico had been running a growing trade deficit, financed by foreign investment, for several years. Mexican inflation was also higher than in its main trading partners. The inflation differential, coupled with the growing trade deficit, led to pressures to devalue the Mexican peso. The government resisted these pressures, attempting to maintain the fixed exchange rate between the peso and the dollar. As scheduled elections approached, one of the presidential candidates was assassinated, and the uncertainty created by this political turmoil and the shaky financial situation of the country caused investors to begin moving their assets from Mexico to other countries. The capital outflow meant that Mexico had to use up much of its foreign exchange reserves to cover the trade deficit. In December 1994, the peso was substantially devalued and subsequently left to float freely.

The financial crisis in Mexico was not caused by NAFTA, although many of those who had opposed the trade agreement tried to claim that there was a link. The impact of the crisis on low-income workers and peasants was severe as continued depreciation of the peso led to increased prices for imported goods, but the crisis was short-lived. Mexican exports increased greatly as a result of the depreciation and the more open trade policies that had been implemented prior to NAFTA as well as the further liberalization carried out in the early years of the FTA. The booming U.S. economy of the 1990s also helped Mexico overcome the financial crisis as U.S. demand for Mexican goods increased. In addition, foreign investment in Mexico returned to previous levels as the financial crisis subsided.

These dramatic financial events make it difficult to assess the impact of NAFTA in the early years following its adoption. For example, the United States has traditionally exported more agricultural goods to Mexico than it imports. This pattern was reversed for about eight months following the peso devaluation. Some commentators who had opposed NAFTA in the first place argued that this reversal showed that NAFTA was a failure. This kind of argument illustrates the common mistake in policy evaluation that has been emphasized in this book. The appropriate counterfactual for assessing the impact of NAFTA on the agricultural trade balance with Mexico is not the way things were before NAFTA took effect but rather the way things would have been at the time the comparison is being made if NAFTA had not been implemented. In fact, the traditional pattern of agricultural trade between Mexico and the United States was reestablished fairly rapidly but at a higher level (more U.S. exports and imports).

Despite the extraordinary modeling effort undertaken to assess the likely impacts of NAFTA prior to its implementation, the use of economy-wide models to conduct ex post evaluations of these policy changes has been quite limited. The Economic Research Service (ERS) of the U.S. Department of Agriculture (USDA) continues to use its partial equilibrium models to monitor and assess the impacts of NAFTA on U.S. agriculture. The ERS researchers note that agricultural trade among the three countries has expanded dramatically compared to what would have occurred if NAFTA had not been adopted.[13] Much other commentary on the impact of NAFTA has relied on ad hoc comparisons that often involve reference to inappropriate counterfactuals. This is true for both antitrade groups and those who believe NAFTA has been beneficial.

Weintraub carried out an ex post analysis of NAFTA after the agreement had been in effect for three years. He devotes one chapter of his book to a discussion of the appropriate way to evaluate NAFTA, severely criticizing most of the public discourse on the question as "one-sided advocacy." Weintraub argues that focusing on the bilateral trade balance between the United States and Mexico or on changes in the numbers of jobs are incorrect criteria for evaluating the agreement. He suggests that the appropriate bases for assessing NAFTA's impact as a trade agreement include the volume of total trade, changes in intraindustry trade and specialization due to economies of scale and increased technical efficiency, productivity changes with their associated impact on wages, and sectoral adjustments due to changes in the competitive position of particular industries. Weintraub also notes that the trade agreement could have repercussions for the environment and for the development of formal and informal institutions and includes these areas in his list of evaluation criteria.[14]

Weintraub finds that the economic effects of NAFTA are largely as predicted by the models used in the ex ante assessments. Trade volumes increased during the first three years, freer trade mitigated the negative impact of the 1994 financial crisis, and there have been sectoral adjustments that, while increasing efficiency and contributing in a very small way to increased productivity in the United States and Mexico, have also led to political opposition from those working in the previously protected industries.

Weintraub believed that it was too early in 1996 to be able to assess the impact of NAFTA and NAAEC on the environment, and that is probably still true today. On institutional and political questions, Weintraub argues that there have been substantial positive changes. Mexican firms and consumers have begun to develop closer ties with their U.S. counterparts, and the further development of Mexico's democracy seems to be under way. It should be noted that in 1999, the Mexican people elected a new president who is a not a member of the Institutional Revolutionary Party (Partido Revolucionario Institucional, or PRI), the party that had ruled Mexico continuously for more than seventy years. It would be wrong to think that these political changes were caused solely by NAFTA, although one might argue that a more open economy can contribute to democratic development.[15]

A more recent assessment of NAFTA can be found in a book by Coffey, Dodds, Lazcano, and Riley, which includes chapters on each of the three member countries.[16] The authors note that the North American Integration and Development Center at UCLA continues to monitor and report on the impacts of NAFTA using the CGE model developed for the ex ante analysis by Hinojosa-Ojeda, Robinson, and others.[17] This work shows that trade and investment patterns between the United States and Mexico had begun to change prior to the implementation of NAFTA as a result of policy reforms undertaken in Mexico in the 1980s. While these trends have continued and perhaps accelerated slightly, NAFTA does not appear to have had a significant impact on trade and employment. The UCLA group estimated in 1997 that 42,000 new jobs had been created and 18,000 lost as a result of NAFTA. The other studies reviewed also support the predictions of the ex ante analyses that found that the short-run economic impact of NAFTA in the United States would be positive but small.

The assessments for Canada and Mexico in the book by Coffey and associates seem to indicate somewhat greater effects than in the case of the United States. Much of the discussion in these countries seems to have shifted, however, from the aggregate impacts on trade, growth, and employment to specific issues such as the U.S.-Canada disputes over trade in wheat and dairy products that have arisen since the implementation of NAFTA. As has always been the case, there may be some discomfort in Mexico and Canada over the increasing influence of the neighboring economic giant not only on economics and business but on cultural life and social and political institutions. Some analysts believe, however, that the greatest long-term benefit of NAFTA will be improved social and political conditions and better relations among the countries and that the economic effects will turn out to be secondary.

THE CASE OF MEXICAN SUBSISTENCE AGRICULTURE AND MIGRATION

As noted earlier, one of the concerns about the impact of NAFTA was the fear that the rural economy in Mexico would be seriously disrupted, leading to increased poverty and migration. Shortly after the adoption of NAFTA, the Mexican government modified one of the centerpieces of the Mexican Revolution, the system of communal land ownership known as the *ejido* system. An *ejido* is a landholding that can be used and passed to one's heirs but not sold, because ownership of the land is vested in the community. Over time, the division of the *ejidos* among the children of the original landholders had resulted in tiny farms which could not be consolidated because of the restrictions on land sales. This system was changed to allow those who wish to sell their holdings and move on to other activities to do so. Many expected that the incentive to leave subsistence farming would become greater as corn markets are fully liberalized. The price for corn in Mexico was two to three times the U.S. price prior to NAFTA. Agricultural productivity was low and the small size of many landholdings made it difficult to introduce improved technologies.

Subsistence farmers produce mainly for home consumption, marketing only whatever small surpluses beyond family needs can be generated. If these farmers opted to sell their land in the face of falling corn prices, they might well end up migrating to Mexico City or the United States. On the other hand, corn markets are to be liberalized gradually over a fifteen-year transition period, and the Mexican government introduced a series of rural policy reforms aimed at easing adjustment in rural areas. Moreover, the situation in rural Mexico is more complicated than indicated by the discussion so far. De Janvry, Sadoulet, and Gordillo de Anda showed that many peasants in Mexico consume almost all of their output, selling little on the market.[18] Such producers will not suffer greatly from falling corn prices. In fact, many of them are actually net purchasers of corn and so will benefit from lower prices when they need to supplement their subsistence production with corn purchased from the market.

This result is also supported by a study based on a village-level CGE model.[19] Recall from Chapter 13 that input-output models are often structured to represent technical relationships in local and regional economies. The same can be done with CGE models in which a general equilibrium model of the local economy is tied to the wider national and international economies through an appropriately specified SAM. The village CGE model in this study is used to assess the impact of NAFTA on agricultural production, rural incomes, and migration out of rural areas. The authors find that liberalization of Mexican corn markets will lead to substantial changes in corn production but will have only limited effects on rural incomes and migration. The reason for this result is that Mexican peasants are highly risk averse, and this has led them to diversify their income sources. As corn production becomes less profitable, the model shows that other income-generating activities increase. The authors also emphasize the benefits to rural households that are net corn purchasers of the lower corn prices under NAFTA.

Because this study was conducted after NAFTA had been adopted, it was possible for the authors to incorporate the details of the trade agreement as well as the new agricultural policies introduced by the Mexican government. This allows them to make realistic predictions about the combined effects of trade liberalization and agricultural policy reform. Based on their results, the authors also propose and evaluate alternatives to the policy reforms adopted in the mid-1990s. The results show that policies to promote technological change can ease the adjustment to new prices without creating distortions in local factor markets, as might occur with the agricultural policy reforms that were actually adopted.

CONCLUSION

While government agencies and policy analysts continue to monitor the impact of NAFTA and critics are still producing anti-NAFTA tracts, the level of analytical activity that prevailed prior to 1994 has not been reached since the adoption of the agreement. One reason, of course, is that the political furor died down somewhat once NAFTA was ratified by the three countries. Another reason is that the provisions of the agreement will not be completely phased in until about 2011, and it may not be possible to assess its full effects until well into the twenty-first century. In the short term, most of the evidence seems to indicate that the predictions of the ex ante studies were largely correct. NAFTA has had small positive impacts on trade and growth, and its modest effects on labor markets have been swamped by the booming economies of all three countries.

Although most of the analytical studies have focused on the economic impacts of the agreement, it is widely recognized that NAFTA is likely to influence political, social, and cultural developments as well. Mexico's democracy seems to have become stronger in the years since NAFTA was adopted, and relations between the United States and Mexico are much improved over earlier periods, which were often characterized by suspicion and distrust. At the same time, there are still many who worry that economic modernization under NAFTA will destroy traditional ways of living in rural Mexico. Those who hold such positions often use the language of ethics and human rights to criticize modernization and globalization. As noted in Chapter 6, ethical judgments cannot be made without a realistic assessment of the relevant facts. Many NAFTA critics assume that the ways of life they wish to protect would be safer if Mexico were insulated from the globalizing world. There is much evidence, however, that local elites insulated from global influences can get away with more extensive human rights abuses than would be the case in more open societies. Amartya Sen has shown that a famine has never occurred in a country with a free press and has also noted, along with many others, that countries with democratic governments have never gone to war with each other.[20] Peasants in the Yucatan have probably suffered more at the hands of Mexican political elites than they are likely to suffer at the hands of global corporations. Still, there is much that remains to be explored in assessing the ethical implications of NAFTA and other changes that lead to greater interaction between countries.

DISCUSSION QUESTIONS

1. What were the main issues related to NAFTA that were debated in the United States prior to the adoption of the agreement? To what extent could accurate information on these issues be produced without the use of analytical models?

2. Would you expect trade liberalization to have a negative impact on the environment? Why or why not?

3. Why were the side agreements to NAFTA on labor and the environment added? Do you think side agreements similar to these two might be useful as models of components that could be added to the multilateral trade agreements negotiated in the WTO?

4. A study showed that with free trade, the United States would import tomatoes from Mexico and the U.S. fresh tomato industry would decline, while with protectionism, fresh tomato production would remain profitable and the industry would continue to hire large numbers of migrant workers from Mexico. In other words, migrant workers and imported tomatoes are substitutes. Which is better for general welfare in the United States and Mexico, protectionism or free trade? Why?

5. If free trade leads to factor-price equalization, then wages for unskilled workers in the United States should fall. What is the likely effect of free trade on employment of unskilled workers? (Hint: With factor-price equalization, the cost of hiring unskilled workers has fallen.)

6. What were the main results of the ex ante CGE analyses of NAFTA? How accurate were the predictions of these analyses?

7. The legal documents defining NAFTA include many hundreds of pages on rules of origin. Why are these rules necessary?

8. Do you think that further development of regional agreements such as NAFTA or Mercosur (Mercado Común del Sur, or Southern Common Market, a South American trade agreement created by Argentina, Brazil, Uruguay, and Paraguay in 1991 and now also including Chile and Bolivia as associate members) will lead to greater trade liberalization in the long run, or are such agreements impediments to multilateral trade liberalization? Explain.

SUGGESTIONS FOR FURTHER READING

Audley, J. J. *Green Politics and Global Trade: NAFTA and the Future of Environmental Politics.* Washington, D.C.: Georgetown University Press, 1997.

Cameron, M. A., and B. W. Tomlin. *The Making of NAFTA: How the Deal Was Done.* Ithaca, N. Y.: Cornell University Press, 2000.

Coffey, P., J. C. Dodds, E. Lazcano, and Robert Riley. *NAFTA: Past, Present and Future.* Boston: Kluwer Academic Publishers, 1999.

Francois, J. F., and C. R. Shiells, eds. *Modeling Trade Policy.* New York: Cambridge University Press, 1994.

Johnson, P. M., and A. Beaulieu. *The Environment and NAFTA.* Washington, D.C.: Island Press, 1996.

McKinney, J. A. *Created from NAFTA: The Structure, Function, and Significance of the Treaty's Related Institutions.* Armonk, New York: M. E. Sharpe, 2000.

NOTES

1. See the discussion of customs unions and free trade agreements in Chapter 5 in the section on the theory of the second best.

2. Nora Lustig, "NAFTA: Potential Impact on Mexico's Economy and Beyond," in *Economic Integration in the Western Hemisphere*, ed. R. Bouzas and J. Ros (Notre Dame, Ind.: University of Notre Dame Press, 1994.)

3. Pat Choate, *Jobs at Risk: Vulnerable U.S. Industries and Jobs under NAFTA-Analyses of the 48 Continental States*, Manufacturing Policy Project (Washington, D.C.: July 1993). A firm's decision to open plants in another country or to move its operations across the border depends on a wide range of specific considerations. There is no evidence that labor costs are the only reason firms make these moves or that the critical labor-cost value is 20 percent of total costs.

4. William J. Baumol, *Environmental Protection, International Spillovers and Trade*, Wicksell Lectures 1971 (Stockholm: Almqvist and Wicksell, 1971).

5. J. F. Francois and C. R. Shiells, "AGE Models of North American Free Trade," in *Modeling Trade Policy*, ed. Francois and Shiells (New York: Cambridge University Press, 1994).

6. Ibid. Francois and Shiells summarize ten such studies. These and other studies are also described in Lustig, "NAFTA," and in N. Lustig, B. P. Bosworth, and R. Z. Lawrence, eds., *North American Free Trade* (Washington, D.C.: Brookings Institution, 1992); and P. J. Kehoe and T. J. Kehoe, eds., *Modeling North American Economic Integration* (Boston: Kluwer Academic Publishers, 1995).

7. The discussion in this section is drawn largely from Francois and Shiells, *Modeling Trade Policy*; and D. K. Brown, A. V. Deardorff, and R. M. Stern, "North American Integration," *Economic Journal* 102 (November 1992): 1507–18.

8. If countries have low trade barriers and already trade extensively with each other, then further liberalization is likely to expand existing trade rather than displace trade from other countries. It is when trade is limited by high barriers among the countries forming the FTA that the removal of the trade barriers is likely to lead to trade diversion.

9. M. Burfisher, S. Robinson, and K. Thierfelder, "Wage Changes in a U.S.-Mexico Free Trade Area," in *Modeling Trade Policy*, ed. J. F. Francois and C. R. Shiells (New York: Cambridge University Press, 1994).

10. I. Trela and J. Whalley, "Trade Liberalization in Quota-Restricted Items: The United States and Mexico in Textiles and Steel," in *Modeling Trade Policy*, ed. J. F. Francois and C. R. Shiells (New York: Cambridge University Press, 1994).

11. S. Levy and S. van Wijnbergen, "Transition Problems in Economic Reform: Agriculture in the North American Free Trade Agreement," *American Economic Review* 85, no. 4 (September 1995): 738–54; and S. Robinson, M. Burfisher, R. Hinojosa-Ojeda, and K. Thierfelder, "Agricultural Policies and Migration in a U.S.-Mexico Free Trade Area: A Computable General Equilibrium Analysis," working paper 617, Department of Agriculture and Natural Resources, University of California-Berkeley, 1991.

12. S. Dessus and M. Bussolo. "Is There a Trade-Off Between Trade Liberalization and Pollution Abatement?" *Journal of Policy Modeling* 20, no. 1 (1998): 11–31.

13. USDA/ERS. "NAFTA: International Agriculture and Trade," ERS-WRS-99-1, Washington, D.C., August 1999. Available at the Internet Web site http://usda.mannlib.cornell.edu/reports/e...ernational/wrs-bb/1999/nafta/wrs99-1s.asc. Accessed on September, 22, 2000.

14. Sidney Weintraub, *NAFTA at Three: A Progress Report* (Washington, D.C.: Center for Strategic and International Studies, 1997). The phrase "one-sided advocacy" appears on page 2.

15. See Paul Krugman, "Mexico's New Deal," *New York Times*, 5 July 2000, p. A21.

16. P. Coffey, J. C. Dodds, E. Lazcano, and Robert Riley. *NAFTA: Past, Present and Future* (Boston: Kluwer Academic Publishers, 1999).

17. See Trela and Whalley, "Trade Liberalization."

18. A. de Janvry, E. Sadoulet, and G. Gordillo de Anda, "NAFTA and Mexico's Maize Producers," *World Development* 23, no. 8 (1995): 1349–62.

19. J. E. Taylor, A. Yunez-Naude, and A. Hampton, "Agricultural Policy Reforms and Village Economies: A Computable General Equilibrium Analysis from Mexico," *Journal of Policy Modeling* 21, no. 4 (1999): 453–80.

20. See Amartya K. Sen, *Development as Freedom* (New York: Alfred A. Knopf, 1999).

15

Conclusion: The Place of Policy Analysis in the Policy Process

The focus of this book has been on the analysis of public policy interventions, that is, the evaluation of the likely consequences of efforts to solve collective action problems. To address this question, it was first necessary to consider the reasons why collective action may be needed. The theoretical concepts presented in Part I help foster an understanding of the circumstances under which governments might elect to intervene to influence economic or social behavior. Policy interventions grounded in the logic of market failures and the desire to improve economic efficiency or to insure distributive justice are justified because they lead to the good consequences of removing inefficiencies and increasing welfare or desired nonconsequentialist outcomes such as establishing and protecting important rights. Analysis of the expected impacts of alternative policy interventions is a critical component in the design and selection of the best policy, where the best policies are those that are the most likely to bring about such outcomes. The methods used to carry out this analytical function are described in Part II.

It seems obvious, however, that a great many public policies are not based on correcting market failures or promoting justice and respect for individual rights. In fact, the model of public policy as a collective effort to solve problems of efficiency, distribution, and justice with a view toward improving the lot of all people appears to be wildly inconsistent with real-world experiences. Lobbyists for special-interest groups help write U.S. legislation on issues of interest to their groups, and political leaders win elections by obtaining pork barrel projects that benefit their districts at the expense of the general public. According to Stiglitz's firsthand account (see Chapter 5), Pareto improvements are often thwarted by individuals and groups seeking to advance their particular interests with little regard for the implications of their actions for others. Public policies made by and for powerful, self-serving individuals driven primarily by avarice or their own special causes are difficult to justify, at least if one believes that the primary rationale for collective action is to make the world a better place for all people and not just for those with the power and means to insure that their interests are advanced. From this perspective, policy analysis as described in this book may be pointless, because public decisions will be made not on the basis of reason and evidence but rather in accordance with the special interests and desires of the rich and powerful.

Despite the problems of rent seeking, special-interest groups, and venal politicians, I believe that there is an important role for good policy analysis as an input into public decision making. In thinking about this role, it may be useful to imagine an idealized process through which public policies might be designed, selected, and implemented. Consider the following:

1. Problem identification.

The first step in determining whether collective action is needed and, if so, what sort of action would be the most effective is to identify a problem. If there is no problem to be solved, there is no need for a collective effort to alter social, economic, cultural, or political relations.

2. Problem diagnosis.

Once a general problem has been identified, it is important to develop a more precise understanding of its nature and scope. Problem diagnosis should include an initial assessment of the social and economic relations attendant to the problem, along with an evaluation of its severity and incidence.

3. Development of alternative solutions.

If a problem has been identified and the initial diagnosis shows that it is indeed a problem meriting some form of collective action, then the next step is to work out one or more procedures that will actually solve the problem. At this stage, it is important to be able to show how a suggested policy will lead to the desired outcome. If such a demonstration is not possible for a proposed policy, that option should be discarded.

4. Policy analysis.

This is the part of the policy process that is the focus of this book. Most policy analysis will begin with an assessment of the good and bad consequences of each alternative, along with some effort to find a common denominator so that the alternatives can be compared and ranked. The comparisons should always include the status quo or an appropriately specified counterfactual so as to keep open the option of doing nothing. Nonconsequentialist considerations should also be brought into the analysis, as illustrated in the case studies and other chapters in Part II.

5. Decision.

The public authorities charged with deciding on public policies then decide which if any of the alternatives should be chosen, taking account of the analytical results, both those that concern consequences and those that do not, along with any other relevant considerations. Policy analysis is an important input into the decision-making process, but it is not the only one.

6. Policy implementation.

The policy chosen as a way to solve the collective action problem is translated into programs, legal instruments, or other measures. Ideally, the particular programs established to implement the policy will accurately reflect the nature and intent of the policy option that was chosen.

7. Ex post evaluation and adjustment.

The effects of the programs that have actually been implemented are monitored with a view toward adjusting and fine-tuning them for greater effectiveness. In many cases, periodic review and assessment will lead to further policy decisions, including, perhaps, the decision to terminate a program if the collective action problem for which it was designed is no longer at issue.

The policy process described in these seven steps is rarely followed for controversial political issues. Consider federal income taxes, which are frequently at the center of U.S. political debates. These debates often proceed without a clear definition of the problem, other than assertions that Americans are overly burdened with income taxes. In some cases, a problem is identified, but there is little or no diagnosis. For example, several U.S. politicians have argued that the tax code is too complex, calling for tax simplification as a remedy. Upon closer examination, however, it turns out that the vast majority of U.S. taxpayers use the short tax form, which is extremely easy to fill out and file. Those who find it in their interest to file the more complex forms generally have wealth and income well above the median and file complicated returns in order to take advantage of a wide range of loopholes. Based on this diagnosis, one might question whether tax simplification is really a significant public policy problem. For many highly politicized public issues, diagnosis, analysis, and reasoned decision making are lost in a haze of hyperbole and misinformation aimed at advancing some set of special interests.

Even in these politicized settings, however, it is possible for policy analysis to play a role in the outcome. Policy analysts and journalists responded to the calls for tax simplification by pointing out that the income tax code was actually fairly simple for most people. Voters recognized that this was not the most pressing problem facing the nation, and the issue of tax simplification faded from the scene. As the nature of the political debate on the income tax shifted from simplification and budget deficits to the disposition of growing federal budget surpluses, political candidates began to propose various tax-cutting schemes, often making claims about the fairness or the distributional impact of their proposals. Many analysts and commentators saw the need for analytical tests of these claims, and numerous articles and opinion pieces assessing the impacts of alternative proposals began appearing on editorial pages and in other outlets.

In some cases, these analytical results were produced by partisan think tanks seeking to exaggerate the social benefits of their preferred tax program. At the same time, however, more objective assessments were offered by academic policy analysts and others who drew on methods similar to those discussed in this book to predict the likely consequences of the different tax proposals. These analytical results are often less prominent than the more simplistic claims that dominate political advertising. Yet they may have an influence in raising issues that journalists or the general public may be able to force into the debate. Thus, although reliable policy analysis usually does not determine the outcome, it may make an important contribution to improving democratic deliberation.

Most public policy issues are far less visible than the highly politicized cases of taxes, gun control, abortion, and so on. For example, while pesticide regulation occasionally becomes a conspicuous public issue (recall the outcry over alar on apples in the 1980s), most legislation on pesticide use is worked out through consultation with experts from universities, the regulatory agencies, or the industry. In many cases, the U.S. Congress has mandated the use of cost-effectiveness or benefit-cost methods to evaluate problems such as those posed by the use and disposal of pesticides and alternative remedies proposed to resolve these problems. In these settings, the policy process may come much closer to following the idealized system outlined in this chapter, allowing policy analysis to play an important role in the resolution of collective action problems. This is true not only in the United States but in other democracies as well. In fact, it is not uncommon for policy analysts to have greater access to decision makers in other countries than they do in the United States. As noted in the case study of NAFTA, for example, Mexican policy analysts probably had much greater influence on the positions taken by Mexican negotiators than their counterparts in the United States had on the U.S. negotiating stance.

The process described in this chapter should not be seen as unidirectional, proceeding linearly from the problem definition to the final implementation of a solution. Diagnosis may reveal that the problem identified in the first step is not really very pressing but that a related issue is. From this diagnosis, it would be sensible to return to the first step and redefine the problem. Likewise, analysis may reveal that none of the proposed policy solutions is efficient but that with some redesign, one or more of the alternatives may lead to welfare increases. Each of the steps in the idealized policy process can be thought of as feeding back on the others. Within this idealized dynamic policy process, policy analysis is critical for its role in the design and identification of beneficial policies.

Although real-world policy making involves conflict, narrow partisanship, and apparent irrationality, the process is often improved through inclusion of the results of reliable policy analysis. Such results can contribute to more enlightened democratic deliberation, improved policy choices, and better policy design and implementation. To play these important roles in real-world public decision making, it is crucial that policy analysis be done in a way that reflects the true social values that are at play. Several of the themes raised in this book are relevant to this question. First, social values may not be accurately reflected in such obvious measures as market prices. In many cases, there are no prices at all attached to important social values. We have shown how divergence between private and social values may arise and noted several conceptual and practical ways to reconcile these differences. At the conceptual level, differences between private and social values reflect market failures such as externalities, which can be remedied with redefinition or reassignment of property rights. Changes in property rights can be made through legal regulations or through the creation of markets in which social values can be discovered. From the Coase theorem, we know that the way in which property rights are assigned has no effect on economic efficiency as long as transaction costs are low. However, property rights assignment has important implications for income distribution, which may be as important a social objective as efficiency.

For practical policy analysis, one cannot assume that externalities have been eliminated or that other types of market failure have been corrected. Analysis of public projects or programs in which price distortions are present requires the use of shadow prices that more accurately reflect social opportunity costs. In many cases, even distorted prices are unavailable, because there is no market and no possibility of creating one. In these cases, the discovery of appropriate social values has to be based on such methods as contingent valuation or hedonic pricing. The purpose of all these methods is to discover measures that accurately reflect the marginal value society attaches to goods and services that may be affected by the public initiative.

There are other market failures that are addressed by public policies. Public goods are likely to be undersupplied by competitive markets due to free riding, and common pool resources can be destroyed by rational individuals unable to establish clear property rights or enforceable contracts. In addition, some markets are not competitive, and the invisible hand often fails in conditions of risk and uncertainty. Many government policies aim to correct these kinds of market failure and thereby to improve the efficiency with which the economy functions. The efficiency gains from such interventions can be substantial, leading to more rapid economic growth and greater general prosperity, but these efficiency gains often require compensation of the losers to satisfy the test of Pareto optimality. Pareto optimality may not be a particularly good guide in deciding on public policies because it tends to protect the status quo. In some instances, social welfare may be increased through redistributive policies or other policies that violate Pareto optimality but for which no compensation is made, particularly if those who would veto these policies would do so in the name of rights that are unwarranted.

The issue of who benefits and who loses from public policies and whether compensation is to be made to the losers brings political and ethical considerations into the analysis. While efficiency is an important social objective, it is simultaneously determined with other important social objectives—including, for example, income distribution—and these other objectives cannot be ignored in policy analysis. This observation is a second theme that has been emphasized in this book. Much policy analysis focuses on finding policies that promote efficiency or, alternatively, on evaluating the costs of policies that are implemented for political or other objectives largely unrelated to efficiency. This is important information for policy makers, but it is not the only kind of information that they may bring to the decision. Many public policy analysts would like to see greater collective rationality and less partisan politics in public decision making. The reality of democratic decision-making, however, as shown in Chapter 5, is likely to be much less neat and precise than the conceptually clear analytical results of benefit-cost analysis. The idea that technical policy analysis is only one input into the decision process represents a recognition that democratic realities may well result in inefficiencies but that the democratic process is still the best way for societies to handle collective action problems.

In addition to recognizing this political dimension, the idea that there is more to policy analysis than economic efficiency also includes recognition that there is an unavoidable ethical component in public policy analysis. The choice of a particular course of action will require value judgments as well as judgments about the nature of important relationships and the factual evidence that has a bearing on the problem. The particular value judgments made in analyzing public policies should be supported by ethical reflection and made clear to the audiences that will use the results of the analysis to make decisions. One might even go so far as to develop a kind of ethical sensitivity analysis designed to show whether the policy recommendation would change if different ethical assumptions were made. For example, consequentialist arguments in favor of the Federal Agriculture Improvement and Reform Act (FAIR) might differ from nonconsequentialist arguments, but if both support the same conclusion, the analysis is not sensitive to the ethical perspective that is chosen.

Another theme that has been emphasized in the chapters on methods and applications is the notion that useful policy analysis must be based on appropriate comparisons. The ideal is to develop a model that allows experiments comparing the evolution of key variables if the policy being analyzed is in place with their evolution if the policy is not implemented. The specification of the appropriate counterfactual is critical if the analysis is to have any meaning at all. Comparing predictions about the state of the world in ten years if a given policy is implemented with the situation today is misleading. Over the next ten years, many things will change whether the policy is implemented or not. Models are used to control for these changes so that the analyst can predict the future with and without the policy change, holding other things constant.

Even if a complex and sophisticated model cannot be built to examine a particular policy issue, it is still important to do one's best to make appropriate comparisons. For example, any casual assessment of FAIR should take account of the fact that prices were extraordinarily high at the time the new farm program was approved. Rational producers responded to these favorable prices by planting more acres, and production increased. At the same time, world demand slumped as the Asian financial crisis of the mid 1990s developed. In the late 1990s, prices plummeted. Many farmers and farm-state legislators blamed the price decline on FAIR and began calling for drastic policy changes. Yet much of the decline in prices would have occurred whether FAIR had been implemented or not. The farm bill did not cause the Asian financial crisis, and farmers would have been able to increase production in response to the high prices in the mid 1990s even if there had been no change in policy. FAIR did allow producers somewhat greater flexibility in their planting decisions, and this change might have been the cause of some of the production increase. The real question is how much of this increase was due to FAIR and how much was due to other causes. Blaming the entire fall in prices on FAIR surely exaggerates its impact.

The methods introduced in Part II of this book have been developed by economists and other policy analysts to handle such issues as the design of appropriate counterfactuals or the discovery of social values when markets are missing or distorted. I have argued that the economic analysis of public policy can be expanded to include political and ethical considerations as well. The conceptual approach to public policy outlined in the first part of the book helps determine when policy interventions may be justified, and the methods described in the second part provide tools for determining whether a particular policy option actually has the potential for resolving a collective action problem in a way that is broadly beneficial. Use of these theoretical concepts and practical evaluation methods can lead to significant improvements in the quality of public decision making and democratic deliberation.

At the same time, it would be an exaggeration to claim that these theories and methods are the only ways to evaluate public policies. In their applied work, political scientists, sociologists, anthropologists, and other social scientists all study and evaluate public polices frequently, using methods that differ significantly from those described in this book. To the extent that these approaches to policy analysis can provide consistent and useful information on the implications of alternative courses of action, such information should be considered by those responsible for choosing and implementing public policies. I have had little to say about these alternative approaches to policy evaluation because I know little about them. In general, I believe that the more analytical approach described in this book holds the greatest promise for insuring that the most important aspects of the problem are systematically included in the information set available to policy makers. On the other hand, I would not want to suggest that scholars and analysts with disciplinary backgrounds in the other social sciences have nothing to contribute to the policy process. The description of such contributions will, however, have to be left to others.

There is one final problem that should be raised before ending this book. Policy analysis can provide important information for the design and selection of public policies, but it is also subject to misuse and abuse. I noted earlier that some of the analytical results on alternative tax-cut proposals were produced by partisan groups seeking to advance particular political agendas. Quantitative analysis generates numerical predictions that often seem to have the authority of scientific pronouncements, even when they are the product of models that are misspecified or analytical procedures that are inappropriate or misleading. It may be extremely time-consuming to take a model apart to discover whether its specification makes sense or to insure that biases have not been introduced either inadvertently or on purpose. Verifying that the analytical framework is appropriate may be beyond the technical competence of the general public and most policy makers. Moreover, it is often difficult, even for experts in policy modeling, to ascertain just how the results being presented were obtained, and it may not be in the interest of partisan commentators to shed light on this question. In this setting, analytical frameworks can take on the appearance of black boxes that generate whatever results are needed to promote a particular point of view.

While these problems are real, they are mitigated by the fact that there are large numbers of more or less objective researchers and analysts in universities, nonpartisan think tanks, and government agencies who not only generate reasonably reliable results but also criticize policy recommendations based on dubious analytical approaches. While there is no guarantee that analytical results designed to serve rent-seeking special-interest groups will always be filtered out before they can do any harm, enough dependable information and correctives for the more biased results are generally placed before the public to insure that reasonable policy choices can be made. It is my hope that this book will contribute to this process by helping students develop a better understanding of public policy and the uses and limitations of policy analysis. With this knowledge, they may be better able to evaluate competing claims about public issues and to improve their own personal decisions on important questions about desirable courses of action. To the extent that this outcome is realized, the general level of democratic discourse will be raised, to the benefit of all. And that would be a good thing.

Index

Page references in *italics* refer to illustrations.

c 12/07